COMPREHENSIVE ANALYTICAL CHEMISTRY

ELSEVIER SCIENTIFIC PUBLISHING COMPANY
335 JAN VAN GALENSTRAAT
P.O. BOX 211, AMSTERDAM, THE NETHERLANDS

Distributors for the United States and Canada:

ELSEVIER NORTH-HOLLAND INC.
52, VANDERBILT AVENUE
NEW YORK, N.Y. 10017

LIBRARY OF CONGRESS CARD NUMBER: 58-10158

ISBN 0-444-41886-5 (Vol. XI)
ISBN 0-444-41735-4 (Series)

WITH 80 ILLUSTRATIONS AND 41 TABLES

PRINTED IN THE NETHERLANDS

COMPREHENSIVE ANALYTICAL CHEMISTRY

ADVISORY BOARD

Contributors to Volume XI

J. Grimshaw, Department of Chemistry, The Queen's University, Belfast, N. Ireland

P. Móritz, Computer and Automation Institute, Hungarian Academy of Sciences, Budapest, Hungary

W.E. van der Linden, Department of Chemistry, The University of Amsterdam, Amsterdam, The Netherlands

Wilson and Wilson's

COMPREHENSIVE ANALYTICAL CHEMISTRY

Edited by

G. SVEHLA, PH.D., D.SC., F.R.I.C.

Reader in Analytical Chemistry
The Queen's University of Belfast

VOLUME XI

The Application of Mathematical Statistics in Analytical Chemistry
Mass Spectrometry
Ion Selective Electrodes

ELSEVIER SCIENTIFIC PUBLISHING COMPANY
AMSTERDAM   OXFORD   NEW YORK
1981

WILSON AND WILSON'S

COMPREHENSIVE ANALYTICAL CHEMISTRY

VOLUMES IN THE SERIES

Vol. IA   Analytical Processes
Gas Analysis
Inorganic Qualitative Analysis
Organic Qualitative Analysis
Inorganic Gravimetric Analysis

Vol. IB   Inorganic Titrimetric Analysis
Organic Quantitative Analysis

Vol. IC   Analytical Chemistry of the Elements

Vol. IIA   Electrochemical Analysis
Electrodeposition
Potentiometric Titrations
Conductometric Titrations
High-frequency Titrations

Vol. IIB   Liquid Chromatography in Columns
Gas Chromatography
Ion Exchangers
Distillation

Vol. IIC   Paper and Thin-Layer Chromatography
Radiochemical Methods
Nuclear Magnetic Resonance and Electron Spin
    Resonance Methods
X-Ray Spectrometry

Vol. IID   Coulometric Analysis

Vol. III   Elemental Analysis with Minute Samples
Standards and Standardization
Separations by Liquid Amalgams
Vacuum Fusion Analysis of Gases in Metals
Electroanalysis in Molten Salts

Preface

In *Comprehensive Analytical Chemistry*, the aim is to provide a work which, in many instances, should be a self-sufficient reference work; but where this is not possible, it should at least be a starting point for any analytical investigation.

It is hoped to include the widest selection of analytical topics that is possible within the compass of the work, and to give material in sufficient detail to allow it to be used directly, not only by professional analytical chemists, but also by those workers whose use of analytical methods is incidental to their work rather than continual. Where it is not possible to give details of methods, full reference to the pertinent original literature is made.

Volume XI contains three longer chapters. The chapter on mathematical statistics builds on the foundations laid down by E.C. Wood in Volume IA, and discusses such important topics as statistical testing, the law of propagation of errors and the setting up of mathematical models in detail. The tailor-made computer programmes make this chapter, despite its highly mathematical content, a very practical contribution indeed. The chapter on mass spectrometry introduces a new field into this series. We hope to return to more specialised aspects of mass spectrometry in a later volume. The last contribution, on ion selective electrodes, describes a rapidly expanding field. Sixteen years ago, when a longer chapter on potentiometric titrations appeared in Volume IIA, these electrodes were just coming into use and merited a mere half page (p. 134) of discussion. Enzyme electrodes were described in Volume VIII and are therefore only briefly treated here.

The three authors come from various European institutions and are well-known experts in their fields. Because of this, and because they are all close personal friends of mine, I very much hope that their contributions will be well received by the public.

Dr. C.L. Graham of the University of Birmingham, England, assisted in the production of the present volume; his contribution is acknowledged with many thanks.

January 1980   G. Svehla

Contents

Chapter 1

The application of mathematical statistics in analytical chemistry

P. MÓRITZ

1. Introduction

The purpose of this contribution is to provide some mathematical-statistical tools for processing results of analytical measurements. It is the aim to describe calculating methods which can be applied directly.

For these purposes, calculation schemes and algorithms are presented keeping in mind the needs of scientists working both with desk-top calculators and with computers. Most of the algorithms are given in the form of Fortran subroutines or functions and their use, together with the explanation of their arguments, is given. A few, more sophisticated algorithms are presented in the form of Algol procedures together with the explanation of their formal parameters. In a sense, this chapter is an extension of the earlier contribution by Wood [1], which appeared in Vol. 1A of this series.

The concepts discussed there will not be dealt with here in detail but will only be referred to briefly where necessary. For more details, appropriate text books should be consulted, e.g. refs. 2—8.

2. Some basic concepts of mathematical statistics

(A) RANDOM ERRORS

Analysts usually carry out several parallel measurements when determining the composition of a sample. The results can be judged from two points of view depending on whether one is interested in their reproducibility (or precision) or their accuracy.

The precision of a method depends on the random errors. If they are high, the reproducibility and the precision are low. The experimental random errors of an analytical method cause variations in the results obtained when successive analyses of the same sample are made in exactly the same way by the same analyst [1].

(B) MEASURES OF CENTRAL TENDENCY

If we wish to express the information contained in a set of results of parallel measurements, we can do this by calculating one of the mean values defined below.

The first is the *arithmetic mean*, often simply called the mean, or average, which is defined by

$$\overline{x} = \frac{1}{n} \sum_{i=1}^{n} x_i$$

where x_i ($i = 1, 2, ..., n$) are the individual analytical results and n the number of parallel results (repetitions). The mean is often denoted by m.

The *geometric mean* is defined by

$$\overline{x}_g = \sqrt[n]{x_1 x, ..., x_n}$$

It is also called the *logarithmic mean* because of the correlation

$$\log \overline{x}_g = \frac{\sum_{i=1}^{n} \log x_i}{n}$$

The geometric mean is sometimes used in statistical calculations. In analytical practice, it is often calculated unintentionally, e.g. if the measured parameter (such as pH) is a logarithmic function of the concentration. A hydrogen ion concentration calculated from the average of the measured pH values is, in fact, the geometric mean of the individual concentrations.

The *harmonic mean* is defined by

$$\overline{x}_h = \frac{n}{\sum_{i=1}^{n} (1/x_i)}$$

2

or

$$\frac{1}{\overline{x}_h} = \frac{\sum\limits_{i=1}^{n} (1/x_i)}{n}$$

In some cases, for example in catalytic methods of analysis, the analytical signal (reaction time) is in reciprocal correlation with the concentration. Thus, averaging parallel values of the measured reaction times, the corresponding concentration value becomes the harmonic mean of the individual concentrations.

Another important quantity which must be mentioned here is the *median*. For its definition, the measured values must be arranged in increasing numerical order, i.e. $x_1 < x_2 \ldots < x_n$. If n is an odd number, then the median is $x_{(n+1)/2}$, i.e. the middle value. If n is an even number, the median is the arithmetic mean of the two middle values, viz.

$$\text{median} = \frac{x_{n/2} + x_{(n/2)+1}}{2}$$

In contrast to the arithmetic mean, the median is insensitive to extreme results. It can be used for characterizing small data sets ($n < 10$) which contain both extremely low and extremely high data.

The *mode* is the most frequently occurring value in a set of data. Whilst it is sometimes used in ordinary statistics to characterize sets, its use in analytical chemistry is not significant.

(C) MEASURES OF DISPERSION

The second quantity which is used to characterize a set of results gives information on the extent to which the individual results differ from the mean. Thus, this parameter gives a measure of the dispersion, or variability, or spread.

(1) Variance and standard deviation

The most important parameter of this kind is the *variance*, s^2, defined by

$$s^2 = \frac{\sum\limits_{i=1}^{n} (x_i - \overline{x})^2}{n-1}$$

The square root of the variance, i.e. s, is the standard deviation *

$$s = \sqrt{\frac{\sum\limits_{i=1}^{n} (x_i - \overline{x})^2}{n-1}}$$

The term $n-1$ in the denominator is called the *degree of freedom* and is sometimes denoted by f. The logic of this nomenclature becomes apparent if one considers that a given $\overline{x}$ can be calculated from $n-1$ freely chosen values of x_i; the nth value however, cannot be chosen freely without altering the value of $\overline{x}$. It must be emphasized that the degree of freedom does not always equal $n-1$ (cf. linear regression analysis).

If results can be grouped, the formula for calculating the standard deviation is different. Suppose that m samples have been analyzed and n parallel determinations carried out on each sample. The results are denoted by x_{ij} where the first index, i, refers to the sample and the second, j, to the jth parallel determination. The mean of the results obtained for the ith sample is $\overline{x}_i$. The standard deviation in this case is calculated as

$$s = \sqrt{\frac{1}{m(n-1)} \sum_{i=1}^{m} \sum_{j=1}^{n} (x_{ij} - \overline{x}_i)^2}$$

Here, the number of degrees of freedom is $m(n-1)$.

(2) Ranges

Further measures of scattering are the range and the semi-inter-quartile range. The *range* is the difference of the greatest and smallest

---

* In the statistical literature, the symbol σ is often used to denote the standard deviation. In the present text, σ will be used for the *population standard deviation*, i.e. a figure obtained from an infinitely large sample. For the standard deviation of a sample of limited size, the symbol s will be used throughout. The sample standard deviation, s, is therefore an estimate of the population standard deviation, σ. Thus, the larger the sample, i.e. the more parallel results available, the better s approximates σ.

A similar distinction will be made between the *population* (theoretical) *mean*, μ, and the sample mean $\overline{x}$.

4

results

$$R = x_n - x_1$$

The range is especially advantageous in characterizing the scattering if the number of results is small ($n < 10$).

In order to define the *semi-interquartile range*, the following quantities will be introduced.

The *quartiles* are the points on the x-scale (scale of results) that divide the results into four equal parts. The lower quartile, Q_1, is the point on the x-scale such that a quarter of the results is less than Q_1. The upper quartile, Q_3, is the point on the x-scale such that three-quarters of the total frequency is below Q_3.

The semi-interquartile range is

$$Q = \frac{Q_3 - Q_1}{2}$$

The greater this range, the greater the dispersion.

(3) The calculation of means and standard deviations by on-line methods

The need of such calculations arises when measurements are carried out with the aid of an instrument controlled by a computer. For example, it can occur that not all measured values should be stored, that only the mean and the standard deviation must be registered continuously or from time to time.

The method of calculation is as follows. Suppose that the mean and the sum of squares of data x_1, x_2, ..., x_{n-1} are known. Let them be m_{n-1}, and z_{n-1}, respectively. Adding the subsequent element x_n to the set, the mean, m_n, and the standard deviation, s_n, are calculated by the iterative formulae

$$m_n = \frac{1}{n}[(n-1)m_{n-1} + x_n]$$

$$s_n = \sqrt{\frac{z_n}{n} - m_n^2}$$

where $z_n = z_{n-1} + x_n^2$.

The subroutine MNON calculates the mean and the standard deviation of measured data got by an on-line method. The results are modified continuously at the arrival of a new measured value. At the first call, some of the formal parameters (N, AM, Z) must be set at zero.

The listing of the subroutine is

```
SUBROUTINE MNON(X,AM,S,N,Z)
U=N+1.0
AM=(AM*(U-1.0)+X)/U
Z=Z+X*X
S=SQRT(Z/U-AM*AM)
N=U
RETJRN
END
```

The parameters are
X   the subsequent element of the set to be evaluated
AM   the mean
S   the standard deviation
N   the number of elements already evaluated
Z   the sum of squares

(D) THE DISTRIBUTION OF ANALYTICAL RESULTS

(1) Histograms

In order to obtain a survey on the distribution of results, they can be tabulated in the form of relative frequencies. In other words, the results are described in terms of grouped data. For this purpose, the range of measurements is partitioned into conveniently chosen, in most cases equidistant, *class intervals*. If the results x are arranged in increasing order $x_1 < x_2 < ... < x_n$ and the number of intervals is N, then the length of one interval is

$$h = \frac{x_n - x_1}{N}$$

The intervals are characterized by the inequalities

$$x_1 + (j - \tfrac{3}{2})h < x < x_1 + (j - \tfrac{1}{2})h$$

where $j = 1, 2, ..., N$. The intervals are also characterized by their central values $X_1, X_2, ..., X_n$ which are expressed in terms of x_i and h as

$$X_1 = x_1, X_2 = x_1 + h, ..., X_j = x_1 + (j-1)h, ..., X_N = x_1 + (N-1)h$$

6

The *class frequency*, n_j is the number of times an x_k falls into the jth class interval. The *relative frequencies* of results in the jth class interval are $\nu_j = n_j/n$.

The cumulative distribution function is a non-decreasing step function

$$F_j = \sum_{i=1}^{j} \nu_i$$

giving the sum of relative frequencies from the first interval up to the jth. It is obvious that $F_1 = \nu_1$ and $F_N = 1$.

The results of 100 parallel determinations of the carbon content of an unknown material in per cent of C, arranged in increasing order is given in Table 1. The frequency distribution of the same results is given in Table 2. The results are represented in graphical form in Fig. 1. Each column is of a width proportional to the class interval and its height is proportional to the results in that class or to the relative frequencies. Such a diagram is called a *histogram*.

The parameters characterizing data sets, mentioned above are calculated for grouped data from the relative frequencies, ν_i in the following way.

TABLE 1

Results of 100 parallel determinations of the carbon content of an unknown material, arranged in increasing order (% C)

32.0	32.2	32.2	32.3	32.3	32.4	32.4
32.4	32.5	32.5	32.5	32.6	32.6	32.6
32.7	32.7	32.7	32.7	32.7	32.8	32.8
32.8	32.8	32.8	32.8	32.8	32.9	32.9
32.9	32.9	32.9	32.9	32.9	32.9	32.9
32.9	32.9	32.9	32.9	33.0	33.0	33.0
33.0	33.0	33.0	33.0	33.0	33.0	33.0
33.0	33.0	33.0	33.0	33.1	33.1	33.1
33.1	33.1	33.1	33.1	33.1	33.1	33.1
33.1	33.1	33.2	33.2	33.2	33.2	33.2
33.2	33.3	33.3	33.3	33.3	33.3	33.3
33.3	33.4	33.4	33.4	33.4	33.5	33.5
33.5	33.5	33.5	33.6	33.6	33.6	33.6
33.6	33.7	33.7	33.7	33.8	33.8	33.9
33.9	34.0					

TABLE 2

The frequency distribution of the results given in Table 1

Carbon content (%) (class)	Tallies	Frequency (F)	Cumulative frequency (ΣF)
32.0	X	0.01	0.01
32.1		0.00	0.01
32.2	X X	0.02	0.03
32.3	X X	0.02	0.05
32.4	X X X	0.03	0.08
32.5	X X X	0.03	0.11
32.6	X X X	0.03	0.14
32.7	X X X X X	0.05	0.19
32.8	X X X X X X X	0.07	0.26
32.9	X X X X X X X X X X X X X	0.13	0.39
33.0	X X X X X X X X X X X X X X	0.14	0.53
33.1	X X X X X X X X X X X X	0.12	0.65
33.2	X X X X X X	0.06	0.71
33.3	X X X X X X X	0.07	0.78
33.4	X X X X	0.04	0.82
33.5	X X X X X	0.05	0.87
33.6	X X X X X	0.05	0.92
33.7	X X X	0.03	0.95
33.8	X X	0.02	0.97
33.9	X X	0.02	0.99
34.0	X	0.01	1.00

The mean value, $\bar{n}$, is given by

$$\bar{x} = \sum_{j=1}^{N} \nu_j X_j$$

and the standard deviation by

$$s = \sqrt{\frac{\sum_{j=1}^{N} \nu_j (X_j - \bar{x})^2}{n-1}}$$

Grouped analytical results can be characterized in addition to the different mean values, which are measures of central tendency, and

8

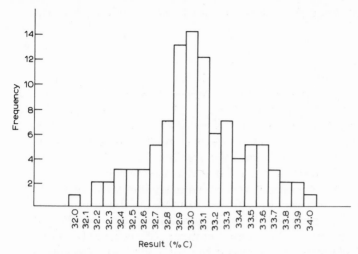

Fig. 1. Histogram of 100 parallel determinations of the carbon content of an unknown material.

the standard deviation which is the measure of dispersion, by two further quantities.

One of them is the measure of *skewness* or asymmetry. The skewness is defined by

$$\rho = \frac{\sum\limits_{j=1}^{N} v_j (X_j - \overline{x})^3}{s^3}$$

If results are distributed symmetrically, then $\rho = 0$. The distribution is asymmetric towards the left-hand side if $\rho > 0$ and towards the right-hand side if $\rho < 0$.

An important constant to the summarized description of grouped data in the immediate neighborhood of the mean is the measure of excess or *curtosis*, which is defined by

$$K = \frac{\sum\limits_{j=1}^{N} v_j (X_j - \overline{x})^4}{s^4} - 3$$

This K is a measure of relative flatness (or peakedness). If $K > 0$, then the relative frequencies near the mean are higher than in the case $K < 0$.

(2) Density functions

The relative frequencies can also be represented in graphical form by the so-called *frequency polygon* (Fig. 2). This is a graph expressing the functionality between the central values of each class interval and the relative frequencies. The cumulative relative frequencies are presented graphically on the *cumulative frequency distribution curve* which is based on the data of Table 2 and is shown in Fig. 3.

If the number of observations was enormously increased, intuition predicts that the staircase-like pattern formed by the tops of the columns in the histogram would become less irregular, especially if advantage were taken of the large number of measurements to diminish the class intervals, increasing the total number of columns accordingly. In the limit, as the number of measurements approaches infinity, the columns become vertical lines and their upper extremities join to form a smooth curve, the theoretical curve fitting the frequency distribution (Fig. 2).

The equation of the curve fitting the frequency distribution is

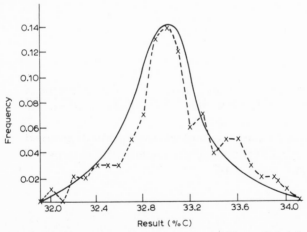

Fig. 2. Frequency distribution curve of 100 parallel determinations of the carbon content of an unknown material. - - - - - -, Frequency polygon; ———, frequency curve of normally distributed results.

10

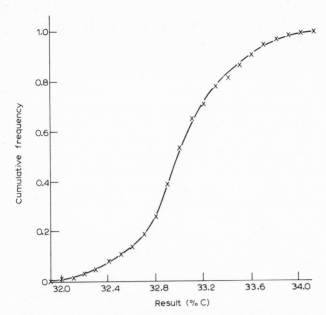

Fig. 3. Cumulative frequency distribution curve of 100 parallel determinations of the carbon content of an unknown material.

called *probability density function* and will be denoted by $\varphi(x)$. The probability that the result of a measurement x does not exceed the value X is

$$P(x \leqslant X) = \int_{-\infty}^{X} \varphi(x)\, dx$$

and the probability that x lies within the interval (x_1, x_2) is

$$P(x_1 \leqslant x \leqslant x_2) = \int_{x_1}^{x_2} \varphi(x)\, dx$$

It is obvious that

$$\int_{-\infty}^{\infty} \varphi(x)\, dx = 1$$

The variable of the probability density functions can be either discrete or continuous. The most important probability density functions will be discussed below.

(E) SOME IMPORTANT PROBABILITY DISTRIBUTIONS

(1) Discrete probability distributions

a. *The binomial distribution.* Let us assume that a number of samples is acceptable or it is not. It is supposed that the probability of acceptance is 1/2 and that it is the same for rejection. If one sample is investigated, $n = 1$, then the overall probability, i.e. the probability of either accepting or rejecting, is

$$P_1 = \tfrac{1}{2} + \tfrac{1}{2} = 1$$

when two samples are investigated, $n = 2$, then there are four possible cases, viz. both are acceptable, only the first is acceptable, only the second is acceptable, or both are rejected. It is obvious that the probability of each case is equal to 1/4. The probabilities that the number of acceptable samples is 2, 1, and 0 are equal to 1/4, 1/2, and 1/4 respectively. (The case in which one sample from the two is acceptable can be realized in two ways.) The overall probability is

$$P_2 = \tfrac{1}{4} + \tfrac{1}{2} + \tfrac{1}{4}$$

or, expressed in powers of 1/2, we obtain

$$P_2 = (\tfrac{1}{2})^2 + 2(\tfrac{1}{2})(\tfrac{1}{2}) + (\tfrac{1}{2})^2 = 1$$

Similarly, for $n = 3$, the probabilities that the number of acceptable samples is 3, 2, 1, and 0 are equal to 1/8, 3/8, 3/8, and 1/8, respectively. The overall probability is

$$P_3 = (\tfrac{1}{2})^3 + 3(\tfrac{1}{2})^2(\tfrac{1}{2}) + 3(\tfrac{1}{2})(\tfrac{1}{2})^2 + (\tfrac{1}{2})^3 = 1$$

In general, the overall probability P_n can be written as

$$P_n = \sum_{i=0}^{n} \binom{n}{i}(\tfrac{1}{2})^{n-i}(\tfrac{1}{2})^i = 1$$

where the symbol $\binom{n}{i}$ denotes the binomial coefficients assuming that n and i are positive integers and $i \leqslant n$.

$$\binom{n}{i} = \frac{n(n-1)\dots(n-i+1)}{1 \times 2 \dots \times i}$$

12

Let us now consider a more general case. An experiment consists of n independent trials in which the probability that any one sample is acceptable is equal to p. The probability that in a series of n samples the number of accepted samples is equal to x is

$$P(x) = \binom{n}{x} p^x q^{n-x} \tag{1}$$

where $x = 0, 1, 2, ..., n$ and $q = 1 - p$

Another example is a radioactive counter. Let us suppose that when it is switched on for a given interval of time, e.g. for a minute, N signals are received. If the probability that $0 \leqslant N < 10$ is denoted by p, the probability that $N \geqslant 10$ is $q = 1 - p$.

The probability that on repeating the experiment n times the number of counts is, in x cases, less than 10 can also be expressed by eqn. (1). Equation (1) is the probability density function of the binomial distribution. The mean and the standard deviation of the binomial distribution are $m = np$ and $s = \sqrt{npq}$.

b. *The Poisson distribution.* The calculation of the probabilities from eqn. (1) according to the binomial distribution is very time consuming when n is a large number. The binomial distribution can be well approximated for $p \leqslant 0.1$ and $n > 30$ by the Poisson distribution. It can be demonstrated that

$$\lim_{n \to \infty} \binom{n}{x} p^x (1-p)^{n-x} = \frac{\lambda^x}{x!} e^{-\lambda}$$

if p/n tends towards 0 in such a manner that np has a finite limit, λ, the parameter of the Poisson distribution. The mean and the standard deviation are $m = \lambda$ and $s = \sqrt{\lambda}$.

The graphs of the Poisson distribution for $\lambda = 0.8$ and $\lambda = 3.5$ are shown in Fig. 4.

The Poisson distribution is known as the law of small numbers.

(2) Continuous distributions

a. *The normal distribution.* The *normal* or *Gaussian* distribution is characterized by the density function

$$\varphi(x) = \frac{1}{\sigma\sqrt{2\pi}} \exp\left(-\frac{(x-\mu)^2}{2\sigma^2}\right) \tag{2}$$

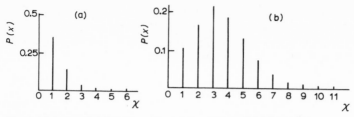

Fig. 4. Poisson distributions for (a) $\lambda = 0.8$ and (b) $\lambda = 3.5$.

In this equation, x is the numerical value of the observation, μ is the mid-value of x (this is both the mean and the median), and the constant or parameter σ is the standard deviation.

The graph of the normal distribution has been shown in Fig. 2 together with the frequency polygon of 100 parallel determinations given in Table 1.

The normal distribution is very important because the distribution of analytical results follows this probability density function in almost every case.

b. The Student or t-distribution. As has been mentioned above, the frequency distribution curve of analytical results can be fitted by the probability density function of the normal distribution only if the number of measurements tends to infinity. In order to make predictions from a finite number of measurements, the *Student* or *t-distribution* must be used.

The probability density function of this distribution is defined by

$$\varphi(t) = \frac{\Gamma[(f+1)/2]}{\Gamma(f/2)\sqrt{\pi f}} \left(1 + \frac{t^2}{f}\right)^{-(f+1)/2} \tag{3}$$

where $t = (x - \overline{x})/s$. The function also depends on a parameter, f, the degrees of freedom. Degrees of freedom can be defined as the number of observations made in excess of the minimum theoretically necessary to estimate a statistical parameter or any unknown quantity. In the case of the t-distribution, the degrees of freedom are one less than the number of measurements, i.e. $f = n - 1$.

The function Γ is the so-called gamma function, defined, in general, by

$$\Gamma(z) = \int_0^\infty e^{-u} u^{z-1} \, du$$

14

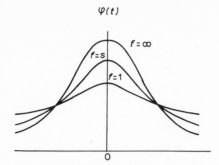

$\varphi(t)$

$f = \infty$

$f = s$

$f = 1$

O

Fig. 5. The Student or t-distribution.

(Special cases of the gamma function are $\Gamma(1) = 1$, $\Gamma(1/2) = \sqrt{\pi}$, $\Gamma(n) = (n-1)!$ where n is a positive integer.) The mean value of the t-distribution is zero; its variance is equal to $f/(f-2)$.

The t-distribution is used for the prediction of the mean from the theoretical mean when the standard deviation is not known and the number of measurements is small. It approaches a normal distribution asymptotically as f tends to infinity. The density plot of the t-distribution is shown for some f values in Fig. 5.

c. *The chi-square, χ^2, distribution.* Let us consider n analytical results, $x_1, x_2, ..., x_n$, being of normal distribution. Transforming these results by the relation

$$u_i = \frac{x_i - \bar{x}}{s}$$

where $i = 1, 2, ..., n$, the mean of the transformed results u_i is zero and their standard deviation is 1. We define the variable χ^2 as the sum of squares of the u_i values.

$$\chi^2 = \sum_{i=1}^{n} u_i^2$$

When taking several data sets and calculating the sum of the squares of the transformed data for each data set, these sums follow the χ^2-distribution.

The density function of the χ^2-distribution is

$$\varphi(\chi^2) = \frac{1}{2^{n/2}\Gamma(n/2)} (\chi^2)^{(n/2)-1} e^{-\chi^2/2} \tag{4}$$

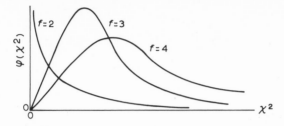

Fig. 6. The χ^2-distribution.

This function is defined for positive χ^2 values. The distribution also depends on the parameter n, the degrees of freedom.

The mean value of the distribution is equal to this parameter and the standard deviation is $\sqrt{2n}$.

The χ^2-distribution is used for determining the standard deviation of the population from which the sample of known variance was taken.

The density plot of the χ^2-distribution is shown on Fig. 6.

d. The Fisher—Snedecor or F-distribution. The probability density function of this distribution is defined by

$$\varphi(F) = \frac{\Gamma[(f_1 + f_2)/2]}{\Gamma(f_1/2)\,\Gamma(f_2/2)} \left(\frac{f_1}{f_2}\right)^{f_1/2} F^{(f_1/2)-1} \left(1 + \frac{f_1}{f_2}F\right)^{-(f_1+f_2)/2} \quad (5)$$

The function $\varphi(F)$ is defined for $F > 0$ and has two degrees of freedom, f_1 and f_2.

This distribution is used for comparing two series of normally distributed analytical results having different standard deviations, s_1 and s_2. The distribution of the variable $F = s_1^2/s_2^2$ is an F-distribution with the degrees of freedom $f_1 = n_1 - 1$ and $f_2 = n_2 - 1$ where n_1

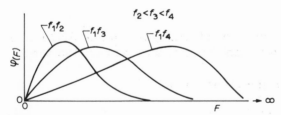

Fig. 7. The F-distribution.

16

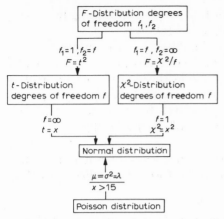

Fig. 8. The connections between some distributions.

and n_2 are the numbers of results in the first and second series, respectively.

The density plot of the F-distribution is shown in Fig. 7.

(3) Relations between distributions

The distributions dealt with above seem to be quite different, but they are connected with each other. The t- and the χ^2-distributions are special cases of the F-distribution and the normal distribution can be regarded as a special case both of the t- and of the χ^2-distributions. The Poisson distribution is a special case of the normal distribution.

A survey of the connections of some of the distributions is shown in Fig. 8.

(4) Confidence levels and confidence intervals

Let us consider the variable x with the density function $\varphi(x)$. The probability, denoted by β, that one measured value of x is within the limits a and b is expressed by the integral of the density function, viz.

$$\beta = P(a \leqslant x \leqslant b) = \int_a^b \varphi(x)\,\mathrm{d}x$$

References pp. 168—169

17

The value $\alpha = 1 - \beta$ is called the confidence level and, using this term, the above statement can be formulated as follows.

The degree of confidence or probability that the interval (a, b), the confidence interval, will contain a random value of x is equal to $1 - \alpha$. The probability that x is outside the confidence interval is equal to α. Otherwise, the value of x lies within the interval (a, b) at level α. The smaller the confidence level for a given confidence interval, the greater the probability that x lies in the confidence interval.

Since the density functions discussed above are not easily integrated, their integrals are published in the form of so-called statistical tables.

a. The normal distribution. Let us first discuss the normal distribution from the point of view of confidence levels and confidence intervals. We want to determine the probability that the value of observation X is greater than b. This probability, denoted by $\alpha = P(X > b)$ is found by integrating the density function (2) from b to ∞.
So

$$\alpha = P(X > b) = \int_{b}^{\infty} \frac{1}{\sigma \sqrt{2\pi}} \exp\left[\frac{(X - \mu)^2}{2\sigma^2} \right] dX \tag{6}$$

This integral is not available in simple form, therefore it would be advantageous to tabulate it. But a separate table would be needed for each (μ, σ) pair. This problem can be overcome by the simple transformation

$$\frac{X - \mu}{\sigma} = x \tag{7}$$

Then $dX = \sigma\, dx$.

Introducing the density function

$$\varphi(x) = \frac{1}{\sqrt{2\pi}} \exp\left(-\frac{x^2}{2}\right)$$

and

$$\frac{b - \mu}{\sigma} = x_\alpha$$

18

eqn. (6) becomes

$$\alpha = P(X > b) = P(x > x_\alpha) = \int\limits_{x_\alpha}^{\infty} \varphi(x)\, dx$$

[The values of α can be calculated by expanding the function $\varphi(x)$ into a power series and integrating term by term.]

The probability α is tabulated versus x_α in Table 3. Defining $x_{\alpha/2}$ by

$$\int\limits_{x_{\alpha/2}}^{\infty} \varphi(x)\, dx = \frac{\alpha}{2}$$

and taking into account that the function $\varphi(x)$ is a symmetrical one, i.e. $\varphi(x) = \varphi(-x)$

$$\int\limits_{-\infty}^{-x_{\alpha/2}} \varphi(x)\, dx = \alpha/2$$

$$P(x < -x_{\alpha/2} \wedge x > x_{\alpha/2}) = \frac{\alpha}{2} + \frac{\alpha}{2} = \alpha$$

The intervals $(0, x_\alpha)$ and $(-x_{\alpha/2}, x_{\alpha/2})$ are called one- and two-sided confidence intervals, respectively, at the level α. It means that the probability that the transformed value of observation x is greater than x_α or is outside the interval $(-x_{\alpha/2}, x_{\alpha/2})$ is equal to α.

Example. What is the probability that the value of observation X lies in the interval $(\mu - k\sigma, \mu + k\sigma)$ where $k = 1-3$?

Making use of transformation (7), the above interval is transformed to $(-k, k)$. The probability $P(-k < x < k)$ is calculated from $P(x > k)$ $P(-k < x < k) = 1 - P(x > k \wedge x < -k) = 1 - 2P(x > k)$ and so $\alpha = 2P(x > k)$.

The results are tabulated on the basis of Table 3 as follows.

k	$P(x > k)$	α	$1 - \alpha$
1	0.1587	0.3174	0.6826
2	0.0228	0.0456	0.9544
3	0.00135	0.0027	0.9973

This means that, within the width of a standard deviation either side

TABLE 3

Normal distribution

x_α	0.00	0.01	0.02	0.03	0.04	0.05	0.06	0.07	0.08	0.09
0.0	0.5000	0.4960	0.4920	0.4880	0.4840	0.4801	0.4761	0.4721	0.4681	0.4641
0.1	0.4602	0.4562	0.4522	0.4483	0.4443	0.4404	0.4364	0.4325	0.4286	0.4247
0.2	0.4207	0.4168	0.4129	0.4090	0.4052	0.4013	0.3974	0.3936	0.3897	0.3859
0.3	0.3821	0.3738	0.3745	0.3707	0.3669	0.3632	0.3594	0.3557	0.3520	0.3483
0.4	0.3446	0.3409	0.3372	0.3336	0.3300	0.3264	0.3228	0.3192	0.3156	0.3121
0.5	0.3085	0.3050	0.3015	0.2981	0.2946	0.2912	0.2877	0.2843	0.2810	0.2776
0.6	0.2743	0.2709	0.2676	0.2643	0.2611	0.2578	0.2546	0.2514	0.2483	0.2451
0.7	0.2420	0.2389	0.2358	0.2327	0.2296	0.2266	0.2236	0.2206	0.2177	0.2148
0.8	0.2119	0.2090	0.2061	0.2033	0.2005	0.1977	0.1949	0.1922	0.1894	0.1867
0.9	0.1841	0.1814	0.1788	0.1762	0.1736	0.1711	0.1685	0.1660	0.1635	0.1611
1.0	0.1587	0.1562	0.1539	0.1515	0.1492	0.1496	0.1446	0.1243	0.1401	0.1379
1.1	0.1357	0.1335	0.1314	0.1292	0.1271	0.1251	0.1230	0.1210	0.1190	0.1170
1.2	0.1151	0.1131	0.1112	0.1093	0.1075	0.1056	0.1038	0.1020	0.1003	0.0985
1.3	0.0968	0.0951	0.0934	0.0918	0.0901	0.0885	0.0869	0.0853	0.0838	0.0823
1.4	0.0808	0.0793	0.0778	0.0764	0.0749	0.0735	0.0721	0.0708	0.0694	0.0681
1.5	0.0668	0.0655	0.0643	0.0630	0.0618	0.0606	0.0594	0.0582	0.0571	0.0559

									0.0188	0.0183
									0.0146	0.0143
									0.0113	0.0110
									0.00866	0.00842
									0.00657	0.00639
									0.00494	0.00480
									0.00368	0.00357
									0.00272	0.00264
									0.00199	0.00193
									0.00144	0.00139
2.0	0.0228	0.0222	0.0217	0.0212	0.0207	0.0202	0.0197	0.0192		
2.1	0.0179	0.0174	0.0170	0.0166	0.0162	0.0158	0.0154	0.0150		
2.2	0.0139	0.0136	0.0132	0.0129	0.0125	0.0122	0.0119	0.0116		
2.3	0.0107	0.0104	0.0102	0.00990	0.00964	0.00939	0.00914	0.00889		
2.4	0.00820	0.00798	0.00776	0.00755	0.00734	0.00714	0.00695	0.00676		
2.5	0.00621	0.00604	0.00587	0.00570	0.00554	0.00539	0.00523	0.00508		
2.6	0.00466	0.00453	0.00440	0.00427	0.00415	0.00402	0.00391	0.00379		
2.7	0.00347	0.00336	0.00326	0.00317	0.00307	0.00298	0.00289	0.00280		
2.8	0.00256	0.00248	0.00240	0.00233	0.00226	0.00219	0.00212	0.00205		
2.9	0.00187	0.00181	0.00175	0.00169	0.00164	0.00159	0.00154	0.00149		
3.0	0.00135									
3.1	0.00099									
3.2	0.00069									
3.3	0.00048									
3.4	0.00034									
3.5	0.00023									
3.6	0.00016									
3.7	0.00011									
3.8	0.00007									
3.9	0.00005									
4.0	0.00003									

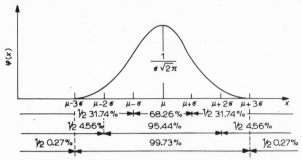

Fig. 9. Confidence intervals of the normal distribution.

of the mean, about $(1 - \alpha)100 = 68\%$ of the observed values are included; if the width is extended to two and three standard deviations either side of the mean, about 95 and 99.7% are included.

The results are also shown in Fig. 9.

b. The t-distribution. When considering the problem of confidence levels and confidence intervals for the *t*-distribution, we suppose that the variable x has been transformed so that its mean value is zero and its standard deviation is unity. The tabulation of $t(\alpha, f)$ for different values of degrees of freedom, f, for some α values is given in Table 4 which is constructed by evaluating the integral

$$\alpha = P[t > t(\alpha, f)] = \int_{t(\alpha, f)}^{\infty} \varphi(x)\,\mathrm{d}x \tag{8}$$

where $\varphi(x)$ is the density function of the *t*-distribution, eqn. (3). When the question what is the probability that the absolute value of the variable t is greater than a given constant, $t(\alpha/2, f)$ is considered, the probability can be expressed as

$$P[|t| > t(\alpha/2, f)] = 2P[t > t(\alpha/2, f)]$$

The probability on the right-hand side is, according to eqn. (8), $\alpha/2$. So

$$P[|t| > t(\alpha/2, f)] = \alpha$$

It follows immediately the probability that t lies within the interval $[-t(\alpha/2, f), t(\alpha/2, f)]$ is $1 - \alpha$. If the two-sided confidence intervals at the level α are to be pointed out, then the t values corresponding to the level $\alpha/2$ must be used.

22

TABLE 4

The critical $t(\alpha, f)$ values of the Student distribution

f	α			
	0.05	0.025	0.01	0.005
1	6.31	12.7	31.8	63.7
2	2.92	4.30	6.97	9.93
3	2.35	3.18	4.54	5.84
4	2.13	2.78	3.75	4.60
5	2.02	2.57	3.37	4.03
6	1.94	2.45	3.14	3.71
7	1.90	2.37	3.00	3.50
8	1.86	2.31	2.90	3.36
9	1.83	2.26	2.82	3.25
10	1.81	2.23	2.76	3.17
11	1.80	2.20	2.72	3.11
12	1.78	2.18	2.68	3.06
13	1.77	2.16	2.65	3.01
14	1.76	2.14	2.62	2.98
15	1.75	2.13	2.60	2.95
16	1.75	2.12	2.58	2.92
17	1.74	2.11	2.57	2.90
18	1.73	2.10	2.55	2.88
19	1.73	2.09	2.54	2.86
20	1.73	2.09	2.53	2.84
21	1.72	2.08	2.52	2.83
22	1.72	2.07	2.51	2.82
23	1.71	2.07	2.50	2.81
24	1.71	2.06	2.49	2.80
25	1.71	2.06	2.49	2.79
26	1.71	2.06	2.48	2.78
27	1.70	2.05	2.47	2.77
28	1.70	2.05	2.47	2.76
29	1.70	2.05	2.46	2.76
30	1.70	2.04	2.46	2.75

Example. How can the results of experiments given below be expressed?

$x_1 = 6.85$ ml	$x_5 = 6.80$ ml	$x_9 = 6.83$ ml
$x_2 = 6.75$ ml	$x_6 = 6.80$ ml	$x_{10} = 6.80$ ml
$x_3 = 6.85$ ml	$x_7 = 6.75$ ml	$x_{11} = 6.75$ ml
$x_4 = 6.80$ ml	$x_8 = 6.70$ ml	$x_{12} = 6.75$ ml

TABLE 5

The critical $\chi^2(\alpha, f)$ values of the χ^2 distribution

f	α		f	α	
	0.05	0.01		0.05	0.01
1	3.84	6.63	16	26.3	32.0
2	5.99	9.21	17	27.6	33.4
3	7.81	11.3	18	28.9	34.8
4	9.49	13.3	19	30.1	36.2
5	11.1	15.1	20	31.4	37.6
6	12.6	16.8	21	32.7	38.9
7	14.1	18.5	22	33.9	40.3
8	15.5	20.1	23	35.2	41.6
9	16.9	21.7	24	36.4	43.0
10	18.3	23.2	25	37.7	44.3
11	19.7	24.7	26	38.9	45.6
12	21.0	26.2	27	40.1	47.0
13	22.4	27.7	28	41.3	48.3
14	23.7	29.1	29	42.6	49.6
15	25.0	30.8	30	43.8	50.9

Here, $\bar{x} = 6.78$ ml, $n = 12$, $f = n - 1 = 11$, $s = 0.047$ ml, $t(0.025, 11) = 2.20$ (Table 4).

$$\bar{x} \pm \frac{t(0.025, 11)}{\sqrt{n}} s = 6.78 \pm \frac{2.20}{\sqrt{12}} 0.047 = 6.78 \pm 0.03 \text{ ml}$$

(The true value lies between 6.75 and 6.81 with 95% probability.)

c. *The χ^2-distribution.* Let us consider the variable χ^2 being of χ^2-distribution. The probability of $\chi^2 > \chi^2(\alpha, f)$, where $\chi^2(\alpha, f)$ is a given constant, is expressed as

$$\alpha = P[\chi^2 > \chi^2(\alpha, f)] = \int_{\chi^2(\alpha, f)}^{\infty} \varphi(\chi^2) \, d\chi^2$$

where $\varphi(\chi^2)$ is the density function of the χ^2-distribution. The tabulation of the values of $\chi^2(\alpha, f)$ for different values of f and for some α values is to be found in Table 5. Note that if $f > 30$, then the approximation

$$\chi^2(\alpha, f) = \tfrac{1}{2}(\sqrt{2f - 1} + x_\alpha)^2$$

24

can be used where x_α is taken from the statistical table of the normal distribution (Table 3).

d. The F-distribution. Suppose that the variable F is of the F-distribution, then the probability of $\alpha = P[F > F(\alpha, f_1, f_2)]$ is given by the integral

$$\alpha = \int_{F(\alpha, f_1, f_2)}^{\infty} \varphi(F) \, \mathrm{d}F$$

where $F(\alpha, f_1, f_2)$ is a fixed constant, f_1 and f_2 are the degrees of freedom and $\varphi(F)$ is the density function of the F-distribution. The critical F values are tabulated versus f_1 and f_2 in terms of the probability α in Table 6.

A very rough approximation of the $F(\alpha, f_1, f_2)$ values is

$$F = \exp\left[\lambda \sqrt{\frac{2(f_1 + f_2)}{f_1 f_2}}\,\right]$$

where $\lambda = 1.645$ for $\alpha = 0.05$, $\lambda = 2.326$ for $\alpha = 0.01$, and $\lambda = 3.090$ for $\alpha = 0.01$.

A better approximation, especially for $f_1 > 24$ and $f_2 > 120$, is

$$F = \exp\left[2\left(\frac{\lambda}{\sqrt{[2f_1 f_2/(f_1 + f_2)]\text{-}\kappa}} - \mu\frac{f_2 - f_1}{f_1 f_2}\right)\right]$$

where λ is the same as above and $\kappa = 1.0$, $\mu = 0.7843$ for $\alpha = 0.05$, $\kappa = 1.4$, $\mu = 1.235$ for $\alpha = 0.01$, and $\kappa = 2.1$, $\mu = 1.925$ for $\alpha = 0.001$ (see ref. 6).

(F) STATISTICAL TESTS

Statistical tests are used to help us to reach some decision about a statistical hypothesis, i.e. a statement about the distribution of parallel analytical results etc. For example, suppose a new analytical method has been introduced for measuring the concentration of an element in a given type of material. It is supposed that the concentration and its standard deviation is already known from prior determinations. We measure the concentration of the element in question by the new method and we wish to test the statements (a) the mean and the standard deviation remain unchanged and (b) the mean remains unchanged while the standard deviation is decreased. Statistical tests are based on calculating some characteristic quantities from

TABLE 6

The critical values of F distribution $F(\alpha, f_1, f_2)$

f_2 \ f_1	3	4	5	6	7	8	10	12	15	20	24	30
$\alpha = 0.05$												
3	9.28	9.12	9.01	8.94	8.89	8.85	8.79	8.74	8.70	8.66	8.64	8.62
4	6.59	6.39	6.26	6.16	6.09	6.04	5.96	5.91	5.86	5.80	5.77	5.75
5	5.41	5.19	5.05	4.95	4.88	4.82	4.74	4.68	4.62	4.56	4.53	4.50
6	4.76	4.53	4.39	4.28	4.21	4.15	4.06	4.00	3.94	3.87	3.84	3.81
7	4.35	4.12	3.97	3.87	3.79	3.73	3.64	3.57	3.51	3.44	3.41	3.38
8	4.07	3.84	3.69	3.58	3.50	3.44	3.35	3.28	3.22	3.15	3.12	3.08
10	3.71	3.48	3.33	3.22	3.14	3.07	2.98	2.91	2.85	2.77	2.74	2.70
12	3.49	3.26	3.11	3.00	2.91	2.85	2.75	2.69	2.62	2.54	2.51	2.47
15	3.29	3.06	2.90	2.79	2.71	2.64	2.54	2.48	2.40	2.33	2.29	2.25
20	3.10	2.87	2.71	2.60	2.51	2.45	2.35	2.28	2.20	2.12	2.08	2.04
24	3.01	2.78	2.62	2.51	2.42	2.36	2.25	2.18	2.11	2.03	1.98	1.94
30	2.92	2.69	2.53	2.42	2.33	2.27	2.16	2.09	2.01	1.93	1.89	1.84
$\alpha = 0.01$												
3	29.46	28.71	28.71	27.91	27.67	27.49	27.23	27.03	26.87	26.69	26.60	26.50
4	16.69	15.98	15.52	15.21	14.98	14.80	14.55	14.37	14.20	14.02	13.93	13.84
5	12.06	11.39	10.97	10.67	10.46	10.29	10.05	9.89	9.72	9.55	9.47	9.38
6	9.78	9.15	8.75	8.47	8.26	8.10	7.87	7.72	7.56	7.40	7.31	7.23
7	8.45	7.85	7.46	7.19	6.99	6.84	6.62	6.47	6.31	6.16	6.07	5.99
8	7.59	7.01	6.63	6.37	6.18	6.03	5.81	5.67	5.52	5.36	5.28	5.20
10	6.55	5.99	5.64	5.39	5.20	5.06	4.85	4.71	4.56	4.41	4.33	4.25
12	5.95	5.41	5.06	4.82	4.64	4.50	4.30	4.16	4.01	3.86	3.78	3.70
15	5.42	4.89	4.56	4.32	4.14	4.00	3.80	3.67	3.52	3.37	3.29	3.21
20	4.94	4.43	4.10	3.87	3.70	3.56	3.37	3.23	3.09	2.94	2.86	2.78
24	4.72	4.22	3.90	3.67	3.50	3.36	3.17	3.03	2.89	2.74	2.66	2.58
30	4.51	4.02	3.70	3.47	3.30	3.17	2.98	2.84	2.70	2.55	2.47	2.39

the measured data and on determining the confidence interval of these quantities at a prescribed confidence level. If the calculated value of this characteristic quantity is outside this interval, then the hypothesis is rejected. It must be emphasized that there do not exist unique conditions of rejecting or accepting a hypothesis. The choice of conditions is, to some extent, subjective. It must be also underlined that statistical tests cannot prove any hypothesis; they can only demonstrate a "lack of disproof".

(1) The most common statistical tests

The most common statistical tests are used to accept or to reject the five hypotheses set out in the following sections.

a. Two results are identical. More precisely, that the difference between the two results, x_1 and x_2, is due to random errors. It is supposed that the standard deviation, s, is known.
The value of u' is calculated from

$$u' = \frac{|x_1 - x_2|}{\sqrt{2}\,s}$$

The hypothesis is accepted if $u' < u$, where $u = 1.96$ if $\alpha = 0.05$, $u = 2.58$ if $\alpha = 0.01$, and $u = 3.00$ if $\alpha = 0.001$.

Note that these u values are the x_α values of the statistical table of normal distribution corresponding to the one-sided level $\alpha/2$ (Table 3).

Example. Are the results $x_1 = 6.85$ and $x_2 = 6.78$ identical at level $\alpha = 0.05$ if $s = 0.047$?

$$u' = \frac{6.85 - 6.78}{\sqrt{2} \times 0.047} = 1.06$$

At level $\alpha/2 = 0.025$, $u = 1.96$. Since $u' < u$, the results can be regarded as identical.

b. The mean is identical with a theoretical mean
The value of t' is calculated from

$$t' = \frac{|\bar{x} - \mu|}{s}\sqrt{n}$$

where s is the standard deviation calculated from the measured data,

$\bar{x}$ is the mean, n is the number of measurements, and μ is the theoretical mean. The hypothesis is accepted if $t' < t$, where $t = t(\alpha, f)$ in which $f = n - 1$ is taken from the statistical table of the t-distribution (Table 4).

Example. The theoretical value of the result of a titration is 6.73 ml. Does the result 6.78 ml determined as the mean of 12 measurements agree with this or is there a significant difference, which indicates a systematic error?

$\bar{x} = 6.78$ ml; $\quad \mu = 6.73$ ml; $\quad s = 0.047$ ml, $\quad n = 12$; $\quad f = n - 1 = 11$; $t(0.025, 11) = 2.20$

Thus

$$t' = \frac{\bar{x} - \mu}{s} \sqrt{n} = \frac{6.78 - 6.73}{0.047} \sqrt{12} = 3.7 > 2.20$$

Since $t' > t$, the difference is significant and there was therefore a positive systematic error committed at each titration.

Example. Supposing the theoretical value is 6.76 ml, how does our conclusion compare with the result given in the previous example?

$\bar{x} = 6.78$ ml; $\quad \mu = 6.76$ ml; $\quad s = 0.047$ ml; $\quad n = 12$; $\quad f = n - 1 = 11$; $t(0.025, 11) = 2.20$.

$$t' = \frac{\bar{x} - \mu}{s} \sqrt{n} = \frac{6.78 - 6.76}{0.047} \sqrt{12} = 1.48 < 2.20$$

Since $t' < t$, the difference is insignificant, and we cannot say that an error was committed when making the measurements.

c. *Two mean values are identical.* Let the two mean values be $\bar{x}_1$ and $\bar{x}_2$ determined from n_1 and n_2 measurements and suppose that the standard deviation of the two sets, s, is identical as proved by the F-test (Sect. e., below).

The value of t' is calculated from

$$t' = \frac{|\bar{x}_1 - \bar{x}_2|}{s} \sqrt{\frac{n_1 n_2}{n_1 + n_2}}$$

and the hypothesis is accepted if $t' < t$ where $t = t(\alpha, f)$ with the degrees of freedom $f = n_1 + n_2 - 2$. This t value is taken from the statistical table of the t-distribution (Table 4).

Example. Two persons obtained the means 18.64 and 18.83, respectively. The number of parallel determinations was $n_1 = 7$ and $n_2 = 13$ and the overall standard deviation, s, was 0.133. Is there a

28

significant difference between the results? $t(0.025, 18) = 2.10$ (taken from Table 4).
We then calculate

$$t' = \frac{|\bar{x}_1 - \bar{x}_2|}{s} \sqrt{\frac{n_1 n_2}{n_1 + n_2}} = \frac{18.83 - 18.64}{0.133} \sqrt{\frac{7 \times 13}{7 + 13}} = 3.2 > 2.10$$

$$f = n_1 + n_2 - 2 = 7 + 13 - 2 = 18$$

Since $t' > t(\alpha, f)$, the two results differ significantly and therefore one of the workers (or both) had systematic errors in his measurements.

Example. Let us examine the question set in the above example but with the values 18.64 and 18.72, respectively. We have the values $n_1 = 7$, $n_2 = 13$, $t(0.025, 18) = 2.10$, $f = n_1 + n_2 - 2 = 18$, and $s = 0.133$.

Thus

$$t' = \frac{|\bar{x}_1 - x_2|}{s} \sqrt{\frac{n_1 n_2}{n_1 + n_2}} = \frac{18.72 - 18.63}{0.133} \sqrt{\frac{7 \times 13}{7 + 13}} = 1.29 < 2.10$$

i.e. there is no significant difference between the two values.
We can now calculate the overall mean.

$$\bar{x} = \frac{n_1 \bar{x}_1 + n_2 \bar{x}_2}{n_1 + n_2} = \frac{7 \times 18.72 + 13 \times 18.64}{20} = \frac{131.04 + 242.32}{20} = 18.67$$

d. The standard deviation is identical with a theoretical standard deviation. If the standard deviation is s, calculated from n measurements, the value of χ'^2 is obtained from

$$\chi'^2 = (n - 1) \frac{s^2}{\sigma^2}$$

where σ is the theoretical standard deviation.
The hypothesis is accepted if $\chi'^2 < \chi^2$, where $\chi^2 = \chi^2(\alpha, f)$, with the degrees of freedom $f = n - 1$. This χ^2 value is taken from the statistical table of the χ^2-distribution (Table 5).

Example. The standard deviation of an analytical method is known to be ± 0.15 mg. Somebody, using the same method, makes 7 determinations and obtains a standard deviation of ± 0.20 mg. Did the person work with an inadequate precision?

We have the values $\sigma = 0.15$, $n = 7$, $\chi^2(0.05, 6) = 12.6$ (taken from Table 5), $s = 0.20$, and $f = n - 1 = 6$.

By calculation

$$\chi'^2 = (n-1)\frac{s^2}{\sigma^2} = 6\frac{0.20^2}{0.15^2} = 10.6 < 12.6$$

Since $\chi'^2 < \chi^2(\alpha, f)$, the precision is acceptable.

Example. Setting the same question as in the above example, but supposing that the standard deviation obtained is ±0.23 mg, we have the values $\sigma = 0.15$, $n = 7$, $s = 0.23$, $f = n - 1 = 6$, and $\chi^2(0.05, 6) = 12.6$.

We then calculate

$$\chi'^2 = (n-1)\frac{s^2}{\sigma^2} = 6\frac{0.23^2}{0.15^2} = 14.1 > 12.6$$

Since $\chi'^2 > \chi^2(\alpha, f)$, the decrease of precision is significant.

e. Two standard deviations are identical. If the two standard deviations are s_1 and s_2, determined from n_1 and n_2 measurements, respectively, the value of F' is calculated from

$$F' = \frac{s_1^2}{s_2^2}$$

The indices 1 and 2 must be chosen so that $F' > 1$.

The hypothesis is accepted if $F' < F$, where $F = F(\alpha, f_1, f_2)$ with $f_1 = n_1 - 1$ and $f_2 = n_2 - 1$. This F value is taken from the statistical table of the F-distribution (Table 6).

Example. One person did 13 parallel measurements and obtained results with a standard deviation of 0.12. Another person did 7 measurements for which the standard deviation was 0.16. Do the two persons work with identical precision, or is the first person's work more reliable than that of the second?

$$s_1 = 0.16 , \quad n_1 = 7 , \quad f_1 = n_1 - 1 = 6 ,$$

$$s_2 = 0.12 , \quad n_2 = 13 , \quad f_2 = n_2 - 1 = 12$$

$F(0.05, 6, 12) = 3.00$ (taken from Table 6)

(N.B. s_1 and s_2 must be chosen so that $F' > 1$.)

$$F' = \frac{s_1^2}{s_2^2} = \frac{0.16^2}{0.12^2} = 1.78 < 3.0$$

30

Thus, $F' < F(\alpha, f_1, f_2)$ and the two standard deviations have to be accepted as identical.

Example. A third person has made 13 parallel determinations with a standard deviation of 0.20. How does this standard deviation compare with that of the first person in the example above?

$s_1 = 0.20$, $n_1 = 13$, $f_1 = 12$,

$F(0.05, 12, 12) = 2.7$ (taken from Table 6) ,

$s_2 = 0.12$, $n_2 = 13$, $f_2 = 12$,

Thus

$$F' = \frac{s_1^2}{s_2^2} = \frac{0.20^2}{0.12^2} = 2.78 > 2.69$$

Since $F' > F(\alpha, f_1, f_2)$, the standard deviation $s = 0.20$ is unacceptable and the third person's work is unreliable.

(2) Processing data for discordant results

When repeating a measurement, it can occur that one result is far removed from the remaining values. Such an extraneous value can occasionally be caused by a random error, but it is better to suppose that this fact cannot be explained. Analysts are, in such cases, of the opinion that this value does not come from the same population and therefore it is called a discordant result.

In such a case, one has to decide whether the extraneous result is caused by random errors or if it really is a discordant one. Discordant results must be removed from the data set or, preferably, they should be replaced by new measured data.

A suitable method for eliminating discordant results has been worked out by Dixon and his coworkers [9—11]. The essential step of the method is to calculate the quotient of some ranges and subranges, of measured data. The quotient is usually denoted by Q and the method is therefore also called the Q test. Let us consider the ordered measured data such as $x_1 < x_2 < ... < x_n$. In the further discussion, the value x_1 is investigated to determine whether it is a discordant result. The range of data is $x_n - x_1$ and sub-ranges are, for example, $x_2 - x_1$, $x_3 - x_1$, $x_{n-1} - x_1$, and $x_{n-2} - x_1$. The ranges and sub-ranges taken into account when calculating the quotient Q is deter-

TABLE 7

Indices to be used in eqn. (9)

n	Indices			
	i	j	k	l
3— 7	2	1	n	1
8—10	2	1	$n - 1$	1
11—13	3	1	$n - 1$	1
14—40	3	1	$n - 1$	1

mined by the number of data. The quotient Q is defined by

$$Q = \frac{x_i - x_j}{x_k - x_l} \tag{9}$$

The indices i, j, k, l are summarized in terms of n in Table 7.

If the value x_n is suspect, then the indices, denoted generally by m, are obtained from $m = n + 1 - p$ where p is the proper index in Table 7.

The ratio Q can be calculated by the function RAT. Its heading is

FUNCTION RAT (I, N, A)

DIMENSION A (N)

and the parameters are

    I = 1 the lowest value is suspect

        2 the greatest value is suspect

    N the number of data

    A the array of data

The function value is the ratio Q defined by eqn. (9) and the listing is

```
FUNCTION RAT(I,N,A)
DIMENSION A(N)
K1=4
IF(N.LT.14) K1=3
IF(N.LT.11) K1=2
IF(N.LT.8) K1=1
K2=2*(K1-1)/3
K3=ISIGN(1,2*K1-5)
IF(I-1)3,2,3
2 J1=2.5+0.5*K3
J2=1
J3=N-K2
J4=1
```

32

```
   GO TO 4
3  J1=N
   J2=N-1.5-0.5*K3
   J3=N
   J4=1+K2
4  AM=A(J3)-A(J4)
   IF(ABS(AM)-1.0E-20)5,6,6
5  RAT=0.0
   RETURN
6  AM=(A(J1)-A(J2))/AM
   RAT=AM
   RETURN
   END
```

Having calculated the ratio Q, the further step is to compare it with the theoretical or critical Q value. The critical Q values referring to the level $\alpha = 0.05$ are summarized in Table 8 (see ref. 11). If the calculated Q value exceeds the tabulated value, the questionable observation must be rejected.

When carrying out the calculations by a computer, it is better not to store the tabulated Q values but to fit them according to Sutarno's idea [12] in the form

$$\ln Q = a + b \ln(\ln n)$$

and to store the coefficients a and b. The coefficients are different for the intervals of n (3, 7), (8, 10), (11, 13), and (14, 40).

The subroutine DIXON has been worked out for calculating the

TABLE 8

Critical Q values

n	Q	n	Q	n	Q
3	0.941	12	0.546	21	0.440
4	0.765	13	0.521	22	0.430
5	0.642	14	0.546	23	0.421
6	0.560	15	0.525	24	0.413
7	0.507	16	0.507	25	0.406
8	0.554	17	0.490	26	0.399
9	0.512	18	0.475	27	0.393
10	0.477	19	0.462	28	0.387
11	0.576	20	0.450	29	0.381
				30	0.376

critical Q values in terms of n. The heading is

SUBROUTINE DIXON (D, N, IE)

The parameters are
D the critical Q value
N the number of data
IE = 1 if N < 3 or N > 40
       0 else
The listing is

```
SUBROUTINE DIXON(D,N,IE)
IE=0
IF(N.LT.3) GO TO 6
IF(N-13)1,1,3
1  IF(N-10)2,2,4
2  IF(N-7) 9,9,8
9  IF(N.EO.3) D=0.941
   IF(N.EO.4) D=0.765
   IF(N.EO.5) D=0.642
   IF(N.EO.6) D=0.560
   IF(N.EO.7) D=0.507
   RETURN
3  A=0.8879695?
   B=-1.531932
   GO TO 5
4  A=0.66539010
   B=-1.3039497
   GO TO 5
8  A=0.41782492
   B=-1.3706695
5  XN=N
   D=B*ALOG(ALOG(XN))+A
   D=EXP(D)
   RETURN
6  IE=1
   RETURN
   END
```

Let us consider the example due to Dean and Dixon [9]. The following series of observations, arranged in order of magnitude, represents calculated percentages of sodium oxide in soda ash. $x_1 = 40.02$, $x_2 = 40.12$, $x_3 = 40.16$, $x_4 = 40.18$, $x_5 = 40.18$, and $x_6 = 40.20$. Supposing that x_1 is questionable, we calculate the ratio

$$Q = \frac{x_2 - x_1}{x_6 - x_1} = \frac{40.12 - 40.02}{40.20 - 40.02} = \frac{0.10}{0.18} = 0.56$$

34

When x_6 is questionable, the ratio is

$$Q = \frac{x_6 - x_5}{x_6 - x_1} = \frac{40.20 - 40.18}{40.20 - 40.02} = \frac{0.02}{0.18} = 0.11$$

The first Q value is equal to the tabulated value of 0.56 and so we reject the value 40.02. The second is essentially lower than the tabulated value and so the value 40.20 is not a discordant result.

(3) Judging data sets for homogeneity

A series of observations is considered as homogeneous if the variance does not exceed a given limit, which depends on the analytical method and on the concentration range. Data sets can be judged for homogeneity by carrying out a test for normality, but such a test is very time-consuming. Instead of this, it is simpler to apply the generalization of the Dixon method [see Sect. 2.(F)(2)] for cancelling discordant results [13].

First, the values x_1 and x_n are investigated. If x_1 or x_n, or both, are discordant results, then they will be cancelled and the corresponding indices recorded. This procedure will be repeated for the remaining subset of data. If neither x_1 nor x_n are discordant results, both are cancelled and the procedure repeated until the number of remaining data exceeds half the number of original results. If the maximum difference of indices recorded is much greater than $n/2$, the the subset of data characterized by these indices can be regarded as a homogeneous one and the mean and standard deviation will be calculated from this subset. The mean and standard deviation calculated in this way characterize the quantity to be measured and its accuracy better than the mean and standard deviation of the whole set of data.

The calculation flow sheet of the method is shown in Fig. 10. The calculation can be carried out by the subroutine DEAN. The declaration of the subroutine is

SUBROUTINE DEAN (N, M, A, IB, IE).

The formal parameters are
N   the number of data
M   the number of discordant results
A(N)   the data array
IB(N) integer array for indicating the discordant results. If A(I) is a discordant result, then IB(I) = 1, else IB(I) = 0.

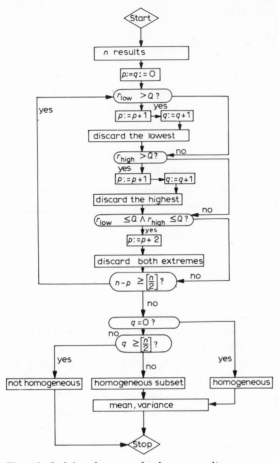

Fig. 10. Judging data sets for homogeneity.

IE   an integer giving the conclusion of the data processing.
IE = 0, the data set is inhomogeneous
IE = 1, the data set is homogeneous
IE = 2, there exists a homogeneous subset
IE = 3, if $N \leqslant 3$ or $N > 40$
The subroutine calls the following subprograms

FUNCTION RAT and SUBROUTINE DIXON

[See Sect. 2.(F)(2)].

36

The listing of the subroutine is

```
 .SUBROUTINE DEAN(N,M,A,IB,IE)
  DIMENSION A(N),IB(N),AA(40),IR(40),IU(40),AR(40),IVA(41),
 XID(41)
  LOGICAL BA,BF
  IF(N.GT.3.AND.N.LE.40) GO TO 1
  IE=3
  RETURN
1 DO 2 I=1,N
2 IB(I)=0
  DO 3 J=1,N
  IR(J)=J
3 AA(J)=A(J)
  N1=N-1
  DO 6 J=1,N1
  JJ=J+1
  DO 5 K=JJ,N
  IF(AA(J)-AA(K))5,5,4
4 AM=AA(J)
  AA(J)=AA(K)
  AA(K)=AM
  I=IR(J)
  IR(J)=IR(K)
  IR(K)=I
5 CONTINUE
6 CONTINUE
  IK=1
  IV=N
  DO 7 J=1,N
7 IU(J)=0
  IU(N+1)=1
  IH=N
  I=0
8 I=I+1
  IF(IH-3)18,18,9
9 IF(2*IH-N) 18,10,10
10 J=0
  DO 11 K=IK,IV
  J=J+1
11 AR(J)=AA(K)
  AM=RAT(1,IH,AR)
  CALL DIXON(0,IH,IX)
  BA=AM.GT.0
  AM=RAT(2,IH,AR)
  BF=AM.GT.0
  IF(BA) IU(IK+1)=1
  IF(BF) IU(IV)=1
  IF(BF) GO TO 13
  IF(BA) GO TO 13
12 IK=IK+1
  GO TO 14
13 IF(BA) GO TO 12
```

```
      CONTINUE
   14 IF(BA) GO TO 16
      IF(BF) GO TO 16
   15 IV=IV-1
      GO TO 17
   16 IF(BF) GO TO 15
   17 IF=IV-IK+1
      GO TO 8
   18 N1=N+1
      DO 19 K=2,N1
   19 IVA(K)=0
      DO 29 K=3,N1
   29 ID(K)=0
      IVA(1)=1
      J=1
      DO 21 K=2,N1
      IF(IU(K))20,21,20
   20 J=J+1
      IVA(J)=K
   21 CONTINUE
      DO 22 K=2,N1
   22 ID(K)=IVA(K)-IVA(K-1)
      IP=-10
      DO 24 K=2,N1
      IV=ID(K)
      IF(IV-IP)24,24,23
   23 IZ=K
      IP=IV
   24 M=N-IP
      IF(2*IP-N) 25,26,26
   25 IE=0
      RETURN
   26 JF=IVA(IZ-1)
      IF(JP.LT.2.AND.(IP+JP).GT.N) GO TO 31
      IE=2
      J1=JP-1
      IF(J1) 32,32,30
   30 DO 27 K=1,J1
      I=IR(K)
   27 IB(I)=1
   32 J2=IP+JP
      IF(N-J2) 33,34,34
   33 RETURN
   34 DO 28 K=J2,N
      I=IR(K)
      IB(I)=1
   28 CONTINUE
      RETURN
   31 IE=1
      RETURN
      END
```

(4) Test for normality

The test for normality is a procedure to determine whether a set of data can be regarded as a population of normal distribution.

Let us suppose that the data x_i ($i = 1, 2, ..., n$) are independent and are arranged in increasing order: $x_1 \leqslant x_2 \leqslant ... \leqslant x_n$. The data are first partitioned into a convenient number of equal intervals. In practice, the number of intervals, N, is chosen so that it is the greatest integer not exceeding $\sqrt{n}$, but is at least 8. The length of one interval is $h = (x_n - x_1)/N$. The end-points of the intervals will be denoted by $X_0, X_1, ..., X_N$ and $X_i = X_0 + jh$ ($j = 1, 2, ..., N$). We calculate the class frequencies, i.e. the numbers ν_j ($j = 1, 2, ..., N$). For a given value of j, ν_j is the number of times an x_i falls into the jth interval. The end points of the intervals X_j are transformed by

$$u_j = \frac{X_j - \bar{x}}{s}$$

where $j = 0, 1, 2, ..., N$, $\bar{x}$ is the arithmetic mean of the data, and s is their standard deviation. These u_j values are substituted into the function $\phi(x)$ which is the integral of the Gaussian probability density function from $-\infty$ to x.

$$\phi(x) = \frac{1}{\sqrt{2\pi}} \int_{-\infty}^{x} \exp(-t^2/2) \, dt \tag{10}$$

These $\phi(x)$ values can be calculated from the α values corresponding to x in Table 3 by $\phi(x) = 1 - \alpha$, or by the function PHI(X) given below. The parameter X corresponds to the variable x in eqn. (10). The listing of the function is

```
      FUNCTION PHI(X)
      IF(X.GE.20.) GO TO 10
      IF(X.LE.-20.) GO TO 11
      P=3.14159265
      AR=SQRT(2*P)
      I=0
      A=X
      PHI=X
12    I=I+1
      C=-X*X*FLOAT(2*I-1)/2/I/(2*I+1)
      A=C*A
      PHI=PHI+A
```

```
        IF(ABS(A).GT.1.E-8) GO TO 12
        PHI=0.5+PHI/AR
        RETURN
10      PHI=1.
        RETURN
11      PHI=0.
        RETURN
        END
```

Then, the differences of these $\phi(u_j)$ values are calculated for each interval.

$$p_j = \phi(u_j) - \phi(u_{j-1})$$

where $j = 1, 2, ..., N$. With the aid of these p_j values, the empirical χ^2 value is calculated.

$$\chi^2 = \sum_{j=1}^{N} \frac{(v_j - np_j)^2}{np_j}$$

This empirical χ^2 value is compared with the theoretical χ^2 values (Table 5) with the degree of freedom $N - 3$ at the prescribed level, e.g. $\alpha = 0.05$.

If the empirical χ^2 is greater than the corresponding theoretical value, then it can be stated with a probability of 95% that the data set can not be regarded as a population of normal distribution.

The calculation outlined above can be performed by the Fortran subroutine ANORM.

Its use is as follows.

CALL ANORM (A, N, N2, AMI, AMA, B)

The arguments and their description are

A(N) array of the x_i data

N the number of data

N2 the number of intervals, calculated by the subroutine

AMI the smallest value of the data set, x_i, to be specified before calling

AMA the greatest value in the data set, x_n, to be specified before calling

B the empirical χ^2 value, calculated by the subroutine

The listing of the subroutine is

```
SUBROUTINE ANORM(A,N,N2,AMI,AMA,B)
DIMENSION A(1000),R(40),PH(40),G(40),RR(40)
AN=FLOAT(N)
```

```
      SN=SQRT(AN)
      N2=INT(SN)
      IF((SN-N2).GT.0.5) N2=N2+1
      IF(N2.LT.8) N2=8
      AM=0.0
      DO 1 I=1,N
1     AM=AM+A(I)/N
      S=J.0
      DO 2 I=1,N
2     S=S+(A(I)-AM)**2
      S=SQRT(S/(N-1))
      H=(AMI-AMI)/N2
      NN=N2+1
      DO 3 I=1,NN
3     R(I)=AMI+(I-1)*H
      DO 4 I=1,40
4     G(I)=0.0
      DO 7 I=1,N
      L=1
      DO 5 J=1,N2
      IF(A(I).LT.AMI+J*H) GO TO 6
      L=L+1
5     CONTINUE
6     G(L)=G(L)+1.0
7     CONTINUE
      SS=0.0
      PH(1)=0.0
      DO 8 I=1,N2
      RR(I)=(R(I+1)-AM)/S
      PH(I+1)=PHI(RR(I))
      H=PH(I+1)-PH(I)
      SS=SS+(G(I)-N*H)**2/N/H
8     CONTINUE
      B=SS
      RETURN
      END
```

N.B. The subroutine calls FUNCTION PHI(X), therefore the program must contain it.

(5) Equality of variances of data groups

Let us suppose that we have n data groups each consisting of m_i $(i = 1, 2, ..., n)$ data with a variance s_i^2 $(i = 1, 2, ..., n)$. The hypothesis is that these variances are equal or, more precisely, the existing small differences are due to random errors, i.e. the variances are homogeneous. We denote the measured data by a_{ij}, where the indices i and j refer to the data group and to the measured value within the group, respectively, and $i = 1, 2, ..., n$ and $j = 1, 2, ..., m_i$. The calculation method, the so-called χ^2 (chi-square) test is performed as follows. We calculate the mean $\bar{a}_i$ of each data group and from that the variances within the data groups

$$s_i^2 = \frac{\sum\limits_{j=1}^{m_i} (a_{ij} - \bar{a}_i)^2}{m_i - 1}$$

and the total variance

$$s^2 = \frac{\sum\limits_{i=1}^{n} \sum\limits_{j=1}^{m_i} (a_{ij} - \bar{a}_i)^2}{m - n}$$

where m is the total number of data

$$m = \sum_{i=1}^{n} m_i$$

The approximate χ^2 value is then calculated from

$$\chi^2 = (m - n) \ln s^2 - \sum_{i=1}^{n} (m_i - 1) \ln s_i^2$$

This value must be compared with the tabulated critical χ_o^2 value respective to $n - 1$ number of degrees. When $\chi^2 < \chi_0^2$, we can accept the hypothesis that there is no evidence of lack of homogeneity among the variances. In the contrary case the corrective factor

$$C = 1 + \frac{\sum\limits_{i=1}^{n} 1/(m_i - 1) - 1/(m - n)}{3(n - 1)}$$

is calculated. When $\chi^2/C > \chi_0^2$, the hypothesis of homogeneity of variances must be rejected; the variances of the individual data groups are significally different. For the purpose of the above calculation, the subroutine BARTLETT has been worked out.

    SUBROUTINE BARTLETT (N, M, A, IE)
    DIMENSION M(N), A(N, 10), AM(10), S(10)

The parameters are

N   number of data groups

M   array of the number of data in the groups

A   the data, a two-dimensional array, the first index refers to data group

IE   IE = 0 the variances are homogeneous

IE = 1 the variances are inhomogeneous

Function required: CHI2.

The listing of the subroutines BARTLETT and CHI2 is

```
SUBROUTINE BARTLETT(N,M,A,IE)
DIMENSION M(N),A(N,10),AM(30),S(30)
MM=0
DO 1 I=1,N
1 MM=MM+M(I)
ST=0.0
DO 2 I=1,N
SS=0.0
MI=M(I)
DO 3 J=1,MI
3 SS=SS+A(I,J)
SS=SS/M(I)
AM(I)=SS
2 CONTINUE
DO 4 I=1,N
SS=0.0
MI=M(I)
DO 5 J=1,MI
SM=A(I,J)-AM(I)
5 SS=SS+SM*SM
ST=ST+SS
S(I)=SS/(M(I)-1)
4 CONTINUE
ST=ST/(MM-N)
SS=0.0
DO 6 I=1,N
6 SS=SS+(M(I)-1)*ALOG(S(I))
CH2=(MM-N)*ALOG(ST)-SS
IE=0
F=CHI2(N-1,IH)
IF(CH2.LE.F) RETURN
ST=0.0
DO 7 I=1,N
7 ST=ST+1/(M(I)-1.)
ST=ST-1/(MM-N)
ST=ST/3/(N-1)+1
CH2=CH2/ST
IF(CH2.GT.F) IE=1
RETURN
END
```

```
      FUNCTION CHI2(N,IE)
      DIMENSION A(30)
      IE=0
      IF(N-30)1,1,2
    1 CONTINUE
      DATA A /3.84,5.99,7.81,9.49,11.1,12.6,14.1,15.5,16.9,18.3,19.7,21.,
     X22.4,23.7,25.,26.3,27.6,28.9,30.1,31.4,32.7,33.9,35.2,36.4,37.7,
     X38.9,40.1,41.3,42.6,43.8/
      CHI2=A(N)
      RETURN
    2 IE=1
      RETURN
      END
```

(6) Test of significance, the F-test

The scope of the method is to compare several groups of measured data to determine whether the mean values of the individual data groups can be regarded as the same or not. In other words, are the differences between the means significant or not? More precisely, let us consider a pattern of n rows and m columns so that each element in the pattern, e.g. the element m_{ij} belonging at the same time to the ith row and the jth column, represents a set of measured data. It is supposed that in moving in the columns from top to bottom and in the rows from left to right the effect of some factor can be recognized. If the mean values are significantly different from each other, the factors causing the differences can be ascertained. Each of the numbers m_{ij} ($i = 1, 2, ..., n; j = 1, 2, ..., m$) is positive and an integer and there is among them at least one which is greater than 1, i.e. $m_{ij} > 0$ and

$$\prod_{i=1}^{n} \prod_{j=1}^{m} m_{ij} > 1$$

The total number of measured data in the ith row and jth column are respectively given by

$$N_i = \sum_{j=1}^{m} m_{ij}$$

and

$$M_j = \sum_{i=1}^{n} m_{ij}$$

The total number of measured data is

$$N = \sum_{i=1}^{n} N_i = \sum_{j=1}^{m} M_j = \sum_{i=1}^{n} \sum_{j=1}^{m} m_{ij}$$

The measured data represented by the element (i, j) of the pattern are denoted by x_{ijk} ($i = 1, 2, ..., n; j = 1, 2, ..., m; k = 1, 2, ..., m_{ij}$). We now calculate the following mean values: the mean of the measured data represented by the element (i, j) of the pattern

$$\overline{a_{ij}} = \sum_{k=1}^{m_{ij}} \frac{x_{ijk}}{m_{ij}}$$

the mean of the elements of the ith row and jth column, respectively

$$\overline{\overline{a_i}} = \sum_{j=1}^{m} \sum_{k=1}^{m_{ij}} \frac{x_{ijk}}{N_i}$$

$$\overline{\overline{a_j}} = \sum_{i=1}^{n} \sum_{k=1}^{m_{ij}} \frac{x_{ijk}}{M_j}$$

and the overall mean

$$\overline{\overline{\overline{a}}} = \sum_{i=1}^{n} \sum_{j=1}^{m} \sum_{k=1}^{m_{ij}} \frac{x_{ijk}}{N}$$

The following sums of squares are then calculated.

$$Q_1 = \sum_{i=1}^{n} N_i (\overline{\overline{a_i}} - \overline{\overline{\overline{a}}})^2$$

$$Q_2 = \sum_{j=1}^{m} M_j (\overline{\overline{a_j}} - \overline{\overline{\overline{a}}})^2$$

$$Q_{12} = \sum_{i=1}^{n} \sum_{j=1}^{m} m_{ij} (\overline{a_{ij}} - \overline{\overline{a_i}} - \overline{\overline{a_j}} + \overline{\overline{\overline{a}}})^2$$

$$Q_r = \sum_{i=1}^{n} \sum_{j=1}^{m} \sum_{k=1}^{m_{ij}} (x_{ijk} - \overline{a_{ij}})^2$$

$$Q_r^* = Q_r + Q_{12}$$

In addition, the following integers, the so-called degrees of freedom are calculated.

$f_1 \ = n - 1$
$f_2 \ = m - 1$
$f_{12} = (n - 1)(m - 1)$
$f_r \ = n - nm$
$f_r^* \ = N - n - m + 1$

Finally, the following quotients, the empirical F values, are obtained.

$$F_1 = \frac{Q_1/f_1}{Q_r/f_r} \qquad F_2 = \frac{Q_2/f_2}{Q_r/f_r} \qquad F_{12} = \frac{Q_{12}/f_{12}}{Q_r/f_r}$$

$$F_1^* = \frac{Q_1/f_1}{Q_r^*/f_r^*} \qquad F_2^* = \frac{Q_2/f_2}{Q_r^*/f_r^*}$$

These F quantities must be compared with the tabulated critical values of the F-test. The basis of comparison is the critical F value denoted, in general, by F_0 belonging to the prescribed level of significance and to the degrees of freedom used in calculating the empirical F values.

If $F_{12} > F_0$, then there is an interaction between the two assumed factors and the mean values of the different data groups differ from each other significantly. On the other hand, if $F_1^* > F_0$ or $F_2^* > F_0$, then there is a significant difference between the mean values of the data in the rows and columns, respectively. The differences between them are not due to random errors.

The above calculation can be carried out by the following Algol procedure.

integer procedure FTEST (r, c, nr, nc, m, x, q, qq, xr, xc, xrc, xm, f, ff);

    name xm;
    integer r, c;
    real xm;
    integer array ff, m, nr, nc;
    array q, qq, f, x, xr, xc, xrc;

The formal parameters are

    integer r the number of rows
    integer c the number of columns
    real xm the overall mean
    integer array nr[1 : r] the number of data in the columns
    integer array nc[1 : c] the number of data in the columns
    integer array m[1 : r, 1 : c] the number of data represented by the element (i, j) of the pattern
    integer array ff [1 : 5] degrees of freedom $f_1, f_2, f_{12}, f_r, f_r^*$
    array q [1 : 5] sum of squares $Q_1, Q_2, Q_{12}, Q_r, Q_r^*$
    array qq [1 : r, 1 : c] the sum of squares of deviations of the data represented by the element (i, j) of the pattern from their mean values
    array f [1 : 5] the empirical F values $F_1, F_2, F_{12}, F_1^*, F_2^*$

TABLE 9

Array elements calculated by the F-test

Value of the procedure	Arrays ff and q					Array f				
	1	2	3	4	5	1	2	3	4	5
0				+						
1		+		+			+			
2	+			+		+				
3	+	+	+	+	+	+	+	+	+	+
4		+		+						+
5	+			+						+
6	+	+		+					+	+

array $x[1:r, 1:c, 1:t]$ the data, where t is the maximum element of array m

array $xr[1:r]$ the mean value of data in the rows

array $xc[1:c]$ the mean value of data in the columns

array $xrc[1:r, 1:c]$ the mean value of data represented by the elements of the pattern

Values must be assigned to the parameters integer r, c; integer array m $[1:r, 1:c]$; array x $[1:r, 1:c, 1:t]$.

The value of the procedure may be 0, 1, 2, 3, 4, 5, or 6. The value 3 means the general case, the values 0, 1, and 2 refer to the special cases r = c = 1, r = 1, and c = 1, respectively. The values 4, 5, and 6 refer to the above special cases r = c = 1, r = 1 and c = 1, respectively, provided that there is only one figure in the pattern at each place.

With regard to the special cases, no value is assigned to some of the elements of the arrays f, ff and q. The elements of the arrays calculated by the procedure are denoted in Table 9 by the + sign.

The listing of the procedure is

```
INTEGER PROCEDURE FTEST   (R, C, NR, NC,
   M, X, Q, QQ, XR, XC, XRC, XM, F, FF);
NAME XM;
INTEGER R, C;
REAL XM;
INTEGER ARRAY FF, M, NR, NC;
ARRAY Q, QQ, F, X, XR, XC, XRC;
BEGIN
INTEGER I, J, K, N, T;
REAL S, S1;
```

```
BOOLEAN B;
BOOLEAN ARRAY BQ[1 : 5];
FOR T: = 1 STEP 1 UNTIL 5 DO
BQ[T]: = FALSE;
FOR T: = 1 STEP 1 UNTIL R DO
BEGIN
N: = 0;
FOR I: = 1 STEP 1 UNTIL C DO
N: = N + M[T, I];
NR[T]: = N;
END;
FOR I: = 1 STEP 1 UNTIL C DO
BEGIN
N: = 0;
FOR T: = 1 STEP 1 UNTIL R DO
N: = M + M[T, I];
NC[I]: = N;
END;
N: = 0;
FOR T: = 1 STEP 1 UNTIL R DO
N: = N + NR[T];
FF[4]: = N - R*C;
B: = FF[4] EQ 0;
XM: = 0;
FOR T: = 1 STEP 1 UNTIL R DO
FOR I: = 1 STEP 1 UNTIL C DO
FOR J: = 1 STEP 1 UNTIL M[T, I] DO
XM: = XM + X[T, I, J];
XM: = XM/N;
IF B THEN
BEGIN
Q[4]: = 0;
GOTO L4;
END;
FOR T: = 1 STEP 1 UNTIL R DO
FOR I: = 1 STEP 1 UNTIL C DO
BEGIN
S: = 0;
FOR J: = 1 STEP 1 UNTIL M[T, I] DO
S: = S + X[T, I, J];
S: = S/M[T, I];
XRC[T, I]: = S;
END;
S: = 0;
FOR T: = 1 STEP 1 UNTIL R DO
FOR I: = 1 STEP 1 UNTIL C DO
BEGIN
S1: = 0;
FOR J: = 1 STEP 1 UNTIL M[T, I] DO
S1: = S1 + (X[T, I, J] - XRC[T, I]) ↑ 2;
QQ[T, I]: = S1;
S: = S + S1;
END;
```

```
Q[4]: = S;
L4:
BQ[4]: = Q[4] LT 10 ↑ (   100);
IF R = 1 THEN
BEGIN
IF C = 1 THEN
BEGIN
K: = 0;
GOTO L1:
END;
K: = 1:
GOTO L2;
END;
K: = IF C = 1 THEN 2 ELSE 3;
FF[1]: = R   1;
FOR T: = 1 STEP 1 UNTIL R DO
BEGIN
S: = 0;
FOR I: = 1 STEP 1 UNTIL C DO
FOR J: = 1 STEP 1 UNTIL M[T, I] DO
S: = S + X[T, I, J];
S: = S/NR[T];
XR[T]: = S;
END;
S: = 0;
FOR T: = 1 STEP 1 UNTIL R DO
S: = S + NR[T] * (XR[T]   XM) ↑ 2;
Q[1]: = S;
BQ[1]: = Q[1] LT 10 ↑ ( 100);
IF K LE 2 THEN
BEGIN
F[1]: = IF BQ[4] THEN 10 ↑ 100 ELSE Q[1] * FF[4]/
   Q[4]/FF[1];
GOTO L1;
END:
L2;
FF[2]: = C   1;
FOR I: = 1 STEP 1 UNTIL C DO
BEGIN
S: = 0;
FOR T: = 1 STEP 1 UNTIL R DO
FOR J: = 1 STEP 1 UNTIL M[T, I] DO
S: = S + X[T, I, J];
S: = S/NC[I];
XC[I]: = S;
END;
S: = 0;
FOR I: = 1 STEP 1 UNTIL C DO
S: = S + NC[I] * (XC[I]   XM) ↑ 2;
Q[2]: = S;
BQ[2]: = Q[2] LT 10 ↑ ( 100);
IF K LE 2 THEN
BEGIN
F[2]: = IF BQ[4] THEN 10 ↑ 100 ELSE
   Q[2] * FF[4]/Q[4]/FF[2];
```

50

```
GOTO L1;
END;
L3;
S: = 0;
FOR T: = 1 STEP 1 UNTIL R DO
FOR I: = 1 STEP 1 UNTIL C DO
S: = S + M[T, I] * (XRC[T, I] - XR[T] - XC[I] +
    + XM) ↑ 2;
Q[3]: = S;
FF[3]: = FF[1] * FF[2];
FF[5]: = FF[3] + FF[4];
Q[5]: = Q[3] + Q[4];
BQ[3]: = Q[3] LT 10 ↑ (-100);
BQ[5]: = Q[5] LT 10 ↑ (- 100);
IF NOT B AND NOT BQ[4] THEN
BEGIN
IF K NE 1 THEN F[1]: = Q[1] * FF[4]/Q[4]/FF[1];
IF K NE 2 THEN F[2]: = Q[2] * FF[4]/Q[4]/FF[2];
F[3]: = Q[3] * FF[4]/Q[4]/FF[3];
END;
IF NOT BQ[5] THEN
BEGIN
F[4]: = Q [1] * FF[5]/Q[5]/FF[1];
F[5]: = Q [2] * FF[5]/Q[5]/FF[2];
END;
IF BQ[4] THEN
FOR I := 1, 2, 3 DO
F[I]: = IF BQ[I] THEN 0 ELSE 10 ↑ 100;
IF BQ[5] THEN
FOR I: = 1, 2 DO
F[I + 3]: = IF BQ[I] THEN 0 ELSE 10 ↑ 100;
L1:
IF B THEN K: = IF K = 0 THEN 0 ELSE K + 3;
FTEST  : = K;
END FTEST;
```

(7) Statistical evaluation of data resulting from several laboratories [14]

This task occurs when evaluating the precision of a standard analytical method. The minimum number of participating laboratories for a precision experiment should be 15. Precision is the closeness of agreement between the results obtained by applying the experimental procedure several times under prescribed conditions.

a. Definitions. Repeatability (permissible tolerance within laboratories), r, is the maximum value below which the absolute difference between two single test (analysis) results on identical material by one operator in a laboratory using the same equipment within a short interval of time, using the standardized (prescribed) test method may be expected to lie with a 95% probability.

$$r = 2.77 \, \sigma_r$$

where σ_r^2 is within-laboratory variance.

Permissible tolerance between laboratories, P, is the maximum value below which the absolute difference between two means of duplicate test results on identical material obtained by operators in different laboratories using the standardized (prescribed) test method may be expected to lie with a 95% probability

$$P = 2.77 \, \sqrt{\frac{\sigma_r^2}{2} + \sigma_L^2}$$

where σ_L^2 is between-laboratories variance.

Note that if one laboratory performs n_1 determinations with a mean value x_1 and the second laboratory performs n_2 determinations with a mean value x_2, then

$$P(n_1, n_2) = 2.77 \, \sqrt{\sigma_L^2 + \frac{\sigma_r^2}{2}\left(\frac{1}{n_1} + \frac{1}{n_2}\right)}$$

where $P(n_1, n_2)$ is the maximum value below which the absolute difference between x_1 and x_2 may be expected to lie with a 95% probability. For $n_1 = n_2 = 2$, $P(n_1, n_2) = P$.

In order that the analytical data can be used most efficiently, and to avoid misunderstanding, clear instructions must be distributed to the participating laboratories. Such instructions are: each laboratory shall make two replicate determinations on each of two bottles of test samples of the same material. The description of the test sample includes the origin of the test sample, the elements content of the sample, and the method of evaluation (estimation) of the elements content. This information includes, if available, the variance and other necessary statistics of the estimate of the elements content.

b. Reporting results by the participating laboratories. All four independent results must be reported separately, i.e. not the average. The minimum number of significant figures must be specified such that the highest decimal unit (i.e. ...; 10; 1; 0.1; 0.01; 0.001; 0.0001; ...) does not exceed the value of b where b is to be computed as follows.

(a) If the within-laboratory standard deviation, σ_r, is known or reasonably well estimated

$$b = \tfrac{1}{2}\sigma_r$$

(b) If σ_r is unknown

$$b = \frac{\overline{R}}{2\sqrt{n}}$$

$$\overline{R} = \frac{1}{k}\sum R_i$$

where R_i is the ith range of n observations. In this case, both n and k are equal to 2, therefore

$$b = \frac{R_1 + R_2}{4\sqrt{2}}$$

where R_1 is the range of the first bottle and R_2 that of the second.

Note that, in general, there is no harm in reporting more significant figures than the prescribed minimum number.

The most important requirement of a test sample is that it must be homogeneous. Two bottles (units) of test samples of the same material shall be randomly distributed to the participating laboratories. One of the simplest ways to randomize the sample distribution is by numbering each package and using uniformly distributed random numbers to allocate the packages.

c. Processing analytical results. The analytical results submitted by the participating laboratories are processed as follows. For the convenience of processing these results, they should be tabulated systematically, and the following intermediate quantities calculated (Table 10).

i laboratory number
j bottle number
v replicate number:
x_{ijv} the vth replicate result of bottle j and reported by laboratory i
n_{ij} number of replicate determinations of bottle j reported by laboratory i; normally $n_{ij} = 2$
n_i number of results reported by laboratory i; normally $n_i = 4$
R_{ij}^2 square of the range of results of bottle j reported by laboratory i
T_{ij} Σx_{ijv}
$\overline{x}_{ij}$ T_{ij}/n_{ij}, mean of bottle j reported by laboratory i

TABLE 10

Data presentation and computation

Lab. no. (i)	Bottle no. (j)	Analytical results x_{ijv}		No. of replicate (n_{ij})	No. of results per lab. (n_i)	n_i^2	$R_{ij}^2 =$ $(x_{ij1} - x_{i..})$
		$v = 1$	$v = 2$				
1	1						
	2						
2	1						
	2						
3	1						
	2						
k	Total	I		I	II	III	

$T_i \quad \overset{\Sigma}{j} T_{ij}$

$\bar{x}_i \quad T_i/n_i$, mean of results reported by laboratory i

$SS_i \quad \overset{\Sigma\Sigma}{j v} x_{ijv}^2$, sum of squares of results reported by laboratory i

$S_i^2 \ [SS_i - (T_i^2/n_i)]/(n_i - 1)$, within-laboratory variance of laboratory i

Confirmation of the homogeneity of the test sample. The purpose of this step of computation is to determine if between-bottles component of variance is statistically significant in comparison with the within-bottles component of variance. Thus it serves as the statistical confirmation of the homogeneity of the test sample.

Bartlett's test for the homogeneity of variance.

$$B = [2k \ln(III/2k) - IV]\bigg/\bigg[1 + \frac{1}{3(2k-1)}\bigg(2k - \frac{1}{2k}\bigg)\bigg]$$

where k is the number of laboratories of which the data are used for Bartlett's test, and III and IV are the sum of the columns 9 and 10 of Table 10, respectively. Note that, because Bartlett's test uses the logarithmic form of within-bottle ranges, results with a zero within-bottle range are excluded from this computation.

Compare the computed value of B with the critical value taken from the χ^2-distribution for $\alpha = 0.05$ (Table 5) for the degrees of freedom of $(2k - 1)$. Then

T_{ij}	T_{ij}^2/n_{ij}	$\bar{x}_{ij}$	T_i	T_i^2/n_i	$\bar{x}_i$	SS_i	Lab. variance (S_i^2)
V	VI		VII	VIII		IX	X

if $B \leqslant \chi^2$ variances are homogeneous, proceed with Doornbos test (omit Cochran's test) and

if $B > \chi^2$ variances are inhomogeneous, proceed with Cochran's test.

Doornbos' test for discordant results. The basic principle of this test is that a bottle mean is considered to be a discordant one if it is too far from the other bottle means. The data that are discordant in accordance to Cochran's test should be temporarily reinstated for the purpose of Doornbos' test.

Arrange the bottle means, $y_h = \bar{x}_{ij}$ in ascending order

$$y_1 \leqslant y_2 \leqslant y_3 \ldots \leqslant y_{(u-1)} \leqslant y_u$$

and compute the quantities

$$L^* = \sum_h^u y_h$$

$$Q^* = \sum_h^u y_h^2$$

where u is the number of bottle means to be tested.

If y_u is suspect	If y_1 is suspect
$L = L^* - y_u$	$L = L^* - y_1$
$Q = Q^* - y_u^2$	$Q = Q^* - y_1^2$

$$t = \frac{|(u-1)\,y_u - L|}{\sqrt{[(u-1)\,Q - L^2]}}\sqrt{\frac{(u-2)}{u}} \qquad t = \frac{|(u-1)\,y_1 - L|}{\sqrt{[(u-1)\,Q - L^2]}}\sqrt{\frac{(u-2)}{u}}$$

$$t_0 = t\left(\frac{\alpha}{2u}, u - 2\right) \qquad\qquad t_0 = t\left(\frac{\alpha}{2u}, u - 2\right)$$

where t_0 is the critical value taken from the Student t-distribution for level $\alpha/2u$ and degrees of freedom $f = u - 2$. These values can be taken from Table 4, but they are tabulated for direct use in Table 11. Compare the computed value of t with the critical value t_0 and reject the suspected results if t is greater than t_0. Repeat this procedure until all the suspected discordant results have been rejected, i.e., $t \leqslant t_0$.

Two-way analysis of variance (Anova). This calculation is based on the mathematical model

$$x_{ijv} = \mu + b_i + a_{ij} + e_{ijv} \tag{11}$$

where x_{ijv} is the vth replicate result of the jth bottle reported by laboratory i, μ is the expected value of x which is estimated by the grand mean $\bar{x}$, b_i is an error term due to the laboratory, i.e. the deviation between the mean of laboratory i from μ, a_{ij} is an error term due to the inhomogeneity of the test sample, i.e. the deviation between the mean of bottle j from the mean of laboratory i, and e_{ijv} is the within-bottles error.

The basic assumption in this calculation is that the errors b_i, a_{ij}, and e_{ijv} are normally distributed with a mean of 0 and variances of σ_L^2, σ_b^2 and σ_r^2, respectively.

Having computed the two-way Anova, the following decision can be made.

(a) If it is found that the between-bottles component of variance is significant, i.e. the test sample is inhomogeneous, and if there were sufficient data used in this computation, then the value of within-bottles variance can be used as the estimate of σ_r^2 and the value of between-laboratories variance can be used as the estimate of σ_L^2.

(b) If it is found that the between-bottles component of variance is insignificant, i.e. the test sample is homogeneous, the results

TABLE 11

Critical values for Doornbos test for discordant results. N = number of observations or means; t_0 = critical values for $\alpha = 0.05$

N	t_0	N	t_0	N	t_0	N	t_0
3	38.188	41	3.4886	81	3.5669	121	3.6340
4	8.860	42	3.4903	82	3.5688	122	3.6355
5	5.841	43	3.492	83	3.571	123	3.637
6	4.851	44	3.494	84	3.573	124	3.638
7	4.382	45	3.496	85	3.574	125	3.640
8	4.115	46	3.497	86	3.576	126	3.641
9	3.947	47	3.499	87	3.578	127	3.643
10	3.833	48	3.501	88	3.580	128	3.644
11	3.751	49	3.503	89	3.582	129	3.646
12	3.691	50	3.505	90	3.583	130	3.647
13	3.646	51	3.507	91	3.585	131	3.648
14	3.611	52	3.509	92	3.587	132	3.650
15	3.584	53	3.511	93	3.589	133	3.651
16	3.562	54	3.513	94	3.591	134	3.653
17	3.545	55	3.515	95	3.592	135	3.654
18	3.531	56	3.517	96	3.594	136	3.655
19	3.519	57	3.519	97	3.596	137	3.657
20	3.510	58	3.521	98	3.597	138	3.658
21	3.503	59	3.523	99	3.599	139	3.659
22	3.497	60	3.525	100	3.601	140	3.661
23	3.492	61	3.527	101	3.602	141	3.662
24	3.488	62	3.529	102	3.604	142	3.663
25	3.485	63	3.531	103	3.606	143	3.665
26	3.483	64	3.533	104	3.607	144	3.666
27	3.481	65	3.536	105	3.609	145	3.667
28	3.480	66	3.538	106	3.611	146	3.669
29	3.479	67	3.540	107	3.612	147	3.670
30	3.479	68	3.542	108	3.614	148	3.671
31	3.479	69	3.544	109	3.616	149	3.673
32	3.479	70	3.546	110	3.617	150	3.674
33	3.479	71	3.548	111	3.619	151	3.675
34	3.480	72	3.550	112	3.620	152	3.676
35	3.481	73	3.552	113	3.622	153	3.678
36	3.482	74	3.553	114	3.623	154	3.679
37	3.483	75	3.555	115	3.625	155	3.680
38	3.484	76	3.557	116	3.626	156	3.681
39	3.486	77	3.559	117	3.628	157	3.683
40	3.487	78	3.561	118	3.630	158	3.684
		79	3.563	119	3.631	159	3.685
		80	3.565	120	3.633	160	3.686

reported by each laboratory are then combined to form a laboratory result.

Analysis of variance table. The most convenient way to compute the variance expressions of precision is by constructing the so-called analysis of variance table based on the model defined by eqn. (11). Numerical entries in this table can be computed from the quantities listed in Table 10. An example of this Anova table is illustrated in Table 12 in which E(M.S.) represents the expected values of the mean squares.

Numerical construction of a two-way Anova table. The two-way Anova table can be constructed numerically from the intermediate quantities found in Table 10. Only results from laboratories that reported results of two bottles with duplicate results per bottle are included in this computation.

$$\text{SSBL} = \text{VIII} - (V^2/4k)$$
$$\text{SSBB} = \text{VI} - \text{VIII}$$
$$\text{SSW} = \text{IX} - \text{VI}$$
$$\text{SST} = \text{IX} - (V^2/4k)$$
$$\text{DFBL} = k - 1$$
$$\text{DFBB} = k$$
$$\text{DFW} = 2k$$
$$\text{MSBL} = \text{SSBL/DFBL}$$
$$\text{MSBB} = \text{SSBB/DFBB}$$
$$\text{MSW} = \text{SSW/DFW}$$
$$F_c = \text{MSBB/MSW}$$

If $F_c \leqslant F_0\,(0.05; k, 2k)$, it means that the test sample is sufficiently homogeneous and if $F_c > F_0\,(0.05; k, 2k)$, the test sample is inhomogeneous. If it is found that the test sample is inhomogeneous, the

TABLE 12

Two-way analysis of variance table

Source of variance	Sums of squares	Degrees of freedom	Mean squares	E(M.S.)
Between laboratories	SSBL	DFBL	MSBL	$\sigma_r^2 + 2\sigma_b^2 + 4\sigma_L^2$
Between bottles	SSBB	DFBB	MSBB	$\sigma_r^2 + 2\sigma_b^2$
Within laboratories	SSW	DFW	MSW	σ_r^2
Total	SST			

various components of variance can be estimated as follows.

$$\sigma_r^2 = \text{MSW} \tag{12}$$

$$\sigma_b^2 = (\text{MSBB} - \text{MSW})/2$$

$$\sigma_L^2 = (\text{MSBL} - \text{MSBB})/4 \tag{13}$$

$$\bar{x} = V/4K$$

$$V[\bar{x} \ldots] = (\sigma_r^2 + 2\sigma_b^2 + 4\sigma_L^2)/4K$$

All the results (from bottle 1 and bottle 2) reported by each laboratory are combined to form "laboratory results" and laboratory means, $\bar{x}_i$, and laboratory variance S_i^2 computed and listed as in Table 10.

Laboratory results with zero variance. No results with zero within-laboratory variance will be used for the computation of Bartlett's test and Cochran's test. Laboratory results without replicate ($n_i = 1$) are excluded from all calculations.

Bartlett's test for variance homogeneity. For this purpose, Table 13 is constructed with the following contents

    column 1 laboratory number
    column 2 $n_i - 1$
    column 3 $1/(n_i - 1)$
    column 4 T^2/n_i (copy from Table 10)
    column 5 SS_i (copy from Table 10)
    column 6 S_i^2 (copy from Table 10)
    column 7 $\ln S_i^2$
    column 8 $(n_i - 1) \ln S_i^2$

TABLE 13

For computation of the 2nd Bartlett test

Lab. no.	$n_i - 1$	$\dfrac{1}{n_i - 1}$	T_i^2/n_i	SS_i	S_i^2	$\ln S_i^2$	$(n_i - 1)\ln S_i^2$
1	$n_1 - 1$	$\dfrac{1}{n_1 - 1}$	T_1^2/n_1	SS_1	S_1^2	$\ln S_1^2$	$(n_1 - 1)\ln S_1^2$
2	$n_2 - 1$						
k							
Total	XI	XII	VIII	IX	X		XIII

k the number of participating laboratories, excluding those with zero variance

XI   the sum of column 2

XII   the sum of column 3

VIII   the sum of column 4

IX   the sum of column 5

X   the sum of column 6

XIII   the sum of column 8

The Bartlett's statistic, B, can be calculated from

$$B = (\text{XI} \ln S^2 - \text{XIII}) / \left[1 + \frac{1}{3(k-1)} \left(\text{XII} - \frac{1}{\text{XI}} \right) \right]$$

where $S^2 = (\text{IX} - \text{VIII})/\text{XI}$

Compare the computed value of B with the critical value taken from the χ^2-distribution for $\alpha = 0.05$ listed in Table 5 for the degrees of freedom of $(k-1)$. Then

if $B \leqslant \chi^2$ variances are homogeneous, proceed with Dixon's test (omit Cochran's test) and

if $B > \chi^2$ variances are inhomogeneous, proceed with Cochran's test.

Cochran's test. The basic principle of this test is that a set of results reported by a laboratory is a discordant result if the within-bottle range is too large in relation to the others. Compute the quantity

$$c = \frac{R_{ij\,\text{max}}^2}{\sum\limits_{i}^{k} \sum\limits_{j}^{2} R_{ij}^2}$$

where $R_{ij\,\text{max}}^2$ is the square of the maximum bottle range (column III, Table 10). Exclude sets of results with zero within-bottle range when computing c. If c is greater than the critical value listed in Table 14 for critical values C, then reject the result with maximum within-bottle range. Repeat the calculation until all the discordant results are rejected.

Dixon's test. There is a possibility that one discordant result reported by a laboratory causes the whole laboratory results to be identified as discordant by Cochran's test. Thus, by rejecting this particular result, the rest of the laboratory results can be used for

60

TABLE 14

Critical values for Cochran's test for $\alpha = 0.05$. N = the number of variances; n = the average number of the replicate determinations

N	n				
	2	3	4	5	6
2		0.975	0.939	0.906	0.877
3	0.967	0.871	0.798	0.746	0.707
4	0.906	0.768	0.684	0.629	0.590
5	0.841	0.684	0.598	0.544	0.506
6	0.781	0.616	0.532	0.480	0.445
7	0.727	0.561	0.480	0.431	0.397
8	0.680	0.516	0.438	0.391	0.360
9	0.638	0.478	0.403	0.358	0.329
10	0.602	0.445	0.373	0.331	0.303
11	0.570	0.417	0.348	0.308	0.281
12	0.541	0.392	0.326	0.288	0.262
13	0.515	0.371	0.371	0.271	0.246
14	0.492	0.352	0.291	0.255	0.232
15	0.471	0.335	0.276	0.242	0.220
16	0.452	0.319	0.262	0.230	0.208
17	0.434	0.305	0.250	0.219	0.198
18	0.418	0.293	0.240	0.209	0.189
19	0.403	0.281	0.230	0.200	0.181
20	0.389	0.270	0.220	0.192	0.174
21	0.377	0.261	0.212	0.185	0.167
22	0.365	0.252	0.204	0.178	0.162
23	0.354	0.243	0.197	0.172	0.155
24	0.343	0.235	0.191	0.166	0.149
25	0.334	0.228	0.185	0.160	0.144
26	0.325	0.221	0.179	0.155	0.140
27	0.316	0.215	0.173	0.150	0.135
28	0.308	0.209	0.168	0.146	0.131
29	0.300	0.203	0.164	0.142	0.127
30	0.293	0.198	0.159	0.138	0.124
31	0.286	0.193	0.155	0.134	0.120
32	0.280	0.188	0.151	0.131	0.117
33	0.273	0.184	0.147	0.127	0.114
34	0.267	0.179	0.144	0.124	0.111
35	0.262	0.175	0.140	0.121	0.108
36	0.256	0.172	0.137	0.111	0.106
37	0.251	0.168	0.134	0.116	0.103

TABLE 14 (continued)

N	n				
	2	3	4	5	6
38	0.246	0.164	0.131	0.113	0.101
39	0.242	0.161	0.129	0.119	0.099
40	0.237	0.158	0.126	0.108	0.097
60	0.174	0.113	0.089	0.076	0.068
120	0.100	0.063	0.049	0.042	0.037
∞	0	0	0	0	0

further computation. To identify this discordant result, these laboratories are subjected to Dixon's test for within-laboratory discordant results. This test is only to be performed if the number of results reported by a particular laboratory is not less than 4.

If a discordant result is detected by this test, it must be rejected and Cochran's test repeated; otherwise, all the results reported by this particular laboratory are rejected. The process is repeated until all the discordant results are rejected.

Dixon's test for between-laboratories discordant results. The basic principle of this test is that a laboratory mean is considered to be a discordant result if it is too far removed from the other laboratory means. The data that have been rejected by Cochran's and within-laboratory. Dixon's tests are temporarily reinstated for the purpose of this test.

Arrange the laboratory means, $y_h = \bar{x}_i$, in ascending order.

$$y_1 \leqslant y_2 \leqslant y_3 \leqslant ... \leqslant y_{k-1} \leqslant y_k$$

Compute the quantity Q defined by eqn. (9) [see Sect. 2.(F)(2)] for the cases y_1 is suspect and y_k is suspect. Compare the computed Q value with the critical value from Table 8, rejecting the suspected results (either y_1 or y_k) if the computed Q is larger than its corresponding critical value. Repeat the above procedure until all the discordant means have been rejected. If the number of laboratories, k, is reduced to 60% of the original number of laboratories and the Dixon's test still shows that there are more discordant means, it indicates that the model is not accurate. In this case, the results rejected by Dixon's and Cochran's criteria shall be reinstated. However, this fact must be reported.

If no discordant results are detected from any of the tests and all data used for the computation of the two-way Anova, proceed to the computation of various precisions. Use the values of σ_r and σ_L computed by eqns. (12) and (13) instead of those computed by eqns. (14) and (15).

When the process of identification of discordant results as described above is complete and it is found that there are some discordant results, and when there are no new results to replace them, reject all the discordant results and proceed with the one-way analysis of variance as follows.

One-way analysis of variance (Anova). The purpose of this part of the computation is to calculate the grand mean $\bar{x} \ldots$, the within-laboratories variance σ_r^2, the between-laboratories variance σ_L^2, the variance of the grand mean σ_m^2, the repeatability r, and the permissible tolerance between laboratories, P.

The analysis of variance shall be performed after the discordant results have been rejected. The number of sets of laboratory results for the analysis of variance must be at least 60% of the original number of participating laboratories.

Mathematical model. This calculation is based on the mathematical model

$$x_{ij} = \mu + b_i + e_{ij}$$

where x_{ij} is the jth replicate result reported by laboratory i, μ is the expected value of x which is estimated by the grand mean $\bar{x} \ldots$, b_i is an error term due to the laboratory, i.e. the deviation between the mean of laboratory i from μ, and e_{ij} is the within-laboratories error. The basic assumption in this calculation is that both b_i and e_{ij} are normally distributed with a mean of 0 and variance of σ_L^2 and σ_r^2, respectively.

Analysis of variance table. The most convenient way to compute the various expressions of precision is by constructing the so-called analysis of variance table based on the above model. The numerical entries in this table can be computed from the quantities listed in Table 10. An example of this Anova table is illustrated in Table 15.

Numerical construction of Anova table. The Anova table can be

TABLE 15

One-way analysis of variance table

Source of variance	Sums of squares	Degrees of freedom	Mean squares	E(M.S.)
Between laboratories	SSB	DFB	MSB	$\sigma_r^2 + \dfrac{1}{k-1}\left(\sum_i^k n_i - \dfrac{\sum_i^k n_i^2}{\sum_i^k n_i}\right)\sigma_L^2$
Within laboratories	SSW	DFW	MSW	σ_r^2
Total	SST			

constructed numerically from the intermediate quantities found in Table 10.

SSB = VIII − VII^2/I
SSW = IX − VIII
SST = IX −VII^2/I
DFB = $k − 1$
DFW = I − k
MSB = SSB/DFB
MSW = SSW/DFW

where I, II, VII, VIII, and IX are the sums of the columns indicated in Table 10. These sums exclude all the discordant results.

d. Computation of precision. Computation of various precisions is carried out on the basis of the Anova table with the variables defined below.

Within-laboratories variance, σ_r^2

$$\sigma_r^2 = \text{MSW} \tag{14}$$

Between-laboratories variance, σ_L^2

$$\sigma_L^2 = \frac{(\text{MSB-MSW})(k − 1)}{I − (II/I)} \tag{15}$$

Repeatability, r

$$r = 2.77\sigma_r$$

Permissible tolerance between laboratories, P

$$P = 2.77 \sqrt{\sigma_L^2 + \frac{\sigma_r^2}{2}}$$

The grand mean, $\bar{x}$ ...

$$\bar{x} \dots = \text{VII/I} \tag{16}$$

The variance of the grand mean, $V[\bar{x} \dots]$

$$V[\bar{x} \dots] = \frac{II}{I^2} \sigma_L^2 + \frac{\sigma_r^2}{I}$$

Note that if σ_L^2 as calculated from eqn. (15) is found to be negative, σ_L^2 must be replaced by zero.

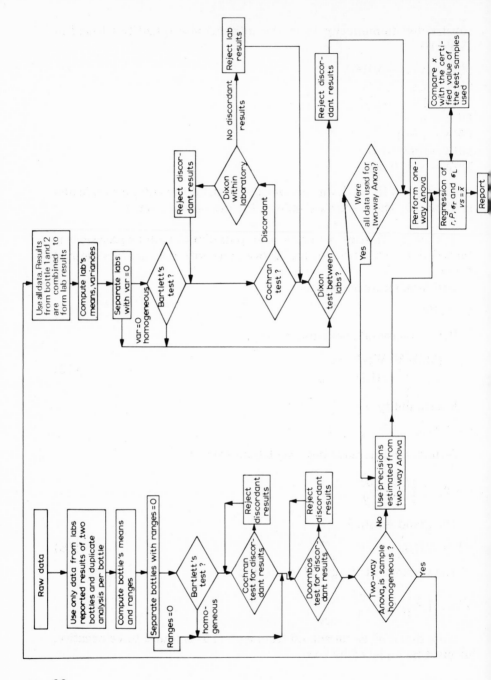

66

e. Functional relations between repeatability, permissible tolerance, within-laboratories standard deviation, between-laboratories standard deviation, and concentration level. If the number of sets of data (the number of levels of concentrations) is small, linear relationships can be set up between repeatability, r, permissible tolerance, P, standard deviations σ_r, σ_L (within- and between-laboratories), and X. This relationship is in the form of

$$y = a + bX$$

where y is either r, P, σ_r, or σ_L, X is the concentration level, as calculated from eqn. (16) $\bar{x}$ . . ., and a and b are constants, which are determined by linear regression [see Sect. 4.(B)(1)]. It is recommended that plots should be made of r, P, σ_r, and σ_L against X and that the regression lines be drawn. If it is found by visual observation that the regression line does not fit the data points, attempts should be made to transform the data so that a better fit can be achieved. An example of this is the transformation into logarithmic form.

The calculation flow sheet of the statistical evaluation of analytical data resulting from several laboratories is shown in Fig. 11.

3. The law of propagation of errors

In some analytical methods, the concentration of the substance is determined by measuring a quantity related to the concentration. Such a method involves carrying out the entire analytical procedure on a weighed amount of sample, denoted by x_1, to obtain a certain value, x_2. The required result, y, is, in the simplest case

$$y = k \frac{x_2}{x_1} \tag{17}$$

where the constant k expresses the proportionality between the measured value and the actual amount of the investigated component. Sometimes k is not a true constant but includes a value, x_3, to be determined experimentally, e.g. the titre of a volumetric solution, so that

$$k = k'x_3$$

where k' is another constant. In some cases, e.g. in the case of standard addition methods, the constant k' is a true constant only over a certain concentration range.

References pp. 168—169   67

In the more general case, in the case of comparative methods, an experimentally determined relationship is used instead of a single value of x_3. This relationship expresses the dependence of the measured value on the concentration of the component to be analyzed and is determined graphically or numerically from the results of a number of experiments. In this case, the result of the analytical procedure is not given by eqn. (17) but by

$$p = \frac{1}{x_1} f(x_2)$$

or

$$y = F(x_1, x_2)$$

In a more general form; this becomes

$$y = F(x_1, x_2, ..., x_n)$$

when n measurements are required for one determination.

The measured quantity, y, being a function of the measured values $x_1, x_2, ..., x_n$, the error of y, denoted by dy, can be expressed in the form of the total differential of the function F, viz.

$$dy = \frac{\partial F}{\partial x_1} dx_1 + \frac{\partial F}{\partial x_2} dx_2 + ... + \frac{\partial F}{\partial x_n} dx_n$$

where $dx_1, dx_2, ..., dx_n$ are the errors of the measured values $x_1, x_2, ..., x_n$.

Example. In the simplest case of eqn. (17) where k is a constant "without error"

$$y = kx$$

$$dy = k \, dx$$

Dividing the second equation by the first

$$\frac{dy}{y} = \frac{dx}{x}$$

This means that the relative error of the quantity to be determined is the same as that of the measured value.

Example.

$$y = kx_1 x_2$$

$$dy = k(x_2 \, dx_1 + x_1 \, dx_2)$$

68

Dividing the second equation by the first

$$\frac{dy}{y} = \frac{dx_1}{x_1} + \frac{dx_2}{x_2}$$

This means that relative errors are additive. The same is true for the quotient.

Example.

$$y = kx^n$$

$$dy = knx^{n-1} \, dx$$

Dividing the second equation by the first

$$\frac{dy}{y} = n \frac{dx}{x}$$

i.e. the relative error of the measured quantity multiplied by the exponent gives the required relative error.

Example.

$$y = k \frac{x}{1-x}$$

$$dy = \frac{k \, dx}{(1-x)^2}$$

The relative error is

$$\frac{dy}{y} = \frac{dx}{(1-x)\, x}$$

This relative error reaches its minimum at $x = 0.5$.

The above examples, together with others, are summarized in Table 16.

The standard deviation, σ, of a result $y = F(x_1, x_2, ..., x_n)$, where the quantities $x_1, x_2, ..., x_n$ have been determined with the standard deviations $\sigma_1, \sigma_2, ..., \sigma_n$, can be calculated by the equation

$$\sigma^2 = \left(\frac{\partial F}{\partial x_1}\right)^2 \sigma_1^2 + \left(\frac{\partial F}{\partial x_2}\right)^2 \sigma_2^2 + ... + \left(\frac{\partial F}{\partial x_n}\right)^2 \sigma_n^2 \qquad (18)$$

(Sometimes, the standard deviation of x_i is denoted by σ_{x_i} or, if there is no ambiguity, by dx_i.) The above equation can be deduced from the definition of the standard deviation assuming that the variables are uncorrelated and their standard deviations are small.

TABLE 16

Error of the final result in terms of the errors of individual measured values

Final result y	Error
$k_1 x_1 + k_2 x_2$	$k_1 dx_1 + k_2 dx_2$
$k x_1 x_2$	$k(x_2 dx_1 + x_1 dx_2)$
$k x^n$	$k n x^{n-1} dx$
$\dfrac{kx}{1-x}$	$\dfrac{k dx}{(1-x)^2}$
$k e^x$	$k e^x dx$
$k \ln x$	$k \dfrac{dx}{x}$

Further, it must be assumed that the partial derivatives, $\partial F / \partial x_i$, are continuous. Some typical examples are given in Table 17.

(A) THE STANDARD DEVIATION OF THE MEAN VALUE

Having measured a quantity n times, the result of the measurement is given as the mean value of the measured data.

TABLE 17

The square of the standard deviation (variance) of the final result y in terms of the standard deviations, σ_i of the partial results x_i (i = 1, 2, ..., n)

Final result y	Variance σ^2
$x_1 \pm x_2$	$\sigma_1^2 + \sigma_2^2$
$x_1 x_2$	$\sigma_1^2 x_2^2 + \sigma_2^2 x_1^2$
$a + k x^n$	$(nkx^{n-1})^2 \sigma^2$
$\dfrac{x_1}{x_2}$	$\left(\dfrac{x_1}{x_2}\right)^2 \left(\dfrac{\sigma_1^2}{x_1^2} + \dfrac{\sigma_2^2}{x_2^2}\right)$
$\ln(x_1 + x_2)$	$\dfrac{\sigma_1^2 + \sigma_2^2}{(x_1 + x_2)^2}$
$\ln \dfrac{x_1}{x_2}$	$\dfrac{\sigma_1^2}{x_1^2} + \dfrac{\sigma_2^2}{x_2^2}$

$$\bar{x} = \frac{1}{n}(x_1 + x_2 + ... + x_n) = \frac{1}{n}\sum_{i=1}^{n} x_i$$

The standard deviation of one measured value is calculated as

$$s_x = \sqrt{\frac{\sum_{i=1}^{n}(\bar{x} - x_i)^2}{n-1}}$$

The accuracy of a single measured value can be characterized by this.

The reproducibility of the mean value is characterized by its standard deviation which is calculated on the basis of the law of propagation of errors. The variance of the mean value is

$$s_{\bar{x}}^2 = \left(\frac{\partial \bar{x}}{\partial x_1} s_{x_1}\right)^2 + \left(\frac{\partial \bar{x}}{\partial x_2} s_{x_2}\right)^2 + ... + \left(\frac{\partial \bar{x}}{\partial x_n} s_{x_n}\right)^2 \qquad (19)$$

Since

$$\frac{\partial \bar{x}}{\partial x_1} = \frac{\partial \bar{x}}{\partial x_2} = ... = \frac{\partial \bar{x}}{\partial x_n} = \frac{1}{n}$$

and

$$s_{x_1} = s_{x_2} = ... = s_{x_n}$$

eqn. (19) can be written

$$s_{\bar{x}}^2 = n\left(\frac{1}{n}s_x\right)^2$$

and the standard deviation of the mean value is

$$s_{\bar{x}} = \frac{s_x}{\sqrt{n}}$$

So, if a lot of parallel measurements have been carried out, the result can be given as

$$\bar{x} \pm \frac{s_x}{\sqrt{n}}$$

If only a few parallel measurements have been carried out, then the result is given by the Student distribution as

$$\bar{x} \pm \frac{t(\alpha, f) s_x}{\sqrt{n}}$$

where $t(\alpha, f)$ is the proportionality factor referring to the level α and to $f = n - 1$ degrees of freedom (Table 4).

Now the interesting question arises, how many parallel measurements is it necessary to carry out? It is obvious that the term $t(\alpha, f)s_x/\sqrt{n}$ containing the number of measurements has an important role to play here. With the aid of the tabulated t-values, e.g. for $\alpha = 0.025$, the following table is constructed.

n	$f = n - 1$	$t(0.025, f)$	$\dfrac{t(0.025, f)}{\sqrt{n}}$	R
1	0	∞		
2	1	12.7	8.98	
				3.62
3	2	4.30	2.48	
				1.56
4	3	3.18	1.59	
				1.28
5	4	2.78	1.24	
				1.18
6	5	2.57	1.05	
				1.13
7	6	2.45	0.93	
				1.12
8	7	2.36	0.83	
				1.08
9	8	2.31	0.77	
				1.08
10	9	2.26	0.71	

The numbers in the last column, denoted by R, are the ratios of pairs of numbers in the penultimate column. It can be concluded that the $t(0.025, f)/\sqrt{n}$ values decrease very quickly to $n = 4$ to but, thereafter, the decrease is slow. This is also demonstrated in Fig. 12. The conclusion can therefore be drawn that it is necessary to carry out

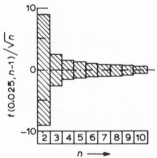

Fig. 12. Confidence limits of the mean.

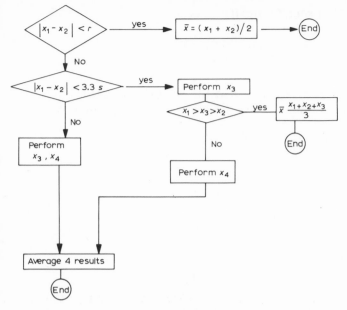

Fig. 13. Necessary repeats of analysis. r = repeatability = 2.77 s; $\overline{x}$ = average.

three or four parallel measurements. Two would not be enough, but with more than four measurements, the accuracy would be practically the same.

The question of how many repeats are necessary in order to have a reliable mean value can also be answered according to Sutarno [12] as follows if an estimate of the standard deviation, s, is available. Two measurements, x_1 and x_2, are first made. If their difference does not exceed $2.77s$, the so-called repeatability, then these two results are averaged and the procedure is finished. On the other hand, if $|x_1 - x_2|$ does not exceed $3.3s$, then a third measurement is made. If this third result, x_3, lies between x_1 and x_2, then the three results are averaged. In all other cases, a fourth measurement must be carried out and all the four results averaged. The flow sheet of the procedure is shown in Fig. 13.

(B) THE APPLICATION OF THE LAW OF PROPAGATION OF ERRORS IN SOME ANALYTICAL METHODS

The propagation of errors will now be discussed in detail from the point of view of some analytical methods.

(1) Measurement of specific gravity

The fundamental equation of this measurement is

$$d = m/v$$

where d is the specific gravity, m is the weight, and v is the volume. Applying eqn. (18), the standard deviation, σ_d, is calculated by

$$\sigma_d^2 = \left(\frac{\partial d}{\partial m}\, \sigma_m\right)^2 + \left(\frac{\partial d}{\partial v}\, \sigma_v\right)^2$$

$$\frac{\partial d}{\partial m} = \frac{\partial(m/v)}{\partial m} = \frac{1}{v}$$

$$\frac{\partial d}{\partial v} = \frac{\partial(m/v)}{\partial v} = -\frac{m}{v^2}$$

$$\sigma_d = \sqrt{\left(\frac{1}{v}\, \sigma_m\right)^2 + \left(-\frac{m}{v^2}\, \sigma_v\right)^2}$$

It must be noted that the standard deviation depends on the measured values (m, v). A generalized form of this statement is: by means of a suitable choice of measured values the inaccuracy can be minimized.
Example 1.

$$s_m = 10^{-4}\ \text{g} \qquad s_v = 10^{-2}\ \text{ml}$$

If $m = 1$ g , $\qquad v = 2$ ml , $\qquad$ and $\qquad d = 0.5$ g ml^{-1}

$$\sigma_d = \sqrt{(\tfrac{1}{2} \times 10^{-4})^2 + (\tfrac{1}{4} \times 10^{-2})^2} = 2.5 \times 10^{-3}\ \text{g ml}^{-1}$$

Example 2.

If $m = 5$ g , $\qquad v = 10$ ml , $\qquad$ and $\qquad d = 0.5$ g ml^{-1}

$$\sigma_d = \sqrt{(\tfrac{1}{5} \times 10^{-4})^2 + (\tfrac{5}{100} \times 10^{-2})^2} = 5 \times 10^{-4}\ \text{g ml}^{-1}$$

In the second case, the standard deviation is essentially smaller.

(2) Gravimetry

In gravimetry, the percentage of the component to be analyzed is calculated by

$$p = 100\ ka/w$$

where k is the stoichiometric conversion factor, a is the weight of the precipitate, and w is the weight of the sample. In order to simplify the application of the law of propagation of errors, let us take the logarithms of both sides

$\ln p = \ln 100 + \ln k + \ln a - \ln w$

Calculating the standard deviation of $\ln p$ in terms of the logarithms on the right-hand side and keeping in mind that the stoichiometric factor k is a constant

$(d \ln p)^2 = (d \ln a)^2 + (d \ln w)^2$

or, taking into account the differential calculus formula

$$\frac{d \ln x}{dx} = \frac{1}{x} \quad \text{i.e.} \quad d(\ln x) = \frac{dx}{x}$$

$$\left(\frac{dp}{p}\right)^2 = \left(\frac{da}{a}\right)^2 + \left(\frac{dw}{w}\right)^2$$

Thus, the relative standard deviation is the sum of the relative standard deviations of the two weights, a and w.

In order to minimize the error of the percentage value, the errors in weighing must be kept as low as possible and the measured values must be high.

(3) Titrimetry

In titrimetry, the percentage of the component to be analyzed is calculated by

$$p = \frac{100kfv}{w}$$

where k is the stoichiometric conversion factor, v is the volume of the titrant, f is its titrimetric factor, and w is the weight of the sample. In order to simplify the application of the law of propagation of errors, let us take the logarithms of both sides

$\ln p = \ln 100 + \ln k + \ln f + \ln v - \ln w$

Calculating the standard deviation of $\ln p$ in terms of the logarithms on the right-hand side, and keeping in mind that the stoichiometric factor k is a constant

$$(\text{d} \ln p)^2 = (\text{d} \ln f)^2 + (\text{d} \ln v)^2 + (\text{d} \ln w)^2$$

or

$$\left(\frac{\text{d}p}{p}\right)^2 = \left(\frac{\text{d}f}{f}\right)^2 + \left(\frac{\text{d}v}{v}\right)^2 + \left(\frac{\text{d}w}{w}\right)^2$$

so the relative standard deviation of the percentage can easily calculated.

It is obvious that the relative variance of the percentage is the sum of three variances and so each of them should be as low as possible. If the condition $\text{d}f/f < 0.001$ or 0.1% is fulfilled, then the first term on the right-hand side can be neglected [7] and the above equation reduces to

$$\left(\frac{\text{d}p}{p}\right)^2 = \left(\frac{\text{d}v}{v}\right)^2 + \left(\frac{\text{d}w}{w}\right)^2$$

The term $\text{d}v/v$ can be kept at a fairly low value by using, for example, a burette of 50 ml with its standard deviation of less than 0.05 ml. By weighing an appropriate quantity, w, so that v will be 30—40 ml, the term $\text{d}v/v$ will be approximately equal to 0.001 and the relative standard deviation practically equal to that of weighing.

(4) The standard addition method [15]

Let us assume that the concentration of an element to be measured is x and the result of some measurement is B. Then a known quantity a of the element in question is added and the value A is measured. The evaluation can be carried out by graphical and numerical methods.

If there exists a linear relationship between the concentration and the measured value, the proportionality

$$\frac{B}{x} = \frac{A}{x + a}$$

is valid (see Fig. 14). After rearranging

$$x = a \frac{B}{A - B}$$

In order to calculate the relative standard deviation of the measured

76

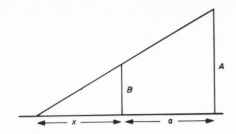

$$\frac{B}{x} = \frac{A}{x+a}$$
$$\therefore x = a\frac{B}{A-B}$$

Fig. 14. Standard addition method.

value, $V_x = s_x/x$, we assume that the relative standard deviation of A and B are equal, viz.

$$\frac{s_A}{A} = \frac{s_B}{B} = V_m$$

By using the law of propagation of errors, [eqn. (18)]

$$s_x^2 = \left(\frac{\partial x}{\partial A} s_A\right)^2 + \left(\frac{\partial x}{\partial B} s_B\right)^2$$

Calculating the partial derivatives

$$\frac{\partial x}{\partial A} = -\frac{aB}{(A-B)^2} = -\frac{x}{(A-B)}$$

$$\frac{\partial x}{\partial B} = \frac{a}{A-B} + \frac{aB}{(A-B)^2} = \frac{aA}{(A-B)^2} = \frac{Ax}{B(A-B)}$$

So

$$s_x^2 = \frac{x^2}{(A-B)^2} s_A^2 + \frac{A^2 x^2}{B^2(A-B)^2} s_B^2$$

Dividing by x^2

$$\frac{s_x^2}{x^2} = \frac{1}{(A-B)^2} s_A^2 + \frac{A^2}{B^2(A-B)^2} s_B^2$$

References pp. 168—169

Then multiplying and dividing the first term by A^2

$$V_x^2 = \frac{s_x^2}{x^2} = \frac{A^2}{(A-B)^2}\frac{s_A^2}{A^2} + \frac{A^2}{(A-B)^2}\frac{s_B^2}{B^2}$$

$$= \frac{A^2}{(A-B)^2}V_m^2 + \frac{A^2}{(A-B)^2}V_m^2 = V_m^2\,2\,\frac{A^2}{(A-B)^2} \qquad (20)$$

It is obvious from Fig. 13 that

$$\frac{A}{x+a} = \frac{A-B}{a}$$

or

$$\frac{A}{A-B} = \frac{x+a}{a}$$

Substituting into eqn. (20).

$$V_x^2 = V_m^2 \cdot 2\left(\frac{x+a}{a}\right)^2$$

or

$$V_x = V_m\sqrt{2}\left(\frac{x+a}{a}\right) = V_m\sqrt{2}\left[\frac{1+(a/x)}{a/x}\right]$$

Constructing the graph of the dependence of V_x/V_m in terms of a/x gives a hyperbola. In order to minimize V_x in terms of a, it is advantageous to choose a value of a/x in the range where the hyperbola approaches its horizontal asymptote. We can say that the value of a/x must be at least 2 or a must be twice the concentration to be measured.

(5) Indirect gravimetric analysis

Indirect gravimetric analysis in the case of two components is based on measuring two different precipitates formed from the same mixture containing the two components. For example, a mixture containing sodium and potassium is first converted into a mixture of NaCl and KCl and weighed. The mixture is then converted into a mixture of Na_2SO_4 and K_2SO_4 and is weighed again.

Denoting the quantity of Na and K by x and y, respectively, and the weight of the two precipitates by A and B, the following equa-

78

tions, expressing the conservation of mass, can be set up.

$$ax + by = A \tag{21}$$

$$cx + fy = B \tag{22}$$

where a, b, c, and f are the stoichiometric factors

$$a = \frac{[\text{NaCl}]}{[\text{Na}]} \qquad b = \frac{[\text{KCl}]}{[\text{K}]}$$

$$c = \frac{[\text{Na}_2\text{SO}_4]}{2[\text{Na}]} \qquad f = \frac{[\text{K}_2\text{SO}_4]}{2[\text{K}]}$$

and the brackets denote molecular or atomic weights.

In order to calculate the standard deviation, s_x and s_y, of the quantities to be measured, let us first solve eqns. (21) and (22).

$$x = \frac{fA - bB}{D} \qquad y = \frac{aB - cA}{D}$$

where $D = af - bc$.

According to the law of propagation of errors, [eqn. (18)]

$$s_x^2 = \left(\frac{\partial x}{\partial A} s_A \right)^2 + \left(\frac{\partial x}{\partial B} s_B \right)^2 \tag{23}$$

and

$$s_y^2 = \left(\frac{\partial y}{\partial A} s_A \right)^2 + \left(\frac{\partial y}{\partial B} s_B \right)^2 \tag{24}$$

We can assume that the standard deviation of the weighing is constant, i.e.

$$s_A = s_B = s_W \tag{25}$$

Calculating the partial derivatives

$$\frac{\partial x}{\partial A} = \frac{f}{D}; \qquad \frac{\partial x}{\partial B} = -\frac{b}{D}; \qquad \frac{\partial y}{\partial A} = \frac{a}{D}; \qquad \frac{\partial y}{\partial B} = -\frac{c}{D}$$

Substituting these derivatives in eqns. (23) and (24) and taking into account eqn. (25), the standard deviations are

$$s_x = \frac{\sqrt{b^2 + f^2}}{D} s_w$$

and

$$s_y = \frac{\sqrt{a^2 + c^2}}{D} s_w$$

The denominator $D = af - bc$ is independent of the concentrations and depends only on the stoichiometry, i.e. on the atomic and molecular weights. The determinant of the system of eqns. (21) and (22)

$$D = \begin{vmatrix} a & b \\ c & f \end{vmatrix} = af - bc$$

is the most important factor in the accuracy of indirect analysis. Its value must be as high as possible. (If the value of D is small, the system of equations is ill-conditioned, i.e. the roots are not well-defined and their accuracy is very low.)

We shall now calculate the determinant D for two fairly general cases of the indirect analysis of two component systems and then we shall draw some conclusions regarding the error of the determination.

We denote the cations and anions by K_i and A_i ($i = 1, 2, 3, ...,$). Let us first assume that two mixed precipitates, $K_1A_1 + K_2A_1$ and $K_1A_2 + K_2A_2$, are formed. Then the system of equations expressing the conservation of mass is

$$\frac{[K_1A_1]}{[K_1]} x + \frac{[K_2A_1]}{[K_2]} y = A$$

$$\frac{[K_1A_2]}{[K_1]} x + \frac{[K_2A_2]}{[K_2]} y = B$$

where the terms in brackets denote the equivalent weights of the ions and compounds.

The determinant of this system of equations is

$$D = \begin{vmatrix} \dfrac{[K_1A_1]}{[K_1]} & \dfrac{[K_2A_1]}{[K_2]} \\ \dfrac{[K_1A_2]}{[K_1]} & \dfrac{[K_2A_2]}{[K_2]} \end{vmatrix} = \frac{1}{[K_1][K_2]} \begin{vmatrix} [K_1 \ A_1] & [K_2 \ A_1] \\ [K_1 \ A_2] & [K_2 \ A_2] \end{vmatrix}$$

Recalling that

$$[K_iA_j] = [K_i] + [A_j] \qquad (i, j = 1, 2)$$

80

TABLE 18

The determinant, D, for two indirect analyses

Ions to be determined		A	B	D
x	y			
K	Na	KCl + NaCl	$K_2SO_4 + Na_2SO_4$	0.2245
Sr	Ca	$SrCO_3 + CaCO_3$	$SrSO_4 + CaSO_4$	0.4883

and the rule of addition of determinants

$$D = \frac{1}{[K_1][K_2]} ([K_1] - [K_2])([A_1] - [A_2])$$

we can see that the accuracy is increased by diminishing the equivalent weights both of the cations and of anions.

Two examples for this type of determination are summarized in Table 18.

In the second case, the two mixed precipitates $K_1A_1 + K_2A_1$ and K_3A_1 or $K_1A_1 + K_1A_2$ and K_1A_3 are formed. The system of equations for the determination of the amounts of K_1 and K_2 is as follows. (Similar equations are valid if the anions A_1 and A_2 must be determined. In this case the K and A must be interchanged.)

$$\frac{[K_1A_1]}{[K_1]} x + \frac{[K_2A_1]}{[K_2]} y = A$$

$$\frac{[K_3A_1]}{[K_1]} x + \frac{[K_3A_1]}{[K_2]} y = B$$

The determinant D is calculated as above.

$$D = \frac{([K_3] - [A_1])([K_1] - [K_2])}{[K_1][K_2]}$$

$$= \frac{1}{[K_1][K_2]} [K_3A_1]([K_1] - [K_2])$$

The accuracy of the determination is increased by increasing the equivalent weight of the second precipitate, K_3A_1, and by increasing the difference in the equivalent weights of the ions to be determined.

TABLE 19

The determinant, D, for some indirect analyses

Ions to be determined		A	B	D
x	y			
K	Na	KCl + NaCl	AgCl	2.5672
K	Na	$K_2SO_4 + Na_2SO_4$	$BaSO_4$	2.0916
Br	Cl	AgBr + AgCl	Ag	1.6929
I	Cl	AgI + AgCl	Ag	2.1928
I	Br	AgI + AgBr	Ag	0.5000
Br	Cl	AgBr + AgCl	AgCl	2.2495
I	Cl	AgI + AgCl	AgCl	2.9139
I	Br	AgI + AgBr	AgCl	0.6643
Br	Cl	AgBr + AgCl	AgBr	2.9472
I	Cl	AgI + AgCl	AgBr	3.8177
I	Br	AgI + AgBr	AgBr	0.8704
Br	Cl	AgBr + AgCl	AgI	3.6848
I	Cl	AgI + AgCl	AgI	4.7731
I	Br	AgI + AgBr	AgI	1.0882

Some examples of this type of indirect analysis are given in Table 19.

From Tables 18 and 19, one can see that, when determining K and Na by indirect analysis, the value of D for the precipitates KCl + NaCl and $K_2SO_4 + Na_2SO_4$ is small enough (Table 18, first line) compared with the precipitates KCl + NaCl and AgCl or $K_2SO_4 + Na_2SO_4$ and $BaSO_4$ (Table 19, first two lines). Therefore, in order to reduce the errors, the latter two methods are recommended instead of the first.

When determining Br and Cl by the same method, it is recommended that it is done by the formation of the precipitates AgBr + AgCl and AgI since the value of D for Br and Cl is, in this case, the highest (Table 19, line 12).

(6) Gas chromatography

The basis of quantitative gas chromatography is as follows. Chromatograms, i.e. curves representing the concentration of the components to be analyzed as a function of the time, are registered.

82

Under given gas chromatographic circumstances, each component has a characteristic peak. The time corresponding to the maximum of the peak is characteristic of the component and the peak area is proportional to its amount.

The most important measurements in gas chromatography are the measurement of retention times and of peak areas. The latter is now measured almost exclusively by integrators, but two approximate methods are still in use. The first of them consists of measuring the peak height and multiplying it by the width at half height (see Fig. 15, BE and HJ).

Another method is to draw tangents to each side of the peak and to construct a triangle with them and the base line. The area of this triangle is approximately proportional to the area of the peak (see Fig. 15, FG and AB).

Having determined the peak area of the ith component of a mixture, A_i, containing n components, its percentage composition, c_i, is calculated in the simplest case by the equation

$$c_i = \frac{a_i A_i}{\sum a_j A_j} \times 100 \tag{26}$$

where A_j (j = 1, 2, ..., n) are the peak areas and a_j are calibration constants relating each peak area to the amount of the corresponding component and which also includes the response factor. The summation in the denominator is extended to all components present in the mixture. (It must be noted that eqn. (26) gives a correct result only if each component has been eluted.)

The relative error in the percentage composition can be calculated

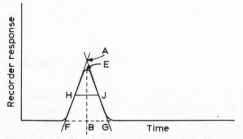

Fig. 15. Idealized chromatogram showing a peak.

by taking logarithms of both sides of eqn. (26).

$$\ln c_i = \ln a_i + \ln A_i - \ln(\sum a_j A_j) + \ln 100 \tag{27}$$

It is assumed that the calibration factors and the peak areas of the components other than number i are error-free constants. In other words, the error in the percentage composition is determined only by the error in the corresponding peak area, A_i. Calculating the derivative of eqn. (27) in terms of A_i gives

$$\frac{d \ln c_i}{d A_i} = \frac{1}{A_i} - \frac{a_i}{\sum a_j A_j}$$

After rearranging

$$\frac{dc_i}{c_i} = \frac{dA_i}{A_i} - \frac{a_i \, dA_i}{\sum a_j A_j} \tag{28}$$

From eqn. (26)

$$\sum a_j A_j = \frac{a_i A_i}{c_i} \times 100$$

Substituting this in eqn. (28)

$$\frac{dc_i}{c_i} = (1 - c_i/100) \frac{dA_i}{A_i}$$

It is obvious that the smaller the relative error of percentage composition, assuming that the relative error of peak area determination is constant, the higher the percentage of the component to be analyzed.

(7) Potentiometry

The basic equation of potentiometric analysis is the Nernst equation, viz.

$$E = E^0 + \frac{RT}{nF} \ln c \tag{29}$$

where E is the measured potential of the electrode, E^0 is the standard electrode potential of the ion to be determined, R is the gas constant, T is the absolute temperature, n is the number of electronic charges of the ion to be determined, F is the Faraday constant, and

84

c is the concentration. Assuming that $T = 298°K$ and taking into account that R and F are universal constants, eqn. (29) can be written as

$$E = E^0 + \frac{0.025}{n} \ln c \tag{30}$$

The concentration c is calculated from eqn. (30) as

$$c = \exp[40.0n(E - E^0)] \tag{31}$$

The variance of c is, according to eqn. (18)

$$s_c^2 = \left(\frac{\partial c}{\partial E} s_E\right)^2$$

and

$$s_c = \frac{\partial c}{\partial E} s_E \tag{32}$$

Differentiating eqn. (31)

$$\frac{\partial c}{\partial E} = 40.0n \exp[40.0n(E - E^0)]$$

and substituting into eqn. (32)

$$s_c = 40n \exp[40n(E - E^0)]s_E$$

The relative standard deviation is

$$\frac{s_c}{c} = 40.0ns_E \tag{33}$$

It can be seen that the relative standard deviation is greater than the standard deviation of the potential measurement.

Example. Calculate the accuracy of the potentiometric analysis. Suppose that $n = 1$ and that the accuracy of the measurement of the potential is almost maximum, $s_E = 10^{-3}$.

From eqn. (33)

$$\frac{s_c}{c} \times 100 = 40.0 \times 1 \times 10^{-3} \times 100 = 4.0\%$$

This value is sufficiently high and therefore acid solutions are, in general, measured by titrimetric methods and not by their pH.

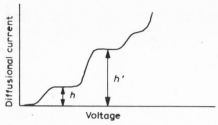

Fig. 16. Idealized polarogram. Standard addition method.

(8) Polarographic analysis

The most important application of polarography in analytical chemistry is the determination of concentrations from polarograms, which are graphs representing the current intensity as a function of the electrode potential (Fig. 16). The concentration is measured by measuring the heights of the polarographic waves on the polarogram since these heights are, under unchanged experimental conditions, proportional to the concentrations. The proportionality factor can be determined by calibration but, in practice, instead of determining the proportionality factor, the calibration is carried out by the standard addition method [see also Sect. 3.(B)(4)].

The polarogram of the solution to be measured is first recorded. Let the concentration of the solution be x, its volume V, and the height of polarographic wave h. A volume A of a solution of known concentration c is then added and the height of the resulting polarographic wave, h' measured (Fig. 16).

The concentration of the solution obtained by adding the solution of known concentration to that to be measured is calculated as follows. The amount of solute in the solution to be measured is xV and that of the solution of known concentration cA. The amount of the solute present in the final solution is therefore $xV + cA$. The volume of the final solution is $V + A$ and so its concentration is $(xV + cA)/(V + A)$.

Because the heights of the polarographic waves are proportional to the concentrations, the proportion of heights and concentrations must be constant.

$$\frac{h}{x} = \frac{h'(V + A)}{xV + cA}$$

86

From this, the concentration is expressed as

$$x = \frac{hc}{h' + (h' - h)(V/A)} \tag{34}$$

When calculating the variance of the measured concentration, it is assumed that the variances of volumes A and V, as well as that of the concentration c, can be neglected in comparison with that of the height of the polarographic waves. This means that the volumes and the concentration of the standard solution can be determined with a greater accuracy than that of the height.

In order to apply the law of propagation of errors, the partial derivatives of eqn. (34), $\partial x/\partial h$ and $\partial x/\partial h'$, must be calculated. With the aid of these derivatives, the standard deviation of the concentration is

$$dx = \frac{c[1 + (V/A)]}{[h' + (h' - h)(V/A)]^2} \sqrt{h^2(dh')^2 + h'^2(dh)^2} \tag{35}$$

Dividing eqn. (35) by eqn. (34) and rearranging

$$\frac{dx}{x} = \frac{h'(V + A)}{h'(V + A) - hV} \sqrt{\left(\frac{dh'}{h'}\right)^2 + \left(\frac{dh}{h}\right)^2} \tag{36}$$

Making use of eqn. (34) again and substituting the denominator by hcA/x

$$\frac{dx}{x} = \frac{h'x}{hc} \frac{A + V}{A} \sqrt{\left(\frac{dh'}{h'}\right)^2 + \left(\frac{dh}{h}\right)^2} \tag{37}$$

By a further transformation, eqn. (37) becomes

$$\frac{dx}{x} = \left[1 + \frac{h/h'}{1 + (A/V) - (h/h')}\right] \sqrt{\left(\frac{dh'}{h'}\right)^2 + \left(\frac{dh}{h}\right)^2} \tag{38}$$

From eqn. (38), it is obvious that, in order to minimize the variance of the concentration at constant relative variances of h' and h, first of all the term A/V must be as high as possible, at least greater than 2 and then the term h/h' must be as small as possible. This can be achieved by increasing the concentration of the standard solution.

(9) Photometry

The basic equation of all photometric measurements is the law of Lambert and Beer.

$$I = I_0 \exp(-\beta l) \tag{39}$$

Equation (39) states that the intensity of light I_0 diminishes to I when passing through a medium (in most cases liquid) of thickness l (cuvette length). The coefficient β is the extinction coefficient. This is related, if there is in the mixture to be analyzed only one light absorbing component, to the molar extinction coefficient a and to the concentration, c by

$$\beta = ca \tag{40}$$

The quantity

$$E = \ln \frac{I_0}{I} \tag{41}$$

is called the extinction of the solution and can be expressed as

$$E = acl \tag{42}$$

or, by using the notation $k = 1/al$

$$c = kE \tag{43}$$

So the concentration is determined by determining E from I and I_0 and then c is calculated from eqn. (43). The proportionality factor, k, must be determined for each cuvette and for each substance by calibration. In order to calculate the variation of the concentration we first calculate the variance of E in terms of dI and dI_0 from eqn. (41).

$$(dE)^2 = \left(\frac{\partial E}{\partial I_0}\right)^2 (dI_0)^2 + \left(\frac{\partial E}{\partial I}\right)^2 (dI)^2 = \frac{(dI_0)^2}{I_0^2} + \frac{(dI)^2}{I^2}$$

It can be assumed that $dI_0 = dI$.

Calculating immediately the relative variance

$$\frac{dE}{E} = \frac{dI}{E} \sqrt{\frac{1}{I_0^2} + \frac{1}{I^2}}$$

Replacing E on the right-land side by $\ln(I_0/I)$ [eqn. (41)] and introducing the notation $x = I/I_0$ gives

$$\frac{dE}{E} = -\frac{dI}{I_0} \frac{\sqrt{x^2 + 1}}{x \ln x} \tag{44}$$

This relative variance of the extinction has its minimum where

$$f(x) = \frac{\sqrt{x^2 + 1}}{x \ln x} \tag{45}$$

is maximum since I_0 and dI are constants.

88

Using differential calculus, it can be shown that at $x = 0.37$ eqn. (45) has a maximum, i.e. eqn. (44) has a minimum. More precisely, the value of x giving an extremum is one root of the equation $x^2 + \ln x + 1 = 0$.

From the above, it is suggested that it is preferable to work at about $I = 0.37\,I_0$. From a more thorough investigation of eqn. (45), it can be concluded that the minimum is very flat and the interval $0.05 I_0 < I < 0.7 I_0$, i.e. $1.3 > E > 0.2$, is quite suitable for the analysis.

Continuing the calculation of the variation of the concentration, we have to calculate the variance of the proportionality factor, k, in eqn. (43). The value of k is determined by calibration using test samples of known concentration and extinction. From eqn. (43)

$$k = \frac{c}{E}$$

therefore the variance of k is

$$\left(\frac{\mathrm{d}k}{k}\right)^2 = \left(\frac{\mathrm{d}c}{c}\right)^2 + \left(\frac{\mathrm{d}E}{E}\right)^2$$

According to experience [7], $(\mathrm{d}c/c)^2$, the variance of concentration, being the sum of the variances of weighing and diluting, can be neglected compared with $(\mathrm{d}E/E)^2$ and so

$$\mathrm{d}k/k \approx \mathrm{d}E/E \qquad .$$

If the value of k is a mean value of n measurements, then its variation is

$$\frac{\mathrm{d}k}{k} = \frac{\mathrm{d}E}{\sqrt{n}E} \tag{46}$$

Now, from eqn. (43) and from the law of propagation of errors, the variance of the measured concentration is

$$\left(\frac{\mathrm{d}c}{c}\right)^2 = \left(\frac{\mathrm{d}k}{k}\right)^2 + \left(\frac{\mathrm{d}E}{E}\right)^2 \qquad \vdots$$

and taking into account eqn. (46)

$$\left(\frac{\mathrm{d}c}{c}\right)^2 = \left(\frac{\mathrm{d}E}{E}\right)^2\left(1 + \frac{1}{n}\right)$$

$$\frac{\mathrm{d}c}{c} = \sqrt{1 + \frac{1}{n}}\,\frac{\mathrm{d}E}{E}$$

References pp. 168—169

This means that the relative variation of the measured concentration tends to that of the extinction by increasing the number of duplicates in the calibration process.

(10) (Non-differential) spectrophotometry [16]

The photometric error can also be derived in another way. For the sake of comparison with differential spectrophotometry [see Sect. 3.(B)(11)], this derivation is described here.

The symbols used in addition to those of the previous section are R, which represents the scale readings, and T, the transmittance, values of which fall between 0 and 1. The correlation between E and T is

$$E = -\ln T$$

The following indices are used: s refers to the solvent, r to a reference solution, x to the unknown solution, and D to total darkness.

Following T and E, some symbols are added in parentheses. Thus, $T(x/s)$ means, for example, the transmittance of the unknown measured against the solvent, the validity of Beer's law is assumed, i.e. the correlation

$$E(x/s) = alc_x = \frac{c_x}{A} = -\ln \frac{I_x}{I_s} \tag{47}$$

holds for each value of c_x, where $A = 1/(al)$. In spectrophotometric work, two adjustments are required before the actual measurement is made. First, the zero point of the transmittance scale must be adjusted. This is done by placing a shutter in the light beam, i.e. the photocell is in complete darkness, and balancing the instrument with resistors (balancing the dark current of the photocell). When this is completed, the instrument indicates 0% transmittance if a completely opaque species is placed in the light beam. The second manipulation brings the adjustment of the 100% transmittance on to the scale, by placing the solvent in the light beam and rebalancing the instrument with a second variable resistor. If a solution is now placed in the light beam, its transmittance, which is equal to the light intensity now falling on the photocell (relative to that passing through the solvent) can be measured. Thus the transmittance scale must be adjusted (calibrated) by two different light intensities, one of them having the value zero, i.e. total darkness.

Calibration of the transmittance scale, however, can also be made

90

by two other light intensities which differ less in value than the two described above. For example, if two reference solutions containing the absorbing species in different concentrations are used for scale calibration, one can adjust the zero transmittance with the more concentrated solution in the light beam and 100% with the other.

The scale calibration is made by an opaque species or shutter (0%) and solvent (100%). Thus, $R_s = 1.00$ and $R_D = 0.00$. Transmittance and extinction readings can be described as

$$T(\text{x/s}) = \frac{I_x}{I_s}$$

and

$$E(\text{x/s}) = -\ln T(\text{x/s})$$

respectively. The concentration of an unknown solution can be expressed as

$$c_x = \frac{E(\text{x/s})}{al} = AE(\text{x/s})$$

which gives a directly proportional relationship between c_x and E. The coefficient of variation of the concentration determination

$$\frac{dc_x}{c_x} = -\frac{1}{e^{-E(\text{x/s})}E(\text{x/s})}\, dT(\text{x/s})$$

is directly proportional to the standard deviation of transmittance readings $dT(\text{x/s})$ and depends on the extinction values themselves.

It can be demonstrated by differentiating the coefficient of $dT(\text{x/s})$ in terms of $E(\text{x/s})$ that the above expression of dc_x/c_x has a minimum at $E(\text{x/s}) = 1.0$ where the most precise results can be expected.

(11) Differential spectrophotometry [16]

Differential spectrophotometric methods were developed to improve the precision of results. This can be achieved by the appropriate expansion of the scale used for the measurement of light intensities. Scale expansion can be made by adjusting the transmittance scale with solutions of suitable concentrations. The transmittance of any solution with a concentration between the two calibrating solutions may be measured with an increased precision, compared with ordinary spectrophotometry, the scale now being more sensitive.

The most simple differential spectrophotometric method is the transmittance method. In this method, 100% transmittance is given by a reference solution containing the absorbing species to be measured in a known concentration, c_r ($R_r = 1.00$). The transmittance and extinction readings in this case are

$$T(x/r) = \frac{I_x}{I_r} \quad \text{and} \quad E(x/r) = -\ln T(x/r) \tag{48}$$

respectively. The name of the method is derived from the fact that the transmittance read in this case can be regarded as the ratio of two transmittances

$$T(x/r) = \frac{T(x/s)}{T(r/s)}$$

where $T(x/s) = I_x/I_s$ and $T(r/s) = I_r/I_s$.

The concentration of the unknown can be expressed as

$$c_x = \frac{E(x/r)}{al} + c_r = AE(x/r) + B \tag{49}$$

yielding a linear (but not directly proportional) correlation between concentration and extinction.

The standard deviation of concentration is calculated in the following way. Applying Beer's law, eqn. (47), to the reference solution

$$c_r = AE(r/s)$$

and substituting this in eqn. (49)

$$c_x = AE(x/r) + AE(r/s)$$

From the law of propagation of errors

$$(dc_x)^2 = A^2 \, [dE(x/r)]^2 + A^2 \, [dE(r/s)]^2 \tag{50}$$

The extinction of the reference solution can be regarded as an error-free constant since it has been determined by a calibration process. Its variance can be neglected in comparison with that of $dE(x/r)$. Equation (50) thus reduces to

$$dc_x = A \, dE(x/r)$$

which is the standard deviation of the concentration determination.

92

The relative standard deviation is

$$\frac{dc_x}{c_x} = \frac{A \, dE(x/r)}{A[E(x/r) + E(r/s)]} \tag{51}$$

From eqn. (48)

$$dE(x/r) = -\frac{dT(x/r)}{T(x/r)} \tag{52}$$

Substituting eqns. (48) and (52) into eqn. (51)

$$\frac{dc_x}{c_x} = -\frac{dT(x/r)}{\exp[-E(x/r)][E(x/r) + E(r/s)]} \tag{53}$$

Studying eqn. (53) more closely, it can be stated that, as the extinction of the reference solution increases, the relative standard deviation of concentration determination decreases.

It can be shown that the coefficient of $dT(x/r)$ in eqn. (53)

$$F(x/r) = -\frac{1}{\exp[-E(x/r)][E(x/r) + E(r/s)]}$$

has a minimum at $E(x/r) = 1 - E(r/s)$ when regarding $E(r/x)$ as variable. This means that the minimum shifts towards lower $E(x/r)$ extinction values, finally disappearing at $E(r/s) = 1$. For $E(r/s) > 1$, the coefficient $F(x/r)$ is a monotonous function of $E(x/r)$.

Because the precision increases with decreasing $F(x/r)$, the following conclusions can be drawn.

(a) An increase of precision can be expected if the extinction of the reference solution (measured against the solvent) increases, i.e. if reference solutions of higher concentration are used.

(b) If $E(r/s) > 1$, then the precision of the measurement increases with decreasing extinction of the unknown solution (measured against the reference), $E(x/r)$, gaining maximum precision if $E(x/r) = 0$, i.e. if the concentration of the unknown solution is equal to that of the reference solution.

(c) The case $F(x/r) < 1$ is especially favourable because the standard deviation of the transmittance readings, $dT(x/r)$, has to be reduced to gain the relative standard deviation of the concentration measurements.

(12) Emission spectrographic analysis [17]

The fundamental equation of the analysis is due to Lomakin and Scheiber.

$$\frac{\Delta S}{\gamma} = b \ln c + \ln a \tag{54}$$

where ΔS is the difference in the blackening of the analytical lines, c is the concentration, γ is the contrast factor of the photographic plate, and b and a are parameters.

In order to calculate the relative standard deviation of the concentration, we first obtain the expression for $\ln c$ from Eqn. (54), viz.

$$\ln c = \frac{1}{b} \frac{\Delta S}{\gamma} - \frac{\ln a}{b}$$

Calculating the total differential of $\ln c$

$$d(\ln c) = \frac{dc}{c} = \frac{d(\Delta S)}{b\gamma} - \frac{\Delta S}{b\gamma^2} d\gamma$$

or

$$dc = c \frac{1}{b\gamma} d \Delta S + c \frac{\Delta S}{b^2} d\gamma$$

Applying the law of propagation of errors

$$\left(\frac{dc}{c}\right)^2 = \left(\frac{1}{b\gamma}\right)^2 \left\{ [d(\Delta S)]^2 + \frac{(\Delta S)^2}{\gamma^2} (d\gamma)^2 \right\}$$

When studying the scattering of analytical results obtained by using one photographic plate, γ is constant and we get

$$\frac{dc}{c} = \frac{d(\Delta S)}{b\gamma} \tag{55}$$

Equation (55) is usually applied in current analytical work for determining the error, expressed as a percentage of the concentration, using experimental $d(\Delta S)$ values. From the above equations, it further follows that to increase the accuracy, γ and b must be increased, i.e. the contrast factor must be as high as possible. This can be provided by using extracted material. The accuracy can also be increased by choosing analytical pairs of lines so that ΔS should be as small as possible in the middle of the range of concentrations.

94

From eqn. (55), it follows that for the analysis of emission spectra the relative standard deviation does not depend on the concentration to be determined. For this reason, emission spectrographic analysis is used particularly extensively in the analysis of low concentrations. In chemical analysis in this range of concentrations, the relative standard increases with decreasing concentration.

Let us now consider a very important method in spectroscopic analysis, the method with two analytical pairs of lines. This method makes it possible to obtain the maximum information from one spectrogram. In this case, the fundamental equation of quantitative spectroscopic analysis is

$$\frac{\Delta S_1 + \Delta S_2}{\gamma} = (b_1 + b_2) \ln c + \ln a_1 + \ln a_2$$

Applying the law of propagation of errors and supposing that the correlation between ΔS_1 and ΔS_2 can be neglected, the following expression can be deduced for the relative standard deviation of the concentration.

$$\left(\frac{dc}{c}\right)^2 = \frac{[d(\Delta S_1)]^2 + [d(\Delta S_2)]^2}{[\gamma(b_1 + b_2)]^2}$$

Further, if it is assumed that $b_1 \approx b_2 = b$ and $d(\Delta S_1) \approx d(\Delta S_2) = d(\Delta S)$ then

$$\frac{dc}{c} = \frac{d(\Delta S)}{\sqrt{2}\gamma b}$$

It means that when applying this method, the relative standard deviation is reduced by $\sqrt{2}$, i.e. by approximately 1.4.

(13) Diffuse reflectance spectroscopy [18]

Reflectance spectroscopy can make a unique contribution when an attempt is made to measure the absorption with turbid or colloidal systems or when one wishes to determine the absorption spectra of substances adsorbed on solid surfaces.

The radiation reflected from a finely divided solid consists of a regular, or specular, reflection component and a diffuse reflection is important when diffuse reflectance is measured.

The most general theory of diffusion reflection is given by Kubelka and Munk. Their equation for an infinite thick opaque layer, when the reflectance of a sample diluted with a non- or low-absorbing powder is measured against the pure powder, may be written in the form

$$\frac{(1-R)^2}{2R} = \frac{2.303\epsilon c}{s} \tag{56}$$

where R is the diffuse reflectance of the layer relative to a non- or low-absorbing standard, s is the scattering coefficient, ϵ is the extinction coefficient, and c is the molar concentration. Under constant experimental conditions, the left-hand side of eqn. (56) is proportional to the molar concentration. Let us denote the proportionality factor by $1/k$. Equation (56) can then be rewritten

$$c = \frac{k(1-R)^2}{2R} \tag{57}$$

where $k = s/2.303\ \epsilon$. So, by measuring R, the concentration c can be calculated.

In order to determine the optimum concentration range for analysis, the relative error of the concentration must be calculated. Calculating the derivative of eqn. (57)

$$\frac{dc}{dR} = \frac{k(R^2-1)}{2R^2}$$

After dividing by eqn. (57) and rearranging

$$\frac{dc}{c} = \frac{R+1}{R(R-1)}\,dR$$

The relative error, when the error in reading the reflectance is constant, is determined by

$$f(R) = \frac{R+1}{R(R-1)} \tag{58}$$

To determine the value of R minimizing the relative error in c, the derivative of eqn. (58), i.e.

$$f'(R) = \frac{-R^2-2R+1}{R^2(R-1)^2}$$

must be equated to zero. This is so if $R^2 + 2R - 1 = 0$. The roots of this equation are $R = -1 \pm \sqrt{2}$. Since negative R values are meaning-

96

less, the minimum relative error in c, or the optimum value for reflectance, occurs at $R = 0.414$.

This result is in accordance with the following practical experiences. With high concentrations, so little radiation is reflected that the sensitivity of the spectrophotometer becomes inadequate. With low concentrations, the error in reading the instrument becomes too large.

When plotting the reflectance against the logarithm of the concentration, the curves obtained have an inflection point at $R = 0.414$. The optimum range for analysis corresponds, in this case, to the section of greatest slope.

The relative accuracy of measurement can be estimated by the use of the identity

$$\frac{dc}{c} = d(\ln c) = \frac{d(\ln c)}{dR} dR = \frac{2.303 \, d(\log c)}{dR} dR$$

Assuming a constant reading error of 1% R, i.e. $dR = 0.01$, the % relative error of concentration is $2.303 \, d(\log c)/dR$.

(14) Mass spectrometry

Analysis by mass spectrometry is based on the fact that the peak height of each component of a mixture is, proportional, under appropriate experimental conditions, to the partial pressure of the component to be analyzed, viz.

$$h_i = c_i p_i$$

where h_i is the peak height, p_i is the partial pressure, and c_i is the sensitivity factor depending on the component.

Since the sensitivity factors are independent of the other components present in the mixture only to a first approximation, the peaks are, in general, linearly superimposed and so the peak heights are expressed by the equation

$$h_j = \sum_{i=1}^{n} c_{ij} p_i \tag{59}$$

where c_{ij} is the sensitivity factor of the ith component related to the jth peak, and n is the number of components in the mixture.

Since, at constant total pressure, the partial pressures are proportional to the percentage molar composition, in order to determine the latter the partial pressures of the components must be calculated.

This is done by setting up at least n linear equations of the type in eqn. (59).

$$h_j = \sum_{j=1}^{n} c_{ij}p_j \qquad j = 1, 2, ..., m \tag{60}$$

where $m \geqslant n$. Suppose that the mutual sensitivity factors c_{ij} are known and the peak heights, h_j, are measured, the partial pressures p_i can be calculated by solving the system of linear equations shown in eqn. (60).

If $m = n$, then writing eqn. (60) in vector form

$$\mathbf{C}\,\mathbf{p} = \mathbf{h}$$

where $\mathbf{C}$ is the matrix of sensitivity factors, having n rows and n columns, $\mathbf{p}$ is the column vector of partial pressures, and $\mathbf{h}$ is the column vector of peak heights, both vectors having n elements. The partial pressures are obtained by calculating the inverse matrix of $\mathbf{C}$, denoted by $\mathbf{C}^{-1}$.

$$\mathbf{p} = \mathbf{C}^{-1}\,\mathbf{h} \tag{61}$$

The inverse matrix can be calculated by partial pivoting when using a computer by the Fortran subroutine INV [see Sect. 4.(B)(6)].

The partial pressure of the ith component is calculated by rewriting eqn. (61) in scalar form

$$p_i = \sum_{j=1}^{n} c'_{ij}h_j \qquad (i = 1, 2, ..., n)$$

where c'_{ij} are the elements of the inverse matrix, $\mathbf{C}^{-1}$.

Suppose that the error of the partial pressure p_i depends only on the error of the corresponding peak height h_i and calculating the derivative of p_i with respect to h_i

$$\frac{\mathrm{d}p_i}{\mathrm{d}h_i} = c'_{ii}$$

or

$$\mathrm{d}p_i = c'_{ii}\,\mathrm{d}h_i$$

98

it can be recognized that the absolute error of the partial pressure is proportional to that of peak height.

If the system of eqn. (60) is set up from more than n equations $(m > n)$, then the determination of partial pressures is a problem of multi-variable linear regression. The calculation method and the estimation of errors is outlined in Sect. 4.(B)(6).

(15) Radio counting methods

The importance of analytical techniques applying counting methods, e.g. radiometry or direct measuring X-ray spectroscopy, is growing. The composition of the test sample is determined in these methods by counting characteristic discrete events (e.g. impulses). In order to be able to compare the results, the number of impulses, x, is referred to unit time, e.g. to a minute. The method is called counting to pre-selected time. The counting process is repeated several times and the mean value, $\bar{x}$, is calculated. The frequency of impulses is defined as the number of impulses $\bar{x}$ divided by the time interval T, i.e.

$$\nu = \bar{x}/T \tag{62}$$

This can be determined either by counting the impulses during a prescribed interval or by measuring the time needed for observing a prescribed number of impulses.

Counting processes, in general, follow the Poisson distribution. For this type of distribution

$$s = \sqrt{\bar{x}} \tag{63}$$

(see Sect. 2.(E)(1)b). The relative error of the measurement is

$$\frac{\mathrm{d}x}{\bar{x}} = \frac{\sqrt{\bar{x}}}{\bar{x}} = \frac{1}{\sqrt{\bar{x}}}$$

The frequency of impulses defined by eqn. (62) is obtained as the difference of the frequencies due to the preparate plus the background and of the frequency due to the background alone

$$\nu = \nu_p - \nu_b$$

where the index p refers to the preparate plus the background and b to the background. Supposed that the time interval of counting, T, is prescribed and constant

$$\nu = \frac{\bar{x}_p}{T} - \frac{\bar{x}_b}{T} = \frac{\bar{x}}{T} \tag{64}$$

and so the relative error of ν is the same as that of $\bar{x}$. Equation (64) can be transformed into

$$\bar{x} = \bar{x}_p - \bar{x}_b$$

From this, taking into account eqn. (63), according to the law of propagation of errors, the variance of countings is

$$(d\bar{x})^2 = (d\bar{x}_p)^2 + (d\bar{x}_b)^2 = \bar{x}_p + \bar{x}_b$$

In radiometric methods, it is assumed that there exists a linear relationship between the frequency of impulses, ν, and the concentration, c, i.e.

$$c = k \nu \tag{65}$$

where k is a constant. This constant must be determined by a calibration process.

$$k = \frac{c_0}{\nu_0}$$

where index 0 refers to a standard sample.

When the counting of impulses is always performed over constant time intervals, then

$$c = c_0 \frac{\bar{x}}{\bar{x}_0}$$

Applying the law of propagation of errors

$$\left(\frac{dc}{c}\right)^2 = \left(\frac{dc_0}{c_0}\right)^2 + \left(\frac{d\bar{x}}{\bar{x}}\right)^2 + \left(\frac{d\bar{x}_0}{\bar{x}_0}\right)^2$$

Assuming that $\bar{x} \approx \bar{x}_0$ and $d\bar{x} \approx d\bar{x}_0$ and taking into account eqn. (63)

$$\left(\frac{dc}{c}\right)^2 = \left(\frac{dc_0}{c_0}\right)^2 + \frac{1}{\bar{x}} + \frac{1}{\bar{x}_0} \tag{66}$$

The total relative variance of the concentration can be reduced by increasing the number of counts and decreasing the relative variance of weighing and dilution. It is recommended that all three terms on the right-hand side of eqn. (66) be of the same order of magnitude.

100

4. Methodology of setting up mathematical models

Most physico-chemical methods used in analytical chemistry utilize the dependence of a certain measurable quantity on the concentration of the substances to be determined in solution. This dependence may be written

$$y = f(x) \qquad (67)$$

where y is the value of the quantity measured and x is the corresponding concentration of the component to be determined. The functional dependence between the measured quantity and concentration of the solution is mostly determined experimentally and the relationship obtained is frequently treated graphically. Such a curve of the relationship (67) between the concentration of the substance determined and the quantity measured is called a *calibration graph* [2]. Although graphical methods have a great advantage in their illustrative character, recently it has become more common to treat the relationship between the measured quantity and concentration numerically.

The determination of the above functional dependence is carried out by regression analysis or, more generally, by setting up a mathematical model.

Before treating some mathematical models in detail, let us discuss the problem of smoothing experimental results.

(A) SMOOTHING EXPERIMENTAL RESULTS

When treating experimental results, it often occurs that some data points are inaccurate and these must be found and rejected. On the other hand, sometimes one is not interested in the functional relationship between the data but only in the tendency of the relationship. In this case, the graph of smoothed points gives a better survey of the tendency than the plot of the measured data. These tasks can be carried out by a suitable smoothing procedure. A simple procedure of smoothing is to substitute each point by its weighted average. Let the observed data be $(x_1, y_1), ..., (x_{i-1}, y_{i-1}), (x_i, y_i), (x_{i+1}, y_{i+1}), ..., (x_n, y_n)$ and the smoothed values of the dependent variable $\bar{y}_1, \bar{y}_2, ..., \bar{y}_i, ..., \bar{y}_n$. The value of $\bar{y}_i$ is most simply calculated by

$$\bar{y}_i = \frac{y_{i-1} + 2y_i + y_{i+1}}{4}$$

where $i = 2, 3, ..., n - 1$

$$\bar{y}_1 = \frac{2y_2 + y_3}{3} \qquad \bar{y}_n = \frac{y_{n-1} + 2y_n}{3}$$

This procedure is repeated several times, up to a maximum of 10, and the discordant results rejected using the inequality

$$|y_i - \bar{y}_i| > f \frac{\sum_{i=1}^{n} |y_i - \bar{y}_i|}{n}$$

A suitable choice for the value of the factor f is 2 or 3. When plotting the observed data and the remaining smoothed points, one can see that the latter lie on the focus line of the observed points.

For this purpose, and for arranging the smoothed values of y and of the remaining x values into new arrays, the subroutine SMOOTH was constructed.

The declaration of the subroutine is

SUBROUTINE SMOOTH (N, X, Y, NF, F)

The parameters are

N the number of input data, at the end the number of remaining data
X the array of abscissae
Y the array of ordinates
NF the number of weightings to be performed by the subroutine CENT
F the factor defining the tolerance

Subroutine required: CENT
The listing of the subroutine is

```
SUBROUTINE SMOOTH(N,X,Y,NF,F)
REAL ME
DIMENSION X(N),Y(N),YR(40),IND(40)
CALL CENT(N,Y,YR,NF,ME,DE)
I=0
TM=ME*F
DO 2 J=1,N
DI=ABS(Y(J)-YR(J))
IF(DI.LE.TM) GC TO 2
I=I+1
IND(I)=J
```

102

```
2 CONTINUE
  IF(I.EQ.0) GO TO 5
  K=IND(1)
  IND(I+1)=N+1
  DO 4 J=1,I
  ID=IND(J+1)-J-1
  DO 3 MM=K,ID
  MJ=MM+J
  X(MM)=X(MJ)
  Y(MM)=Y(MJ)
  YR(MM)=YR(MJ)
3 CONTINUE
4 K=ID+1
5 L=I
  N=N-L
  CALL CENT(N,Y,YR,NF,ME,DE)
  RETURN
  END
```

The declaration of the subroutine CENT is

SUBROUTINE CENT (N, Y, YR, NF, ME, DE)

The parameters are

N the number of measured data
Y the array of ordinates
YR an auxiliary array for storing the input ordinates
NF the number of weighting
ME the mean deviation of the original and weighted ordinates
DE the maximum deviation of the original and weighted ordinates.

The listing of the subroutine is

```
  SUBROUTINE CENT(N,Y,YR,NF,ME,DE)
  REAL ME
  DIMENSION Y(N),YR(N)
  DO 1 I=1,N
1 YR(I)=Y(I)
  M=N-1
  DO 3 I=1,NF
  T1=Y(1)
  Y(1)=(T1*2.0+Y(2))/3.0
  DO 2 J=2,M
  T2=T1
  T1=Y(J)
2 Y(J)=(T2+Y(J)*2.0+Y(J+1))/4.0
  T2=T1
  Y(N)=(T2+Y(N)*2.0)/3.0
3 CONTINUE
  ME=0.0
```

```
      DE=0.0
      DO 5 I=1,N
      DI=ABS (Y(I)-YR(I))
      ME=ME+DI
      IF (DI-DE)5,5,4
4     DE=DI
5     CONTINUE
      U=N
      ME=ME/U
      RETURN
      END
```

(B) LINEAR MODELS

(1) Linear regression

The most simple statistical model applicable to cases in which two variables x and y are measured simultaneously is the linear model. It is assumed that a linear relationship exists between the variables, viz.

$$y = a + bx \tag{68}$$

In other words the measured values of the variables x and y are best fitted by a linear function. The process of finding the equation of best fit is called regression analysis. The values of the parameters a and b are determined, in most cases, by the method of least squares. This means that the sum of squares of the deviations of the individual measured y values from their calculated values $(a + bx)$ is a minimum compared with any other pairs of (a, b).

Let us suppose that n points are measured, i.e. $(x_1, y_1), (x_2, y_2),...,$ (x_n, y_n). It is assumed that the variable x is, for practical purposes, error-free. Then the method of least squares is applied to eqn. (68) as follows.

$$S = \sum_{i=1}^{n} (y_i - a - bx_i)^2 \tag{69}$$

Here, S is the sum of squares and the constants, a and b must be determined so that S is a minimum. The sum S reaches its minimum if

$$\frac{\partial S}{\partial a} = \frac{\partial S}{\partial b} = 0$$

Having calculated the partial differential quotients of eqn. (69), equating them to zero and rearranging gives

$$\sum_{i=1}^{n} y_i x_i = a \sum_{i=1}^{n} x_i + b \sum_{i=1}^{n} x_i^2$$

$$\sum_{i=1}^{n} y_i = an + b \sum_{i=1}^{n} x_i$$

The constants are now calculated from this system of linear equations. The practical calculation is carried out using the following scheme which contains the measured data and some auxiliary quantities. The symbols $\bar{x}$, $\bar{y}$, and $\bar{T}$ denote the average of the x_i, y_i, and $T_i = x_i + y_i$ values. In the following, $\Sigma_{i=1}^{n}$ is simply written Σ.

Measured values

x	y	T $(= x + y)$	x^2	y^2	T^2	xy
.	.	.	.	.	.	.
.	.	.	.	.	.	.
x_i	y_i	T_i	x_i^2	y_i^2	T_i^2	$x_i y_i$
.	.	.	.	.	.	.
.	.	.	.	.	.	.
Σx_i	Σy_i	ΣT_i	Σx_i^2	Σy_i^2	ΣT_i^2	$\Sigma x_i y_i$

The steps in the calculation are as follows.
(1) Tabulate x_i, y_i, and T_i.
(2) Calculate the sums Σx_i, Σy_i, and ΣT_i.
　　Check: $\Sigma T_i = \Sigma x_i + \Sigma y_i$.
(3) Calculate the averages $\bar{x}$, $\bar{y}$, and $\bar{T}$.
(4) Calculate the squares x_i^2, y_i^2, T_i^2 and the products $x_i y_i$.
(5) Calculate the sums Σx_i^2, Σy_i^2, ΣT_i^2 and $\Sigma x_i y_i$.
　　Check: $\Sigma T_i^2 = \Sigma x_i^2 + 2\Sigma x_i y_i + \Sigma y_i^2$.
　　If there is an error, check each set of data (horizontally): $T_i^2 = x_i^2 + 2x_i y_i + y_i^2$.
(6) Calculate the sums of the square deviations.
　　$SSD(x^2) = \Sigma x_i^2 - \bar{x}\Sigma x_i$
　　$SSD(y^2) = \Sigma y_i^2 - \bar{y}\Sigma y_i$
　　$SSD(T^2) = \Sigma T_i^2 - \bar{T}\Sigma T_i$

$$\text{SSD}(xy) = \Sigma y_i^2 - \bar{y}\Sigma x_i$$

Check: $\text{SSD}(T^2) = \text{SSD}(x^2) + 2\,\text{SSD}(xy) + \text{SSD}(y^2)$

(7) Calculate the constants.

$$b = \frac{\text{SSD}(xy)}{\text{SSD}(x^2)} \tag{70}$$

and

$$a = \bar{y} - b\bar{x} \tag{71}$$

(8) Calculate the standard deviation

$$S_{y/x} = \sqrt{\frac{\text{SSD}(y^2) - [\text{SSD}(xy)^2/\text{SSD}(x^2)]}{n - 2}} \tag{72}$$

Example

x_i	y_i	T_i	x_i^2	y_i^2	T_i^2	$x_i y_i$
0.1	0.09	0.19	0.01	0.0081	0.0361	0.009
0.2	0.14	0.34	0.04	0.0196	0.1156	0.028
0.3	0.19	0.49	0.09	0.0361	0.2401	0.057
0.4	0.24	0.64	0.16	0.0576	0.4096	0.096
0.5	0.31	0.81	0.25	0.0961	0.6561	0.155
0.6	0.37	0.97	0.36	0.1369	0.9409	0.222
0.7	0.40	1.10	0.49	0.1600	1.2100	0.280
0.8	0.47	1.27	0.64	0.2209	1.6129	0.376
0.9	0.55	1.45	0.81	0.3025	2.1025	0.495
1.0	0.63	1.63	1.00	0.3969	2.6569	0.630
Totals 5.5	3.39	8.89	3.85	1.4347	9.9807	2.348

$\bar{x} = 0.55$, $\bar{y} = 0.339$, $\bar{T} = 0.889$, $n = 10$.

Checking

$\Sigma x_i + \Sigma y_i = \Sigma T_i$	$5.5 + 3.39 = 8.89$
$\Sigma x_i^2 + 2\Sigma x_i y_i + \Sigma y_i^2 = \Sigma T_i^2$	$3.85 + 2 \times 2.348 + 1.4347 = 9.9807$

Calculation of SSD values

$\text{SSD}(x^2) = \Sigma x_i^2 - \bar{x}\Sigma x_i$	$3.85 - 0.55 \times 5.5 = 0.825$
$\text{SSD}(y^2) = \Sigma y_i^2 - \bar{y}\Sigma y_i$	$1.4347 - 0.339 \times 3.39 = 0.28549$
$\text{SSD}(T^2) = \Sigma T_i^2 - \bar{T}\Sigma T_i$	$9.9807 - 0.889 \times 8.89 = 2.07749$
$\text{SSD}(xy) = \Sigma x_i y_i - \bar{y}\Sigma x_i$	$2.348 - 0.339 \times 5.5 = 0.4835$

Checking

$$\text{SSD}(x^2) + 2\,\text{SSD}(xy) + \text{SSD}(y^2) = \text{SSD}(T^2)$$

$$0.825 + 2 \times 0.4835 + 0.28549 = 2.07749$$

Calculation of the constants from eqns. (70) and (71)

$$b = \frac{\mathrm{SSD}(xy)}{\mathrm{SSD}(x^2)} = \frac{0.4835}{0.825} = 0.58606$$

$$a = \bar{y} - b\bar{x} = 0.339 - 0.58606 \times 0.55 = 0.0167$$

Calculation of the standard deviation from eqn. (72)

$$S_{y/x} = \sqrt{\frac{\mathrm{SSD}(y^2) - [\mathrm{SSD}(xy)^2/\mathrm{SSD}(x^2)]}{n-2}}$$

$$= \sqrt{\frac{0.28549 - (0.4835^2/0.825)}{8}} = 0.00213$$

(2) Using calibration graphs

The use of calibration graphs will be outlined below for linear relationships, i.e. for linear models. This explanation can be generalized for more complicated models.

Let us suppose that n pairs of data have been measured, viz. $(x_1, y_1), (x_2, y_2), ..., (x_i, y_i), ..., (x_n, y_n)$. These points are indicated on Fig. 17.

The model describing the functional relationship between the con-

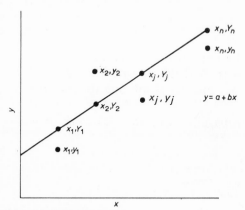

Fig. 17. Calibration line.

References pp. 168—169

centration x and the measured quantity y is

$$y = a + bx \tag{68}$$

The constants a and b are calculated from eqns. (70) and (71).

On Fig. 17, there is also indicated the calibration line and the points on the line belonging to x_i are denoted by Y_i ($i = 1, 2, ..., n$), i.e. $Y_i = a + bx_i$.

The standard deviation of the points around the line is calculated by

$$s = \sqrt{\frac{\sum(y_i - Y_i)^2}{n-2}} = \sqrt{\frac{\sum y_i^2 - a\sum y_i - b\sum x_i^2}{n-2}}$$

$$= \sqrt{\frac{\sum y_i^2 - \dfrac{\sum y_i}{n} - \dfrac{[\sum x_i y_i - (\sum x_i \sum y_i)/n]^2}{\sum x_i^2 - [(\sum x_i)^2/n]}}{n-2}} \tag{73}$$

The standard deviations of the constants a and b, denoted by s_a and s_b, are

$$s_b = \sqrt{\frac{ns^2}{n\sum x_i^2 - (\sum x_i)^2}}$$

and

$$s_a = s_b \sqrt{\frac{\sum x_i^2}{n}}$$

Calibration lines can be used by two different ways.

(i) Sometimes, the problem is to find the y_k value belonging to a given value of x, denoted by x_k, and to point out its confidence interval.

On the basis of the law of propagation of errors, it can be demonstrated that

$$y_k = a + bx_k \pm t(\alpha, f)\, s \sqrt{\frac{1}{n} + \frac{\sum n(x_k - x_M)^2}{n\sum x_i^2 - (\sum x_i)^2}}$$

where x_M is the arithmetic mean of the observed x_i values, t is the

108

Student factor at levels α and degrees of freedom, f, where $f = n - 2$.

It can be recognized that the confidence interval depends first of all on the difference $x_k - x_M$, that is the accuracy of the calibration line diminishes when moving away from the mean value. Therefore, it is not recommended that such lines are used for extrapolation.

(ii) Calibration graphs are used in analytical chemistry more frequently in the following way. The analysis is carried out m times and from these measured data the mean value $\bar{y}_A$ is calculated. The concentration to be determined, x_A, and its confidence interval is calculated by the equation

$$x_A = \frac{\bar{y}_A - a}{b} \pm t(\alpha, f)\, \frac{s}{b}\, \sqrt{\frac{1}{m} + \frac{1}{n} + \frac{\sum n(\bar{y}_A - y_M)^2}{b^2[m \sum x_i^2 - (\sum x_i)^2]}}$$

where $f = n - 2$, s is the standard deviation defined by eqn. (73), and y_M is the arithmetic mean of all observed y_i values used in setting up the calibration graph.

The width of the confidence interval depends on the slope, b, the number of measurements both in setting up the calibration curve and in carrying out the determination, n and m, respectively, and the difference $y_M - \bar{y}_A$. It is therefore advantageous if the calibration line is steep. It is recommended that the calibration line be determined from as many points as possible.

(3) Linear regression if both of the variables are subject to experimental error [19]

In the preceding method, it was assumed that one of the variables, x, is practically error-free. If this cannot be assumed, i.e. both x and y are subject to experimental errors, then the following method is suggested.

Calculate s_x and s_y, the standard deviations of the variables x and y from

$$s_x = \sqrt{\frac{1}{n}\sum_{i=1}^{n} x_i^2 - \bar{x}^2}$$

$$s_y = \sqrt{\frac{1}{n}\sum_{i=1}^{n} y_i^2 - \bar{y}^2}$$

where $\bar{x}$ and $\bar{y}$ are arithmetic mean values. Then carry out the linear regression as outlined above and determine the constants of the equation

$$y = a_1 + b_1 x_1 \tag{74}$$

In this step, it is assumed that x is error-free. In the second step of calculation, y is regarded as error-free and the constants of the equation

$$x = a_2' + b_2' y \tag{75}$$

are determined by another regression.

Rearranging eqn. (75)

$$y = a_2 + b_2 x \tag{76}$$

and this equation is, in general, different from eqn. (74). The constants in eqn. (76) can be calculated from

$$b_2 = b_1/r^2$$

and

$$a_2 = \overline{y} - b_2 \overline{x}$$

Here, r is the correlation coefficient given by

$$r = \frac{\dfrac{1}{n} \displaystyle\sum_{i=1}^{n} x_i y_i - \overline{x}\,\overline{y}}{s_x s_y}$$

It can be demonstrated that $|r| < 1$. If the correlation between the variables is close, then $|r|$ is approximately equal to unity; if there is no practical correlation between the variables, then $|r|$ is almost zero.

The graphs of eqns. (74) and (76) are different straight lines, having a common intercept at $x = \overline{x}$ and $y = \overline{y}$ (Fig. 18, lines 1 and 2).

If the angle, α, between these lines is small, then the correlation is close, $b_1 \approx b_2$. If, on the other hand, α is large, the two lines of regression differ significantly from each other and the constants b_1 and b_2 are very different.

If x is determined without error, then eqn. (74) would be the best fit according to the least squares principle and, similarly, if y determined without error, then eqn. (76) would be the best fit. If both variables are subject to experimental error, then the line [Fig. 18,

110

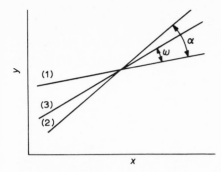

Fig. 18. Regression lines. (1) $y(x)$; (2) $x(y)$; (3) the line of best fit.

line (3)] giving the best fit lies between the lines (1) and (2). The angle between lines (1) and (2) is denoted by α and that between lines (1) and (3) by ω. These angles are calculated by

$$\tan \alpha = \frac{b_2 - b_1}{1 + b_1 b_2} = \frac{1 - r^2}{b_1^2 + r^2} b_1$$

and

$$\omega = \frac{s_x^2}{s_x^2 + s_y^2} \alpha$$

If the linear equation of best fit [Fig. 18, line (3)] is written as

$$y = a_3 + b_3 x$$

then its constants are calculated by

$$b_3 = \frac{b_1 + \tan \omega}{1 - b_1 \tan \omega}$$

and

$$a_3 = \bar{y} - b_3 \bar{x}$$

Example

x	y	x	y	x	y
66.8	24.2	54.7	24.4	49.28	31.4
56.6	24.4	54.0	23.2	47.32	33.6
56.2	23.8	54.0	23.0	44.6	35.6
55.5	25.6	50.12	30.2	43.20	36.0

The mean values of the variables are

$$\frac{\sum\limits_{i=1}^{n} x_i}{n} = \bar{x} = 52.693$$

$$\frac{\sum\limits_{i=1}^{n} y_i}{n} = \bar{y} = 27.950$$

The standard deviations are calculated to be $s_x = 6.0872$ and $s_y = 4.8416$ and the correlation coefficient $r = -0.83124$.

Carrying out a linear regression assuming that x is error-free

$$y = -0.661x + 62.79$$

and the constants in eqn. (76) are calculated as $b_2 = -1.045$ and $a_2 = 27.950 - (-1.045) \, 52.693 = 81.90$.

The angles α and ω and their tangents are

$$\tan \alpha = \frac{1 - r^2}{b_1^2 + r^2} \, b_1 = \frac{1 - (-0.83124)^2}{(-0.6611)^2 + (-0.83124)^2} \times (-0.6611) = -0.1811$$

$$\alpha = -0.1792$$

$$\omega = \frac{6.090^2}{6.090^2 + 4.8416^2} \, (-0.1792) = -0.09801$$

$$\tan \omega = -0.10013$$

The constants of the equation of best fit are

$$b_3 = \frac{-0.6611 - (-0.10013)}{1 - (-0.6611)(-0.10013)} = -0.81525$$

$$a_3 = 27.950 - (-0.81525) \, 52.693 = 70.908$$

So the equation of best fit is

$$y = 70.908 - 0.81525x \tag{77}$$

or

$$x = 86.977 - 1.2266y \tag{78}$$

The calculated values of the variables x' and y' using eqns. (77) and

112

(78) and the deviations $\delta x = x' - x$ and $\delta y = y' - y$ are summarized in the following table.

y'	x'	δy	δx
16.21	57.20	−7.99	−9.60
24.70	56.96	0.30	0.36
25.03	57.68	1.23	1.48
30.09	49.99	−0.11	−0.13
30.76	48.55	−0.61	−0.73
32.42	45.90	−1.18	−1.42
35.85	43.02	−0.15	−0.18
26.28	56.96	1.88	2.26
25.61	55.52	0.01	0.02
34.68	43.50	−0.92	−1.10
26.86	58.40	3.66	4.40
26.86	58.64	3.86	4.64

The standard deviations are $s_{y/x}$ = 3.16 and $s_{x/y}$ = 3.80.

(4) Two intersecting straight lines [20]

For experimental data which appear to behave according to two different distinct linear relationships, a general model can be set up which allows for a smooth transition from one linear régime to the other. The transition is accomplished by a curve incorporating a transition parameter.

A model in which different straight line relationships are assumed to the left and right of some unknown point of intersection x_0 can be constructed by parameterizing it in terms of the average slope, a_1, and the average difference in slope, a_2, that is

$$y = a_0 + a_1(x - x_0) + a_2(x - x_0) \operatorname{sgn}(x - x_0) \tag{79}$$

where

$$\operatorname{sgn}(s) = \begin{cases} -1 & (s < 0) \\ 0 & (s = 0) \\ +1 & (s > 0) \end{cases}$$

A more general model can be set up which permits a smooth transition from one régime to another by replacing the sign function sgn$(x - x_0)$ in eqn. (79) by the transition function trn$\{(x - x_0)/\gamma\}$. By varying the transition or scale parameter, trn(s/γ) produces abrupt transition $(\gamma \approx 0)$ or very gradual transition $(\gamma > 1)$. Model (79) can be written in the more general form

$$y = a_0 + a_1(x - x_0) + a_2(x - x_0)\, \text{trn}(x - x_0) \tag{80}$$

The transition function should satisfy the following conditions.

$\lim\limits_{s \to \infty} \text{trn}(|s|/\gamma) = 1$

$\text{trn}(0) = 0$

$\lim\limits_{\gamma \to 0} \text{trn}(s/\gamma) = \text{sgn }(s)$

$\lim\limits_{s \to \infty} \text{trn}(s/\gamma) = s$

There are many transition functions satisfying the above conditions, but eqn. (80) is insensitive to the particular form of the transition function. So only one transition function, trn(s/γ) = tanh(s/γ), is recommended. The general eqn. (80) thus becomes

$$y = a_0 + a_1(x - x_0) + a_2(x - x_0) \tanh\left(\frac{x - x_0}{\gamma}\right)$$

(see Fig. 19). For any symmetrical transition function, the radius of

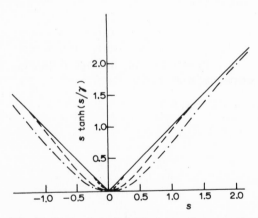

Fig. 19. Plots of $s$tanh(s/γ) versus s for several values of γ. ———, $\gamma = 0$; ······, $\gamma = 0.1$; — — —, $\gamma = 0.5$; — . — ., $\gamma = 1$.

114

curvature, R, at the point of intersection x_0 is

$$R = \frac{\gamma}{2|a_2|} (1 + a_1^2)^{3/2}$$

The parameters are estimated by an iterative method, each iteration consisting of two steps. Initial values must be provided for x_0 and γ as well as for their increments. The constants a_0, a_1, a_2, are then determined by linear regression. After that, the constants x_0 and γ are changed in order to minimize the sum of squares of deviations. This can be done by one of the optimization techniques. Before calculating the new sum of squares of deviations, the costants a_0, a_1, a_2 must be again determined by linear regression.

The subject is further discussed by Ahsanullah and Ehsanes Saleh [21].

(5) Families of straight lines [21]

Data relating to numerous physico-chemical phenomena can be presented in the form of families of parallel straight lines or lines having common intercepts. Methods developed for treating such data statistically involve the simultaneous regression of all lines of the family by the use of the principle of least squares.

a. Straight lines having equal slopes. A family of straight lines having equal slopes can be expressed as

$$y_i = A_i + BX_i \ (i = 1, 2, ..., n)$$

where subscript i denotes any of the individual lines, X_i is the measured value of the independent variable on line i, y_i is the corresponding calculated value of ordinate, B is the common slope, and A_i is the intercept of line i (see, for example, Fig. 20). The deviation of the calculated results, y, from the observed results, Y, for any point on line i can be expressed as

$$\Delta_i = y_i - Y_i = A_i + BX_i - Y_i$$

where Δ_i is the deviation or the residual. Then, $\Sigma \Delta_i$ represents the sum of the residuals of all the points on line i and $\Sigma \Delta_i^2$ represents the sum of the squares of the residuals. If the sums of the squares of all residuals are simultaneously kept at a minimum, parallel regression lines will be obtained. As a result of minimizing this sum of the

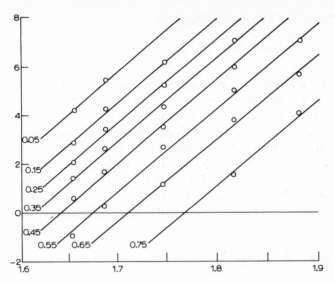

Fig. 20. Straight lines having equal slopes.

squares, the constants A_i and B are

$$B = \frac{\sum\limits_{a}^{n} (\sum X_i Y_i - \overline{X}_i \sum Y_i)}{\sum\limits_{a}^{n} (\sum X_i^2 - \overline{X}_i \sum X_i)}$$

$$A_i = \overline{Y}_i - B\overline{X}_i$$

where $\overline{X}_i = X_i/N_i$, $\overline{Y}_i = Y_i/N_i$, N_i is the number of points on line i, and $\sum\limits_{a}^{n}$ represents the summation for each line.

The common variance, S^2, can be obtained by summing the squares of the deviations of Y from its calculated value y and dividing by $\sum\limits_{a}^{n} (N_i - 1) - 1$, i.e.

$$S^2 = \frac{\sum\limits_{a}^{n} (\sum \Delta_i^2)}{\sum\limits_{a}^{n} (N_i - 1) - 1}$$

116

The variance of the common slope is given by

$$S_B^2 = \frac{\displaystyle\sum_a^n [\sum (X_i - \overline{X}_i)^2]}{\displaystyle\sum_a^n [\sum (X_i - \overline{X}_i)^2]^2} S^2$$

and the variances of intercepts by

$$S_{A_i}^2 = \frac{S^2}{N_i} \frac{1 + \overline{X}_i \sum X_i}{\displaystyle\sum_a^n (\sum X_i^2 - \overline{X}_i \sum X_i)}$$

b. *Straight lines having a common intercept.* A family of straight lines having a common intercept can be expressed as

$$y_i = A + B_i X_i$$

where X_i is the value of the independent fixed variable, y_i is the corresponding calculated value of the dependent variable, A is the common intercept, B_i is the slope of any line, and subscript i denotes

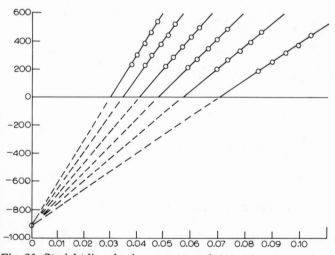

Fig. 21. Straight lines having a common intercept.

References pp. 168—169

117

any of the individual lines (see, for example, Fig. 21). The deviation of the calculated result, y, from the observed result, Y for any point on line i can be expressed as

$$\Delta_i = y_i - Y_i = A + B_i X_i - Y_i$$

Then, Δ_i represents the sum of the residuals of all points on the line i and $\Sigma \Delta_i^2$ represents the sum of the squares of the residuals. If, the sums of the squares of all residuals are simultaneously kept at a minimum, regression lines with common intercepts will be obtained. As a result of minimizing this sum of squares, the constants are

$$A = \frac{\displaystyle\sum_a^n N_i(\overline{Y}_i - \overline{X}_i \sum X_i Y_i / \sum \overline{X}_i^2)}{\displaystyle\sum_a^n N_i(1 - \overline{X}_i \sum X_i / \sum X_i^2)}$$

$$B_i = \frac{\sum X_i Y_i - A \sum X_i}{X_i^2}$$

The common variance can be obtained as above.
The variance of the common intercept is given by

$$S_A^2 = \frac{\displaystyle\sum_a^n N_i S^2 - (\sum X_i / \sum X_i^2)^2 (\sum X_i^2) S^2}{[\displaystyle\sum_a^n N_i \sum (X_i - \overline{X}_i)^2 / \sum X_i^2]^2}$$

and the variances of the slopes by

$$S_{B_i}^2 = \left[\frac{1}{X_i^2} + \left(\frac{\sum X_i}{\sum X_i^2} \right)^2 / \sum_a^n N_i(1 - \overline{X}_i \sum X_i / \sum X_i^2) \right] S^2$$

(6) Multi-variable linear regression [23—26]

The method outlined in Sect. 4.(B)(1) can be extended so that it finds a linear approximation to a function of several variables. A

118

linear function of n variables may be written as

$$y = a_0 + a_1 x_1 + a_2 x_2 + \ldots + a_n x_n = \sum_{j=0}^{n} a_j x_j$$

The mean value, $\overline{x}_j$, of each x_j is first calculated together with the mean, $\overline{y}$, of the y values. The linear approximation is then written

$$y = \overline{y} + a_1(x_1 - \overline{x}_1) + a_2(x_2 - \overline{x}_2) + \ldots + a_n(x_n - \overline{x}_n)$$

$$= \overline{y} + \sum_{j=1}^{n} (x_j - x_j)$$

Applying the principle of least squares

$$S = \sum_{i=1}^{m} [y_i - \overline{y} - \sum_{j=1}^{n} a_j(x_{ij} - \overline{x}_j)]^2$$

where m is the number of measured $(x_1, x_2, \ldots, x_n, y)$ points, $(x_{i1}, x_{i2}, \ldots, x_{in}, y_i)$, $(i = 1, 2, \ldots, m)$. The constants $a_1, a_2, \ldots, a_n$ are determined by putting the partial derivatives of S equal to zero.

$$\frac{\partial S}{\partial a_k} = -2 \sum_{i=1}^{m} [y - \overline{y} - \sum_{j=1}^{n} a_j(x - \overline{x}_j)] (x - \overline{x}_k) \qquad (k = 1, 2, \ldots, n)$$

After rearranging

$$\sum_{j=1}^{n} [a_j \sum_{i=1}^{m} (x_i - \overline{x}_j)(x_i - \overline{x}_k)] = \sum_{i=1}^{m} (y_i - \overline{y})(x_i - \overline{x}_k)$$

The constants $a_1, a_2, \ldots, a_n$ are determined by solving this system of equations. Making use of the notations

$$SXX_{jk} = \sum_{i=1}^{m} (x_i - \overline{x}_j)(x_i - \overline{x}_k)$$

$$SXY_k = \sum_{i=1}^{m} (y_i - \overline{y})(x_i - \overline{x}_k)$$

the system of linear equations to be solved is

$$SXX_{11}a_1 + SXX_{12}a_2 + ... + SXX_{1n}a_n = SXY_1$$
$$SXX_{21}a_1 + SXX_{22}a_2 + ... + SXX_{2n}a_n = SXY_2$$
$$\vdots$$
$$SXX_{n1}a_1 + SXX_{n2}a_2 + ... + SXX_{nn}a_n = SXY_n$$

This system of equations can be written in matrix form, viz.

SXX a = SXY

The vector **a** of the constants can be expressed by means of the inverse matrix SXX^{-1}

a = SXX⁻¹ SXY

Having determined the constants $a_1, a_2, ..., a_n$, we can calculate the y values at each point. Let us denote them by $Y_1, Y_2, ..., Y_n$. The standard deviation, S, is given by

$$S = \sqrt{\frac{\sum_{i=1}^{m} (Y_i - y_i)^2}{m - n - 1}}$$

The error in the coefficients, $da_1, da_2, ..., da_n$ is

$$da_j = S \sqrt{SXX_{jj}}$$

The calculation of the inverse matrix may be carried out by the Fortran subroutine INV. The basis of the calculation is the so-called Gauss elimination with partial pivoting. Further, the subroutine performs a suitable exchange of rows and columns in the matrix, so that we always have the biggest possible element on the diagonal. If a diagonal element becomes less than or equal to a reasonably small quantity, say 10^{-8}, then we are close to having no solution, i.e. the two equations are nearly identical or nearly proportional.

The subroutine INV calculates the inverse matrix corresponding to a given matrix T(N, N). The original matrix will be replaced by the inverse matrix. Thus, the original matrix is lost.

The declaration is: SUBROUTINE INV (N, A, EPS, IE)

The parameters are

N the number of rows and columns

A the matrix to be inverted

EPS   prescribed limit for the diagonal element

IE   logical parameter. Its actual value is normally 0; it becomes 1 only if a diagonal element becomes less than or equal to EPS.

The listing of the subroutine INV is as follows.

```
      SUBROUTINE INV(N,A,EPS,IE)
      DIMENSION A(N,N),B(10),C(10),IP(10),IQ(10)
      IE=0
      DO 12 I=1,N
      PIV=0.0
      DO 1 J=I,N
      DO 1 K=I,N
      IF(ABS(A(J,K)).LE.ABS(PIV)) GO TO 1
      PIV=A(J,K)
      IP(I)=J
      IQ(I)=K
    1 CONTINUE
      IF(ABS(PIV).GT.EPS) GO TO 2
      IE=1
      RETURN
    2 IF(IP(I).EQ.I) GO TO 4
      DO 3 K=1,N
      IZ=IP(I)
      AZ=A(IZ,K)
      A(IZ,K)=A(I,K)
    3 A(I,K)=AZ
    4 IF(IQ(I).EQ.I) GO TO 6
      DO 5 J=1,N
      IZ=IQ(I)
      AZ=A(J,IZ)
      A(J,IZ)=A(J,I)
    5 A(J,I)=AZ
    6 DO 10 K=1,N
      IF(K-I) 8,7,8
    7 B(K)=1.0/PIV
      C(K)=1.0
      GO TO 9
    8 B(K)=-A(I,K)/PIV
      C(K)=A(K,I)
    9 A(I,K)=0.0
      A(K,I)=0.0
   10 CONTINUE
      DO 11 J=1,N
      DO 11 K=1,N
   11 A(J,K)=A(J,K)+C(J)*B(K)
   12 CONTINUE
      DO 18 IK=1,N
      I=N+1-IK
```

References pp. 168—169

```
      IF(IP(I)-I)  13,15,13
13 DO 14 J=1,N
      IZ=IP(I)
      AZ=A(J,IZ)
      A(J,IZ)=A(J,I)
14 A(J,I)=AZ
15 IF(IQ(I)-I)  16,18,16
16 DO 17 K=1,N
      IZ=IQ(I)
      AZ=A(IZ,K)
      A(IZ,K)=A(I,K)
17 A(I,K)=AZ
18 CONTINUE
      RETURN
      END
```

(7) Polynomials of a single variable

The linear model of several variables [Sect. 4.(B)(6)] can be used for the approximation by polynomials.

$$y = a_0 + a_1 x + a_2 x^2 + \dots + a_n x^n$$

This may be done by interpretating the higher powers of one variable as other independent variables.

$$x_1 = x$$
$$x_2 = x^2$$
$$\vdots$$
$$x_n = x^n$$

When the degree of the required polynomial in x becomes high, then the inversion of matrix **SXX** becomes inaccurate. In such cases, the approximation by orthogonal polynomials or by orthogonal moments is recommended.

a. Approximation of measured data by orthogonal polynomials. Let $(x_i y_i)$ $(i = 1, 2, \dots, n)$ be a set of measured data which we want to approximate by the polynomial of degree $m - 1$

$$y = \sum_{j=0}^{m-1} a_j x^j$$

122

The degree of the polynomial is assumed to be at least 2. The coefficients $a_0, a_1, ..., a_{m-1}$ can be determined easily by using orthogonal polynomials.

Two polynomials, $P_j(x)$ and $P_k(x)$, are called orthogonal if

$$\sum_{i=1}^{n} P_j(x_i) P_k(x_i) = 0 \qquad (81)$$

if $j \neq k$. The $P_j(x)$ polynomials with respect to x_i ($i = 1, 2, ..., n$) can be generated by the recurrent relationship

$$P_1(x) = 1$$

$$P_2(x) = x - \alpha_1$$

$$\vdots$$

$$P_k(x) = (x - \alpha_{k-1}) P_{k-1}(x) - \beta_{k-2} P_{k-2}(x)$$

where

$$\alpha_k = \frac{\sum_{i=1}^{n} x_i [P_k(x_i)]^2}{\sum_{i=1}^{n} [P_k(x_i)]^2} \qquad (82)$$

$$\beta_{k-1} = \frac{\sum_{i=1}^{n} x_i P_k(x_i) P_{k-1}(x_i)}{\sum_{i=1}^{n} P_{k-1}^2(x_i)} \qquad (83)$$

According to our assumption, the measured data are expressed as linear combinations of orthogonal polynomials.

$$y = \sum_{j=1}^{m} c_j P_j(x) \qquad (84)$$

The approximation is, in the sense of least squares, the best one if the sum

$$S = \sum_{i=1}^{n} [y_i - \sum_{j=1}^{m} c_j P_j(x)]^2$$

reaches its minimum. Calculating the derivative of S with respect to the constant c_k and putting it equal to zero

$$\sum_{i=1}^{n} y_i P_k(x_i) = \sum_{i=1}^{n} \sum_{j=1}^{m} c_j P_j(x_i) P_k(x_i) \qquad (k = 1, 2, ..., m)$$

After rearranging and taking into account eqn. (81)

$$c_j = \frac{\sum_{i=1}^{n} y_i P_j(x_i)}{\sum_{i=1}^{n} [P_j(x_i)]^2}$$

The advantage of this method is that each coefficient is obtained by two summations in a division and each coefficient is determined independently of the others. If the accuracy of the approximation is not sufficient, only one further coefficient must be calculated in order to achieve a better approximation.

The approximation can be carried out by a computer using the Fortran subroutine ORTHP.

Its use is as follows.

CALL ORTHP (X, Y, N, A, B, C, M, P1, P2)

The arguments and their description are

X (N)	array of the x_i data
Y (N)	array of the y_i data
N	number of data pairs
A (M)	the coefficients α_i, defined by eqn. (82)
B (M)	the coefficients β_i, defined by eqn. (83)
C (M)	the coefficients of the approximation in terms of orthogonals polynomials, eqn. (84)
M	number of polynomials (the degree of the resulting approximation is $M - 1$)
P1 (N), P2 (N)	working arrays. It is not necessary to assign any value to their elements

The listing of the subroutine is

```
SUBROUTINE ORTHP(X,Y,N,A,B,C,M,P1,P2)
DIMENSIONX(N),Y(N),P1(N),P2(N),A(M),B(M),C(M)
DOUBLE PRECISION SA,SB,SC,DN
```

```
        SA=0.0
        SC=0.0
        BB=0.0
        DO 1 I=1,N
        P1(I)=0.0
        P2(I)=1.0
        SA=SA+X(I)
1       SC=SC+Y(I)
        DN=FLOAT(N)
        C(1)=SC/DN
        ASSIGN 2 TO L
        DO 6 K=2,M
        GO TO L,(2,3)
2       ASSIGN 3 TO L
        GO TO 4
3       BB=SB/D
        B(K-2)=BB
4       D=DN
        AA=SA/DN
        A(K-1)=AA
        DN=0.0
        SA=0.0
        SB=0.0
        SC=0.0
        DO 5 I=1,N
        H=P2(I)
        XI=X(I)
        P=(XI-AA)*H-BB*P1(I)
        P1(I)=H
        P2(I)=P
        SC=SC+P*Y(I)
        PP=P*P
        DN=DN+PP
        SA=SA+XI*PP
5       SB=SB+XI*P*H
6       C(K)=SC/DN
        RETURN
        END
```

Having calculated the coefficients of the approximating function defined by eqn. (84) as described above, the values of this function can be calculated by the Fortran function APPR.

Its use is as follows.

Y = APPR (A, B, C, M, X)

Here

A(M), B(M), C(M) are the same arrays as in the subroutine ORTHP
M   is the number of used polynomials
X   is the input argument to be substituted in eqn. (84).

The listing of the function is

```
FUNCTION APPR(A,B,C,M,X)
DIMENSION A(M),B(M),C(M)
XX=X
PK=1.0
PL=XX-A(1)
Y=C(1)+C(2)*PL
DO 1 K=3,M
H=PL
PL=(XX-A(K-1))*PL-B(K-2)*PK
PK=H
Y=Y+C(K)*PL
1    CONTINUE
APPR=Y
RETURN
END
```

Example. In a potentiometric titration, equal volumes of the titrant have been added to the solution containing the element to be determined. The volumes, x ml, and the measured potential differences, y mV, are

x (ml)	y (mV)	x (ml)	y (mV)
13.0	333	14.0	103
13.2	314	14.2	63
13.4	288	14.4	41
13.6	242	14.6	18
13.8	184	14.8	2

We want to approximate the data by means of a polynomial of the third degree using orthogonal polynomials. The task can be carried out by the Fortran programme on p. 127.

The constants of the approximating function calculated by the subroutine ORTHP and the values of the approximating function calculated by the function APPR are summarized in the output of the programme on p. 128.

The results of the approximation can be interpreted as follows.

$$y = 158.8\,P_1(x) - 209.21\,P_2(x) + 26.705\,P_3(x) + 131.31\,P_4(x)$$

where $P_1(x) = 1$
$\qquad P_2(x) = x - 13.9$

```
DIMENSION X(10),Y(10),A(3),B(3),C(4),P1(10),P2(10)
DATA Y/333.0,314.0,288.0,242.0,184.0,103.0,63.0,41.0,18.0,2.0/
J=12.8
DO 1 I=1,10
U=U+0.2
X(I)=U
CALL ORTHP(X,Y,10,A,B,C,4,P1,P2)
PRINT 100
PRINT 2
PRINT 12
DO 3 J=1,3
PRINT 8,A(J),B(J)
PRINT 12
PRINT 5,(C(J),J=1,4)
PRINT 12
PRINT 9
S=0.0
DO 4 J=1,10
XX=X(J)
YY=APPR(A,B,C,4,XX)
Z=YY-Y(J)
S=S+Z*Z
PRINT 6,XX,YY,Y(J),Z
S=SQRT(S/6)
PRINT 12
PRINT 7,S
FORMAT(7X,21HCONSTANTS ALPHA,BETA:)
FORMAT(14H COEFFICIENTS:,4E16.5)
FORMAT(2X,F8.1,F9.2,F9.1,F9.2)
FORMAT(20H STANDARD DEVIATION=,F8.2)
FORMAT(2X,2E16.5)
FORMAT(2X,35H    X      Y CALC.    Y OBS.    DEV.)
FORMAT(/)
FORMAT(1H1)
STOP
END
```

References pp. 168—169

CONSTANTS ALPHA,BETA

0.13900E+02   0.33000E+00
0.13900E+02   0.25600E+00
0.13900E+02   0.00000E+00

COEFFICIENTS: 0.15880E+03   -0.20921E+03   0.20705E+02   0.13131E+03

X	Y CALC.	Y OBS.	DEV.
13.0	333.44	333.0	0.44
13.2	318.35	314.0	4.35
13.4	283.33	288.0	-4.67
13.6	234.69	242.0	-7.31
13.3	178.74	184.0	-5.26
14.0	121.77	103.0	18.77
14.2	70.09	63.0	7.09
14.4	30.00	41.0	-11.00
14.6	7.80	13.0	-10.20
14.8	9.30	2.0	7.80

STANDARD DEVIATION=   11.62

128

$$P_3(x) = (x - 13.9)^2 - 0.33$$
$$P_4(x) = (x - 13.9)^3 - 0.586\,(x - 13.9)$$

It can be observed that, if the independent variable is equidistant, then

$$\alpha_k = \sum_{i=1}^{n} x_i/n \qquad (k = 1, 2, ..., m-1)$$

b. *Approximation by orthogonal moments [27–29].* Given the values of $y_0, y_1, ..., y_{N-1}$ corresponding to $u_0, u_1, ..., u_{N-1}$, an approximation by a parabola of degree n, $F = F(u)$ is required such that, according to the principle of least squares, the sum of the squares of the deviations $F(u) - y$ shall be a minimum, the values of the variable u being equidistant $u_{i+1} - u_i = h$.

The shortest way to reach this is to first introduce a variable x by $x = (u - u_0)/h$; then x will take the values 0, 1, 2, ..., $N - 1$. Let us denote the approximating function by $f(x)$. Then, the function $f(x)$ is expanded into a series of orthogonal polynomials. The polynomial $U_m(x) = U_m$ of degree m will be termed orthogonal with respect to $x = 0, 1, 2, ..., N - 1$ if

$$\sum_{x=0}^{N-1} U_m U_n = 0 \tag{85}$$

for all values of m different from n.

The expansion of $f(x)$ may be written

$$f(x) = a_0 U_0 + a_1 U_1 + ... + a_n U_n \tag{86}$$

where the coefficients a_m are to be determined so that

$$S = \sum_{x=0}^{N-1} [f(x) - y]^2$$

is a minimum.

Putting $dS/da_m = 0$, we get, in consequence of orthogonality

$$\sum_{x=0}^{N-1} U_m y = a_m \sum_{x=0}^{N-1} U_m^2 \tag{87}$$

It follows that a_m is independent of the degree u of the function $f(x)$. This is important.

TABLE 20

The $\beta_{m\nu}$ numbers

m	ν					
	0	1	2	3	4	5
1	−1	1				
2	1	−3	2			
3	−1	6	−10	5		
4	1	−10	30	−35	14	
5	−1	15	−70	140	−126	42

The first member is equal to Θ_m, the orthogonal moment of order m of the observations. This will be computed later. The values of U_m and ΣU_m^2 may be deduced starting from definition (85).

We find U_m given by its Newton expansion

$$U_m(x) = C_m \sum_{\nu=0}^{m} (-1)^{m+\nu} \binom{m+\nu}{m}\binom{N-\nu-1}{m-\nu}\binom{x}{\nu} \tag{88}$$

where C_m is an arbitrary constant. The value of C_m has no great importance considering that in eqn. (86) $a_m U_m$ is independent of C_m and therefore f(x) too. Nevertheless, to simplify our formulae, we shall write C_m as

$$C_m = \frac{1}{(m+1)\binom{N}{m+1}} \tag{89}$$

From eqn. (85) it follows, moreover, that

$$\sum_{x=0}^{N-1} U_m^2 = C_m^2 \binom{2m}{m}\binom{N+m}{2m+1} \tag{90}$$

To shorten the work of computation, tables giving $\beta_{m\nu}$ (Table 20) and $C_{m\nu}$ (Table 21) may be used.

$$\beta_{m\nu} = (-1)^{m+\nu}\binom{m+\nu}{m}\binom{m}{\nu}\frac{1}{\nu+1} \tag{91}$$

130

TABLE 21

The $C_{m\nu}$ numbers

N	C_{10}	C_{11}	C_{20}	C_{21}
3	−1.5	1.5	0.5	−1.5
4	−1.8	1.2	1	−2
5	−2	1	1.428 571 429	−2.142 857 143
6	−2.142 857 143	0.857 142 857 1	1.785 714 286	−2.142 857 143
7	−2.25	0.75	2.083 333 333	−2.083 333 333
8	−2.333 333 333	0.666 666 6667	2.333 333 333	−2
9	−2.4	0.6	2.545 454 545	−1.909 090 900
10	−2.454 545 455	0.545 454 545 5	2.727 272 727	−1.818 181 818
11	−2.5	0.5	2.884 615 385	−1.730 769 231
12	−2.538 461 538	0.461 538 461 5	3.021 978 021	−1.648 351 648
13	−2.571 428 571	0.428 571 428 6	3.142 857 143	−1.571 428 571
14	−2.6	0.4	3.25	−1.5
15	−2.625	0.375	3.345 588 235	−1.433 823 529
16	−2.647 058 823	0.352 941 176 4	3.431 372 549	−1.372 549 020
17	−2.666 666 667	0.333 333 333 3	3.508 771 930	−1.315 789 474
18	−2.684 210 526	0.315 789 473 6	3.578 947 368	−1.263 157 894
19	−2.7	0.3	3.642 857 143	−1.214 285 714
20	−2.714 285 714	0.285 714 285 7	3.701 298 702	−1.168 831 169
21	−2.727 272 727	0.272 727 2727	3.754 940 711	−1.126 482 213
22	−2.739 130 436	0.260 869 565 2	3.804 347 826	−1.086 956 522
23	−2.75	0.25	3.85	−1.05
24	−2.76	0.24	3.892 307 692	−1.015 384 615
25	−2.769 230 769	0.230 769 230 8	3.931 623 933	−0.982 905 982 9
26	−2.777 777 778	0.222 222 222 2	3.968 253 969	−0.952 380 952 4
27	−2.785 714 286	0.214 285 714 2	4.002 463 055	−0.923 645 320 5
28	−2.793 103 447	0.206 896 551 6	4.034 482 738	−0.896 551 723 9
29	−2.8	0.2	4.046 516 128	−0.870 967 7418
30	−2.806 451 614	0.193 548 387 1	4.092 741 935	−0.846 774 193 4

TABLE 21 (continued)

N	C_{22}	C_{30}	C_{31}	C_{32}	C_{33}
3	3	−0.2	0.8	−0.2	4
4	2	−0.5	1.5	−2.5	2.5
5	1.428 571 429	−0.833 333 333 3	2	−2.5	1.666 666 667
6	1.071 428 572	−1.166 666 667	2.333 333 333	−2.333 333 333	1.166 666 667
7	0.833 333 333 3	−1.484 848 485	2.545 454 545	−2.121 212 121	0.848 484 848 5
8	0.666 666 666 7	−1.781 818 182	2.672 727 273	−1.909 090 909	0.636 363 636 4
9	0.545 454 545 5	−2.055 944 056	2.741 258 742	−1.713 286 714	0.489 510 489 6
10	0.454 545 454 5	−2.307 692 308	2.769 230 769	−1.538 461 538	0.384 615 384 6
11	0.384 615 384 6	−2.538 461 538	2.769 230 769	−1.384 615 384	0.307 692 307 6
12	0.329 670 329 7	−2.75	2.75	−1.25	0.25
13	0.285 714 285 7	−2.944 117 648	2.717 647 060	−1.132 352 941	0.205 882 353 0
14	0.25	−3.122 549 020	2.676 470 588	−1.029 411 765	0.171 568 627 5
15	0.220 588 235 2	−3.286 893 705	2.629 514 964	−0.939 112 487 0	0.144 478 844 2
16	0.196 078 431 4	−3.438 569 491	2.578 947 368	−0.859 649 122 8	0.122 807 017 5
17	0.175 438 569 5	−3.578 947 369	2.526 315 790	−0.789 473 684 2	0.105 263 157 9
18	0.157 894 736 8	−3.709 090 909	2.472 727 273	−0.727 272 727 3	0.090 909 090 91
19	0.142 857 142 9	−3.830 039 526	2.418 972 332	−0.671 936 758 9	0.079 051 383 40
20	0.129 870 129 9	−3.942 687 747	2.365 612 648	−0.622 529 644 3	0.069 169 960 48
21	0.118 577 075 1	−4.047 826 087	2.313 043 478	−0.578 260 869 6	0.060 869 565 22
22	0.108 695 652 2	−4.146 153 846	2.261 538 461	−0.538 461 538 4	0.053 846 153 84
23	0.1	−4.238 290 598	2.211 282 051	−0.502 564 102 5	0.047 863 247 86
24	0.092 307 692 28	−4.324 786 325	2.162 393 163	−0.470 085 470 1	0.042 735 042 74
25	0.085 470 085 50	−4.406 130 268	2.114 942 528	−0.440 613 026 8	0.038 314 176 24
26	0.079 365 079 38	−4.482 758 621	2.068 965 517	−0.413 793 103 4	0.034 482 758 62
27	0.073 891 625 64	−4.555 061 179	2.024 471 635	−0.389 321 468 2	0.031 145 717 46
28	0.068 965 517 22	−4.623 387 098	1.981 451 613	−0.366 935 484 0	0.028 225 806 46
29	0.064 516 129 02	−4.688 049 852	1.939 882 697	−0.346 407 624 5	0.025 659 824 04
30	0.060 483 870 96				

$$C_{mv} = (-1)^{m+v}(2m+1)\binom{m+v}{m}\frac{\binom{N-v-1}{m-v}}{\binom{N+m}{m}} \tag{92}$$

With the aid of these values and eqn. (80), eqn. (88) of the orthogonal polynomial becomes

$$U_m = \sum_{v=0}^{m} \beta_{mv}\frac{\binom{x}{v}}{\binom{N}{v+1}} \tag{93}$$

Moreover, from eqn. (90) we obtain

$$\sum_{x=0}^{N-1} U_m^2 = \frac{1}{N|C_{m0}|} \tag{94}$$

and from eqn. (87)

$$a_m = N|C_{m0}|\theta_m \tag{95}$$

Finally, the approximating function (86) will be

$$f(x) = N\sum_{m=0}^{n} |C_{m0}|\theta_m \sum_{v=0}^{m} \beta_{mv}\frac{\binom{x}{v}}{\binom{N}{v+1}}$$

or

$$f(x) = \sum_{m=0}^{n}\sum_{v=0}^{m} C_{mv}\,\theta_m\binom{x}{v} \tag{96}$$

since

$$C_{mv} = \beta_{mv}|C_{m0}|N/\binom{N}{v+1}$$

This formula may be useful if one is working beyond the ranges of the C_{mv} tables.

Thus, the approximation function is easily obtained in its most favourable form (Newton expansion); only the computation of the orthogonal moments Θ_0, Θ_1, ..., Θ_n is needed.

From eqn. (96), we immediately obtain

$$\Delta^\nu f(0) = \sum_{m=\nu}^{n} C_{m\nu}\theta_m$$

and the approximating function will be

$$f(x) = \sum_{\nu=0}^{n} \binom{x}{\nu} \Delta^\nu f(0)$$

Starting from the difference $\Delta^\nu(f(0))$, a table of the approximate values of $f(x)$ and of their differences may be computed by the method of the addition of differences. If $f(x)$ is of degree n, then $\Delta^n f(x)$ is a constant. Into the first line of a table are written the numbers $\Delta^n f(0)$, $\Delta^{n-1} f(0)$, ..., $\Delta f(0)$, $f(0)$. In addition, the numbers $\Delta^n f(0)$, $\Delta^n f(1)$, $\Delta^n f(2)$, ... are placed in the first column and since $\Delta^n f(x)$ is constant, they will be all equal. In the rest of the table, each number shall be equal to the sum of the number above it and that which preceeds the latter in its line. Hence

$$\Delta^{m-1} f(x) = \Delta^{m-1} f(x-1) + \Delta^m f(x-1)$$

This equation follows directly from the definition of the differences. The last column will contain the approximate values, the column last but one their first differences, and so on.

The measure of the precision obtained is defined in this method by

$$\sigma_n^2 = \frac{1}{N} \sum_{x=0}^{N-1} [f(x) - y]^2 = \frac{1}{N} [\sum_{x=0}^{N-1} y^2$$

$$+ \sum_{x=0}^{N-1} \sum_{m=0}^{n} a_m^2 U_m^2 - 2 \sum_{x=0}^{N-1} \sum_{m=0}^{n} a_m U_m y]$$

and in consequence of eqns. (93)—(95)

$$\sigma_n^2 = \frac{1}{N} \sum_{x=0}^{N-1} y^2 - \sum_{m=0}^{n} |C_{m0}|\theta_m^2 \qquad (97)$$

Therefore, to determine σ_n^2, the computation of Σy^2 is necessary in addition to the orthogonal moments.

134

Should the obtained precision be insufficient, we have only to compute Θ_{n+1} and use eqns. (96) and (97). The work previously done would not be lost, as it would when working by other methods.

In order to compute the orthogonal moments, the binomial moments B_0, B_1, ..., B_n are first determined.

$$B_m = \sum_{x=0}^{N-1} \binom{x}{m} y$$

A table is constructed in the following way. In the first line we put $y(N-1)$ and $n+1$ zeros and in the ξth line of the first column the number $y(N-\xi)$. Starting from these initial conditions, we determine the number $f(\xi, \eta)$ of line ξ and column η with the aid of the equation

$$f(\xi, \eta) = f(\xi - 1, \eta - 1) + f(\xi - 1, \eta) \tag{98}$$

It can be shown that the solution of eqn. (98) is

$$f(\xi, \eta) = \sum_{\nu=1}^{\xi-1} \binom{\xi - \nu - 1}{\eta - 2} y(N - \nu)$$

This will give, for $\xi = N + 1$

$$f(N + 1, \eta) = \sum_{\nu=1}^{N} \binom{N - \nu}{\eta - 2} y(N - \nu) = B_{\eta - 2}$$

That is, the row $N+1$ will contain the binomial moments B_0, B_1, ..., B_n; the moment B_m will be in the column $m + 2$.

In this way, we also obtain, by simple addition without any multiplication, every binomial moment needed.

If N is large, the moments grow rapidly with their order and become very large numbers. This is an inconvenience and to obviate it we introduce the mean binomial moments by

$$T_m = \sum_{x=0}^{N-1} \binom{x}{m} y / \binom{N}{m+1} = B_m / \binom{N}{m+1}$$

these are always of the same order of magnitude as y, whatever the order m may be.

Multiplying eqn. (93) by y and summing from $x = 0$ to $x = N$, we get

$$\theta_m = \sum_{\nu=0}^{m} \beta_{m\nu} T_\nu \qquad (99)$$

The orthogonal moments are determined from this equation.

Summing up, the only real work to be done is the computation of the binomial moments given above.

Starting from eqn. (96), we obtain a table of the approximate values $f(x)$ and of their first n differences by the method of the addition of differences.

Example. Let us consider the example already dealt with on p. 126. The following table summarizes the measured volumes, u, the potential differences, y, and the transformed volumes, x. The variable x was introduced by the transformation $x = (u - 13.0)/0.20$.

u	x	y	u	x	y
13.00	0	333	14.00	5	103
13.20	1	314	14.20	6	63
13.40	2	288	14.40	7	41
13.60	3	242	14.60	8	18
13.80	4	184	14.80	9	2

We note that the number of measured points, N, is 10 and we want to approximate these measured data by a polynomial of the third degree.

Calculating the binomial moments according to eqn. (98)

y	B_0	B_1	B_2	B_3
2	0	0	0	0
18	2	0	0	0
41	20	2	0	0
63	61	22	2	0
103	124	83	24	2
184	227	207	107	26
242	411	434	314	133
288	653	845	748	447
314	941	1498	1593	1195
333	1255	2439	3091	2788
	1588	3694	5530	5879

The mean binomial moments are

$$T_0 = \frac{B_0}{\binom{N}{1}} = \frac{1588}{\binom{10}{1}} = 158.8$$

$$T_1 = \frac{B_1}{\binom{N}{2}} = \frac{3694}{\binom{10}{2}} = \frac{3694}{45} = 82.09$$

$$T_2 = \frac{B_2}{\binom{N}{3}} = \frac{5530}{\binom{10}{3}} = \frac{5530}{120} = 46.08$$

$$T_3 = \frac{B_3}{\binom{N}{4}} = \frac{5879}{\binom{10}{4}} = \frac{5879}{210} = 28.00$$

The orthogonal moments are calculated from eqn. (99) using Table 20 for the $\beta_{m\nu}$.

$\theta_0 = T_0 = 158.8000$ $\qquad\qquad$ $\theta_0^2 = 25217.44$

$\theta_1 = T_1 - T_0 = -76.7111$ $\qquad\qquad$ $\theta_1^2 = 5884.59$

$\theta_2 = 2T_2 - 3T_1 + T_0 = 4.7000$ $\qquad\qquad$ $\theta_2^2 = 22.09$

$\theta_3 = 5T_3 - 10T_2 + 6T_1 - T_0 = 12.8762$ $\quad$ $\theta_3^2 = 165.80$

The coefficients $C_{m\nu}$ from Table 21 are

$C_{10} = -2.454$ $\qquad$ $C_{20} = 2.727$ $\qquad$ $C_{30} = -2.056$

$C_{11} = 0.545$ $\qquad$ $C_{21} = 1.818$ $\qquad$ $C_{31} = 2.741$

$\qquad\qquad\qquad$ $C_{22} = 0.454$ $\qquad$ $C_{32} = -1.715$

$\qquad\qquad\qquad\qquad\qquad$ $C_{33} = 0.490$

The precision can be calculated from eqn. (97) without calculating the coefficients.

$$\sigma_3^2 = \frac{\sum\limits_{x=0}^{N-1} y^2}{10} - \sum\limits_{m=0}^{n} |C_{m0}| \theta_m^2 = 40143.60 - 40062.55 = 81.04$$

$\sigma_3 = 9.00$

The approximating function is

$$f(x) = b_0 + b_1\binom{x}{1} + b_2\binom{x}{2} + b_3\binom{x}{3}$$

and the coefficients are

$$b_0 = \theta_0 + C_{10}\theta_1 + C_{20}\theta_2 + C_{30}\theta_3 = 333.4364$$

$$b_1 = C_{11}\theta_1 + C_{21}\theta_2 + C_{31}\theta_3 = -15.0909$$

$$b_2 = C_{22}\theta_2 + C_{32}\theta_3 = -19.9242$$

$$b_3 = C_{33}\theta_3 = 6.3030$$

The table of the approximate values of $f(x)$ are computed by the method of the addition of differences. The differences between calculated and measured data $\epsilon = f(x) - y$ are also included.

$b_3 = \Delta^3 f(x)$	$b_2 = \Delta^2 f(x)$	$b_1 = \Delta f(x)$	$b_0 = f(x)$	y	ϵ	ϵ^2
6.30	−19.92	−15.09	333.44	333	0.44	0.19
6.30	−13.62	−35.01	318.34	314	4.34	18.88
6.30	−7.32	−48.63	283.33	288	−4.67	21.81
6.30	−1.02	−55.95	234.69	242	−7.31	53.38
6.30	5.28	−56.97	178.74	184	−5.26	27.67
6.30	11.58	−61.69	121.77	103	18.77	352.30
6.30	17.88	−40.11	70.09	63	7.09	50.24
	24.18	−22.23	30.00	41	−11.00	121.07
		1.95	7.80	18	−10.20	104.04
			9.80	2	7.80	60.84
						810.42

The measure of precision, the mean of square deviations is

$$\sigma_3^2 = \frac{\sum\limits_{x=0}^{N-1} \epsilon^2}{N} = \frac{810.42}{10} = 81.04$$

and

$$\sigma_3 = 9.00$$

which is the same as the value obtained earlier using only the orthogonal moments.

(C) NON-LINEAR MODELS

Models are regarded as non-linear if there is a non-linearity with respect to the constants. Some special cases will first be discussed followed by the general problem.

(1) Regression analysis by linearizing transformations

In many situations, the relationship between two variables is not a linear one. In order to apply the principle of least squares easily to these relationships, the equations should be transformed into a linear form (see, for example, refs. 30, 31). Let us assume, for example, that there exists an exponential relationship between the variables, viz.

$$y = ae^{bx}$$

This can be transformed by taking logarithms of both sides to

$$\ln y = \ln a + bx$$

or

$$Y = a_0 + bx$$

where $Y = \ln y$ and $a_0 = \ln a$. Several types of easily transformable functions and their linearized forms are shown in Table 22. Using

TABLE 22

Linearizing transformations

Non-linear equation	Linearized equation	Linearized variables	
		Y	X
$y = a + bx^p$	$y = a + bX$	y	x^p
$y = \sqrt[p]{a + bx}$	$Y = a + bx$	y^p	x
$y = \dfrac{x}{a + bx}$	$Y = a + bx$	$\dfrac{x}{y}$	x
$y = ae^{bx}$	$Y = \ln a + bx$	$\ln y$	x
$e^y = ax^b$	$y = \ln a + b \ln x$	y	$\ln x$
$y = ax^b$	$Y = \ln a + bX$	$\ln y$	$\ln x$

these transformations, there exists a linear relationship between the linearized variables

$$Y = a + bX$$

If it assumed that the relationship between the variables is

$$y = ax^n e^{-bx^2}$$

then taking logarithms of both sides

$$\ln y = \ln a + n \ln x - bx^2$$

The constants $\ln a$, n, and b are determined by a two independent variables linear regression, the variables being $\ln x$ and x^2.

In the case of some transformations, a fixed, reliable, practically "error-free" data pair $(x_0 y_0)$ is needed. It is advisable to take it from the middle of the data set and to correct it by linear or cubic interpolation. These transformations are shown in Table 23. In this case, the independent variable will never be transformed, i.e. $X = x$ and the

TABLE 23

Linearizing transformations using a fixed data pair

Non-linear equation	Linearized variable, Y
$y = a + bx + cx^2$	$\dfrac{y - y_0}{x - x_0}$
$y = \dfrac{a + bx}{c + dx}$	$\dfrac{x - x_0}{y - y_0}$
$y = \dfrac{1}{ax^2 + bx + c}$	$\dfrac{y_0 - y}{yy_0(x - x_0)}$
$y = \dfrac{ax^2 + bx + c}{x^2}$	$\dfrac{yx^2 - y_0x_0^2}{x - x_0}$
$y = \dfrac{x}{ax^2 + bx + c}$	$\dfrac{xy_0 - x_0y}{y(x - x_0)}$
$y = +\sqrt{ax^2 + bx + c}$	$\dfrac{y_2 - y_0^2}{x - x_0}$
$y = +\dfrac{1}{\sqrt{ax^2 + bx + c}}$	$\dfrac{y^2 - y_0^2}{y^2y_0^2(x - x_0)}$

linearized equation is always

$$Y = a + bx$$

(2) Approximation by exponential functions

a. *Approximation by one exponential function.* When starting with exponential functions, the Algol procedure "expfit" worked out by Späth [32] will be presented for determining the constants in the model

$$y = a + be^{-cx}$$

Further explanation is given in the comment on the procedure. The listing of the procedure is

procedure expfit (x, y, p, n, ca, ce, eps, a, b, c, s, fx, exit);
    value n, ca, ce, eps; **integer** n; **real** ca, ce, eps, a, b, c, s;
    label exit; **array** x, y, p, fx;
comment If the method of least squares is used to determine the parameters a, b, c of a curve $f(x) = a + be^{-cx}$ which approximates n data points (x_i, y_i) with associated weights p_i, then

$$s(a, b, c) = \sum_{i=1}^{n} p_i [y_i - f(x_i)]^2 \tag{I}$$

must be a minimum. A necessary condition for this is that

$$\frac{\partial s}{\partial a} = \frac{\partial s}{\partial b} = \frac{\partial s}{\partial c} = 0 \tag{II}$$

Usually (see ref. 33) it is attempted to solve this system of non-linear equations by an iterative method which is based upon the linearization of f in (II) and the convergence of which depends on the given starting values for a, b, c.

A simpler and more effective way which can always be chosen if there is only one non-linear parameter in f is the following. It is always possible to eliminate $a = a(c)$ and $b = b(c)$ from the equations $\partial s/\partial a = 0$ and $\partial s/\partial b = 0$ and to put these expressions into $\partial s/\partial c = 0$. This gives only one equation in one variable

$$F(c) := \frac{\partial s}{\partial c} [a(c), b(c), c] = 0$$

If a value c' is calculated with $F(c') = 0$ then the corresponding values of a and b are obtained from $a' = a(c')$ and $b' = b(c')$.

The following procedure is based upon this idea which is fully treated in ref. 34. It allows a triple (a, b, c) to be found which solves (II) if one makes available a non-local procedure Rootfinder which is able to get a zero c of a function $F(c)$ in the interval $[ca, ce]$ with the relative accuracy eps, if $\text{sign}[F(ca)] \neq \text{sign}[F(ce)]$ otherwise leaving to the global label exit. As the above $F(c)$ is discontinuous at $c = 0$, $[ca, ce]$ must not contain 0. The speed and efficiency of the algorithm depend on the choice of the procedure Rootfinder.

Most of the symbols are self-explanatory. The array fx finally contains the values $a + be^{-cx_i}$;

```
begin integer i; real t, u, v, w, fc, h0, h1, h2, h3, h4, h5, h6, h7;
procedure fronc (c, fc); value c; real c, fc;
comment computes for a given c the value fc = F(c) and a = a(c),
b = b(c);
begin h0 := h1 := h2 : = h3 := h4 := h5 := h6 := h7 := 0;
for i := 1 step 1 until n do
begin
    t := x[i]; u := exp(−c × t); v := p[i]; w := y[i];
    h0 := h0 + v; h1 := h1 + u × v; h2 := h2 + u × u × v;
    h3 := h3 + v × w; h4 := h4 + u × v × w;
    h5 := h5 + t × u × v;
    h6 := h6 + t × u × u × v; h7 := h7 − u × v × w × t
end i;
t := 1.0/(h0 × h2 − h1 × h1); a := t × (h2 × h3 − h1 × h4);
b := t × (h0 × h4 − h1 × h3); fc := h7 + (h5 × a + h6 × b)
end fronc;
Rootfinder (fronc, ca, ce, eps, c, exit); t := 0;
for i := 1 step 1 until n do
begin
    v := fx[i] := a + b × exp(−c × x[i]); v := v − y[i];
    t := t + p[i] × v × v
end i;
s := t
end expfit;
```

b. Approximation by a sum of exponential functions [35]. Let us assume that the approximated data (x_i, y_i) $(i = 0, 1, ..., N − 1)$ have the equidistant argument with a step h and the number of measured pairs of data is N.

We want to approximate them by a function

$$y = P_1 \exp(\alpha_1 x) + P_2 \exp(\alpha_2 x) + \ldots + P_n \exp(a_n x) \tag{100}$$

where $\alpha_1 \neq \alpha_2 \neq \ldots \neq \alpha_n$ and $n \leqslant N/2$.
The case $n = 2$.

Fixing three equidistant arguments x_0, $x_0 + h$, and $x_0 + 2h$ and denoting the corresponding y values by y_0, y_1, and y_2, one can verify by substitution

$$\exp[\alpha_1 + \alpha_2)h]y_2 - [\exp(\alpha_1 h) + \exp(\alpha_2 h)]y_1 + y_0 = 0 \tag{101}$$

Introducing the following notations

$$\beta_2 = \exp[(\alpha_1 + \alpha_2)h] \tag{102}$$

$$\beta_1 = \exp(\alpha_1 h) + \exp(\alpha_2 h) \tag{103}$$

from eqn. (101) we obtain

$$\beta_2 y_2 - \beta_1 y_1 + y_0 = 0$$

Since β_1 and β_2 are constants, they can be determined from the experimental data by linear regression taking, for example, y_1 and y_2 as independent and y_0 as dependent variables. The constants α_1 and α_2 of the original eqn. (100) are calculated from eqns. (102) and (103) and the constants P_1 and P_2 are determined again by linear regression assuming that α_1 and α_2 are already constants.
The case $n > 2$.

At the assumed equidistant arguments, the function (100) may be written

$$u_i = M_1 m_1^i + M_2 m_2^i + \ldots + M_n m_n^i \tag{104}$$

where we have introduced $x = x_0 + ih$ $(i = 0, 1, \ldots, N - 1)$ and where $M_j = P_j \exp(\alpha_j x_0)$ and $m_j = \exp(\alpha_j h)$ $(j = 1, 2, \ldots, n)$. The constants M_j and m_j will be determined by an iteration procedure and the constants P_i and α_i are obtained from the above equations.

The choice of the first estimate of the coefficient m_j and of α_j $(j = 1, 2, \ldots, n)$ in eqns. (100) or (104) is made on the basis of a selected subset $2n < N$ of (x_i, y_i) pairs.

The calculation of approximate values m_j is performed according to Mička and Schmidt [36] (an alternative method is given in ref. 37) so that from the N experimental values $2n$ values are chosen with the equidistant argument H. From these values, denoted by

$z_0, z_1, ..., z_{2n-1}$, there are set up n linear equations

$$z_n + A_1 z_{n-1} + ... + A_n z_0 = 0$$

$$z_{n-1} + A_1 z_n + ... + A_n z_1 = 0$$

...

$$z_{2n-1} + A_1 z_{2n-2} + ... + A_n z_{n-1} = 0 \qquad (105)$$

which are solved and from the calculated roots $A_1, ..., A_n$ an algebraic equation of the nth order is formed

$$q^n + A_1 q^{n-1} + ... + A_{n-1} q + A_n = 0 \qquad (106)$$

having n roots, q_j ($j = 1, 2, ..., n$). From these the coefficients m_j in eqn. (104) are

$$m_j = q_j^{h/H} \qquad (107)$$

If negative or complex roots are obtained, then the iteration method must start with completely arbitrary values.

When the first approximate values $\alpha_1, \alpha_2, ..., \alpha_n$ and $m_1, m_2, ..., m_n$ in eqns. (100) and (104), respectively, are calculated in any way, the coefficients $M_1, M_2, ..., M_n$ will be determined by the least square method; these coefficients should minimize the sum

$$S = \sum_{i=0}^{N-1} (u_i - y_i)^2 = \sum_{i=0}^{N-1} (M_1 m_1^i + M_2 m_2^i + ... + M_n m_n^i - y_i)^2 \qquad (108)$$

which is, for given m_j values, a function of $M_1, ..., M_n$. The necessary conditions for the minimum S are the equations

$$\frac{\partial S}{\partial M_1} = \frac{\partial S}{\partial M_2} = ... = \frac{\partial S}{\partial M_n} = 0 \qquad (109)$$

from which the system of normal equations for M_j

$$M_1[m_1^i m_1^i] + M_2[m_2^i m_1^i] + ... + M_n[m_1^i m_n^i] = [y_i m_1^i]$$

$$M_1[m_2^i m_1^i] + M_2[m_2^i m_2^i] + ... + M_n[m_2^i m_n^i] = [y_i m_2^i]$$

...

$$M_1[m_n^i m_1^i] + M_2[m_n^i m_2^i] + ... + M_n[m_n^i m_n^i] = [y_i m_n^i] \qquad (110)$$

is obtained in which the term $[m_j^i m_k^i]$ is an abbreviation for

$$\sum_{i=0}^{N-1} m_j^i m_k^i \text{ and } [y_i m_j] \text{ is } \sum_{i=0}^{N-1} y_i m_j^i.$$

144

It is clear that the set m_j determined by eqn. (107) is not the only one possible and it may be that for another choice of m_j the sum of squares S [eqn. (108)] will be smaller. S is a function of m_j, viz.

$$S = S(m_1, m_2, ..., m_n) \tag{111}$$

To find the values $m_1, m_2, ..., m_n$ minimizing the function (111), the following iteration procedure is used: let us consider a_j, a value in the vicinity of point m_j and that

$$a_j = m_j (1 + x_j)$$

where $j = 1, 2, ..., n$ and x_j are small compared with 1. The sum of squares defined by eqn. (108) may be considered a function of x_j at given m_j $(j = 1, 2, ..., n)$ and its minimum leads to a series of equations for x_j

$$\frac{\partial S}{\partial x_j} = 0 \qquad \text{for} \qquad j = 1, 2, ..., n \tag{112}$$

For the numerical solution of this system, all powers $a_j^i = m_j^i (1 + x_j)^i$ are expanded into binomial series. From these expansions, only the first-order term is used; then we substitute

$$a_j^i = m_j^i \left[1 + \binom{i}{1} x_j \right] \tag{113}$$

While this equation is accurate for $i = 1$, it is only approximate for $i = 2, 3, ..., N - 1$. We assume that the neglected terms are sufficiently small.

Substituting m_j in eqn. (108) by a_j as defined by eqn. (113) and evaluating eqn. (112), the following system of linear equations is obtained for x_j.

$$M_1[\mu_1\mu_1] x_1 + M_2[\mu_1\mu_2] x_2 + ... + M_n[\mu_1\mu_n] x_n + [\mu_1\epsilon] = 0$$

$$M_1[\mu_1\mu_2] x_1 + M_2[\mu_2\mu_2] x_2 + ... + M_n[\mu_2\mu_n] x_n + [\mu_2\epsilon] = 0$$

$$\vdots$$

$$M_1[\mu_1\mu_n] x_1 + M_2[\mu_2\mu_n] x_2 + ... + M_n[\mu_n\mu_n] x_n + [\mu_n\epsilon] = 0$$

where $\mu_j = im_j^i$, and $\epsilon_i = u_i - y_i$. (The term $[\mu_j\mu_k]$ is an abbreviation of $\sum_{i=0}^{N-1} i^2 m_j^i m_k^i$, $[\mu_j\epsilon] = \sum_{i=0}^{N-1} im_j^i\epsilon_i$.)

Having x_j, the values found by the first iteration step $m_j = m_j^0$ are corrected according to the equation

$$m_j^{(1)} = m_j^{(0)} (1 + x_j)$$

where $j = 1, 2, ..., n$. By this, one step of the iteration procedure is completed and we start again from the system of equations (110).

The end of the process is bound, for example, to the condition of obtaining the minimum $S^{(r)}$, or for finding the coefficient $m_j^{(r)}$ with a prescribed accuracy.

Note that, for numerical purposes, the equation for S [eqn. (108)]

$$S = \sum_{i=0}^{N-1} (M_1 m_1^i + M_2 m_2^i + ... + M_n m_n^i - y_i)^2$$

may be simplified by the use of normal equations (110). Let us multiply the rows of these equations by the coefficients $M_1, M_2, ..., M_n$. After rearrangement, we obtain

$$M_1^2 [m_1^i m_1^i] + M_1 M_2 [m_1^i m_2^i] + ... + M_1 M_n [m_1^i m_n^i] - M_1 [y_1 m_1^i] = 0$$

$$M_2 M_1 [m_2^i m_1^i] + M_2^2 [m_2^i m_2^i] + ... + M_2 M_n [m_2^i m_n^i] - M_2 [y_i m_2^i] = 0$$

$$\vdots \qquad\qquad \vdots \qquad\qquad \vdots \qquad\qquad \vdots$$

$$M_n M_1 [m_n^i m_1^i] + M_n M_2 [m_n^i m_2^i] + ... + M_n^2 [m_n^i m_n^i] - M_n [y_i m_n^i] = 0$$

If all these zero-valued terms are subtracted from the sum of squares, eqn. (108), we obtain a more simple expression

$$S = [y_i y_i] - M_1 [y_i m_1^i] - M_2 [y_i m_2^i] - ... - M_n [y_i m_n^i]$$

Since the term $[y_i y_i]$ in S represents a constant, it is sufficient to investigate in the course of iteration the term

$$[y_i y_i] - S = \sum_{j=1}^{n} M_j [y_i m_j^i]$$

where $[y_i y_i] = \sum_{i=0}^{N-1} y_i^2$. (See also ref. 38.)

(3) The general case of non-linear models

Suppose we have a random observable dependent variable y_i $(i = 1, 2, ..., m)$ and several non-random, independent variables, x_k $(k = 1, 2, ..., n)$. Let a_j $(j = 1, 2, ..., q)$ be the parameters of the model

$$y = f (x_1, x_2, ..., x_n, a_1, a_2, ..., a_q)$$

146

In order to obtain the desired parameter estimates according to the least square principle, the sum of squares of deviations

$$S = \sum_{i=1}^{m} [y_i - f(x_{1i}, x_{2i}, ..., x_{ni}, a_1, a_2, ..., a_q)]^2$$

must be minimized. There exist many techniques for solving this problem. The more effective ones are the sequential simplex method, the direct search method, and the linearization of the model in a truncated Taylor series. These are discussed in detail by Himmelblau [39]. Of these, we shall discuss the first and the third methods.

   a. Minimizing functions by a modified sequential simplex method. The scope of the method is to find the minimum point of the function of n variables

$$y = F(x_1, x_2, ..., x_n)$$

or

$$y = F(\mathbf{x})$$

where $\mathbf{x}$ is a vector in the n dimensional space.

   The basis of the calculation is the so-called sequential simplex method [40—42]. The first step of the calculation is the construction of a regular simplex (in the case of two or three independent variables the construction of a regular triangle or tetrahedron, respectively) in the space of the independent variables. The method uses a succession of calculations by constructing the mirror image of one vertex of the simplex in the sense that the reflection must always lead to a more favourable region. If this reflection leads to a more favourable region, then the reflected point is rejected and replaced by its mirror image. In the other case, another vertex will be reflected.

   The path taking from the starting point will zig-zag about the line of steepest ascent. This construction process is repeated until the minimum point is reached. This means that, when reflecting all the vertices, none of reflections leads to a more favourable region then the calculation is finished and the minimum point is found with an accuracy determined by the step length.

   The construction of the simplex and the reflection is shown for $n = 2$ in Fig. 22.

   Let us now discuss the definition and properties of the n-dimen-

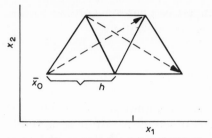

Fig. 22. The principle of the modified sequential simplex method.

sional regular simplex. Let us consider $r + 1$ points in the n-dimensional space where $r \leqslant n$. Let their position vectors be $x_0, x_1, x_2, ...,$ x_r and let the distances $|x_i - x_j|$ $(i, j = 1, 2, ..., 3; i \neq j)$, the so-called edge lengths, be constants. The set of points defined by the sums

$$\lambda_0 x_0 + \lambda_1 x_1 + ... + \lambda_r x_r$$

where $\Sigma_{i=0}^n \lambda_i = 1$ is called an r-dimensional regular simplex. The points defined by $x_0, x_1, x_2, ..., x_r$ are the vertices of the simplex. The lines joining each pair of them are the edges. Further, it will be supposed that $r = n$.

Let us suppose that one edge of the simplex and all the altitudes perpendicular to this edge, are parallel to the axes of coordinates. Let us define the vectors a and b in the form of coordinates as

$$a_i = x_{0i} + h \sqrt{\frac{i+1}{2i}} \qquad (i = 1, 2, ..., n) \tag{114}$$

$$b_i = x_{0i} + h \sqrt{\frac{1}{2i(i+1)}} \tag{115}$$

It can be shown that the vectors $x_1, x_2, ..., x_n$ defined by the relations

$$x_{ij} = x_{0j} \text{ if } j > i \tag{116}$$

$$x_{ij} = a_j \quad \text{if } j = i \tag{117}$$

$$x_{ij} = b_j \quad \text{if } j < i \tag{118}$$

represent a regular simplex in the n-dimensional space. (The step length need not be the same in all directions of the n-dimensional space. We may also introduce a vector h defining the step length in

148

the different directions of the space and then we must replace h in eqns. (114) and (115) by h_i. In this case, the simplex will not be a regular one, but nevertheless, it can be used for the further calculations.)

Let us denote the values of the function to be minimized in the point x_i as

$$y_i = F(x_i) \ (i = 1, 2, ..., n)$$

Rearranging the values of y_i in the decreasing order

$$y_0 > y_1 > ... > y_n$$

and reflecting the vector x_0 corresponding to y_0 by the relation

$$x_0^* = \frac{2}{n} \sum_{i=0}^{n} x_i - \frac{n+2}{n} x_0$$

we obtain a regular simplex. If

$$y_0^* < y_1 \tag{119}$$

is valid, where $y^* = F(x_0^*)$, and x_0^* is in the permitted domain as well, then y_0 and x_0 must be replaced by y_0^* and x_0^*. Should eqn. (119) not be valid or x_0 be outside of the permitted domain, vectors $x_1, x_2, ..., x_n$ must be reflected until the first of the relations

$$y_i^* < y_{i+1} \ (i = 1, 2, ..., n-1) \tag{120}$$

$$y_n^* < y_n \tag{121}$$

becomes valid and the corresponding $x_1^*, x_2^*, ..., x_n^*$ vectors are in the permitted domain. If none of the relations (119)—(121) becomes valid for any y_i^* then the calculation is finished and the minimum point is found with an accuracy corresponding to the length of an edge of the simplex.

If this accuracy is not sufficient, then the calculation must be repeated starting from the minimum point found previously but dividing the starting step length by two. Constraints of any arbitrary form may be taken into account. If the prescribed starting point is outside the permitted domain, then an appropriate starting point lying within the permitted domain can be found.

In order to take into account the constraints, let us suppose that there are m constraints formulated as inequalitites reduced to zero.

This means that if one of the inequalities

$f_1(x) > 0$

$f_2(x) > 0$

$\vdots$

$f_m(x) > 0$

holds for the given point, then the point is outside the permitted domain. Let us introduce the function G with the aid of the following set of functions.

$f_i^*(x) = f_i(x)$

if $f_i(x) > 0$ otherwise it is 0 where $i = 1, 2, ..., m$

$$G(x) = \sum_{i=1}^{m} f_i(x) \tag{122}$$

The value of this function is zero if the point x is within the permitted domain and positive if one of the inequalities does not hold. If the given starting point lies outside the permitted domain, we must find an adequate starting point within the permitted domain. For this purpose, we minimize function (122).

It is supposed that function G is monotonous in the sense that it has a minimum point only if $G(x) = 0$. If $G(x) \neq 0$, we cannot find the minimum point of $F(x)$ starting from the given point and we must start from another one.

In order to calculate the partial difference quotients in the vicinity of the given starting point, or of an arbitrary point in the n-dimensional space, let us define a set of vectors x_i' ($i = 0, 1, 2, ..., n$) as follows.

$x_{ij}' = x_{0j}$

if $j \geqslant i$, otherwise

$x_{ij}' = b_j$

The partial difference quotients are

$$\frac{\Delta F}{\Delta x_i} = \frac{F(x_i) - F(x_i')}{x_{ii} - x_{ii}'} \qquad (i = 1, 2, ..., n)$$

150

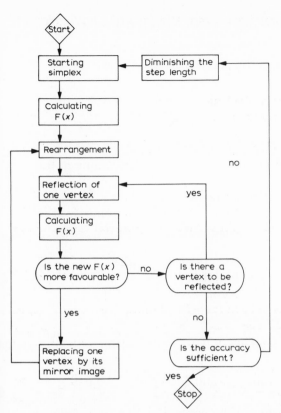

Fig. 23. Minimizing by the sequential simplex method.

This calculation may be carried out from time to time during the optimization and the values of the partial difference quotients may also serve as a criterion for stopping. The calculation flow sheet is shown in Fig. 23.

The calculation method outlined above has been programmed for the most simple case in Fortran. The step length is supposed to be the same in each direction, constraints cannot be taken into account. The declaration of the subroutine OPTSIMP worked out for this case is

SUBROUTINE OPTSIMP (N, N1, K, KM, ST, XN, XO, X, F, Y, YN, EPS, IE)

DIMENSION XN(N), XO(N1, N), X(N), Y(N1), A(6), B(6), R(6)

The parameters are

N number of independent variables
N1 N + 1
K counter for successful reflections
KM specified maximum value of K
ST step length
XN coordinates of the starting point (starting values of the independent variables) before the first call, and the minimum at the end of the calculation
XO coordinates of the vertices of the actual simplex
X the independent variable of the function to be minimized
F external function, the function to be minimized, given in the form F(X, N).
Y the function values at the vertices of the simplex
YN the minimum value of the function
EPS the maximum permitted error in YN
IE if the counter K exceeds KM then IE =1 else IE = 0.

Subroutine required

CHANGE (A, B)

This subroutine puts the values A and B in place of B and A respectively.

Note that the subroutine can be applied in this form for a maximum of 6 independent variables. When it must be applied to a function with more than 6 variables, then the number 6 is to be changed in the declaration part to the actual number of variables.

The listing of the subroutine is

```
      SUBROUTINE OPTSIMP(N,N1,K,KM,ST,XN,XO,X,F,Y,YN,EPS,IE)
      DIMENSION XN(N),XO(N1,N),X(N),Y(N1),A(6),B(6),R(6)
      IE=0
      K=0
    2 DO 3 I=1,N
      A(I)=XN(I)+ST*SQRT((I+1)/2.0/I)
      B(I)=XN(I)+ST*SQRT(1.0/2.0/I/(I+1))
    3 CONTINUE
      DO 5 I=1,N1
      DO 4 J=1,N
      IF(I-J-1)6,7,8
    6 XO(I,J)=XN(J)
```

152

```
       GO TO 4
 7  XO(I,J)=A(J)
       GO TO 4
 8  XO(I,J)=B(J)
 4  CONTINUE
 5  CONTINUE
       DO 10 I=1,N1
       DO 9 J=1,N
 9  X(J)=XO(I,J)
       Y(I)=F(X, I)
10  CONTINUE
11  DO 13 I=1,N
       I1=I+1
       DO 12 J=I1,N1
       IF(Y(I)-Y(J)) 14,12,12
14  CALL CHANGE(Y(I),Y(J))
       DO 15 L=1,N
15  CALL CHANGE (XO(I,L),XO(J,L))
12  CONTINUE
13  CONTINUE
       YN=Y(N1)
       IF(K-KM) 17,17,16
16  IE=1
       RETURN
17  L=1
19  DC 18 J=1,N
       R(J)=0.0
       DO 20 I=1,N1
20  R(J)=XO(I,J)+R(J)
       X(J)=2*R(J)/N-(N+2)*XO(L,J)/N
18  CONTINUE
       W=F(X,N)
       IS=L+1
       IF(L.EQ.(N+1)) IS=L
       IF(W-Y(IS)) 21,22,22
21  CONTINUE
       DO 23 I=1,N
23  XO(L,I)=X(I)
       Y(L)=W
       K=K+1
       GO TO 11
22  L=L+1
       IF(L-N-1) 19,19,24
24  DO 27 I=1,N
27  XN(I)=XO(N1,I)
       YN=Y( I1)
       IF(ABS(Y(N)-Y(N1))-EPS) 25,26,26
25  CONTINUE
       RETURN
```

```
26 ST=ST/2
   GO TO 2
   END


   SUBROUTINE CHANGE (A,B)
   C=A
   A=B
   B=C
   RETURN
   END
```

The above method for minimizing functions of several variables has been programed for the case when taking into account constraints in Algol. The description of the Algol procedure OPTSIMP3 is as follows.

The procedure finds the minimum point of a function $y = F(x)$ by the modified simplex method described above, taking into account constraints.

The most important characteristics of the procedure are

(a) The step length need not be the same in all directions of the n-dimensional space. The basic step lengths are given as an array.

(b) If the starting point is within the permitted domain but some of the vertices of the first simplex would be outside, the procedure tries to construct systematically other simplexes around the given starting point with the specified step length. For this purpose, the sign of the elements of the array step are changed systematically. If it would be impossible to find a simplex with vertices all within the permitted domain, the procedure diminishes the step length in all directions.

(c) The constraints must be given in the form of a real expression of the elements of array x, depending on the value of the formal parameter iconstr, so that, if a point lies outside the permitted domain, the sign of the expression is negative.

(d) If the given starting point is outside the permitted domain, the procedure tries to find a point lying within it.

The procedure heading is

procedure OPTSIMP3 (var, count, cmax, step, xnew, xold, x, function, y, ymin, eps, nconstr, constr, iconstr, ERROR);

value   var, cmax, eps, nconstr;

integer   var, count, cmax, nconstr, iconstr;

real   function, ymin, eps, constr;

array   xnew, xold, x, y, step;

154

label ERROR;
The formal parameters are

integer var	number of independent variables.
integer count	counter. Must be 0 at first call. Recalling the procedure its last value must be stored.
integer cmax	if count becomes higher than cmax, the procedure makes an exit to the label ERROR.
array step [1 : var]	its ith element is the step length in the ith direction in the var dimensional space.
array xnew [1 : var]	coordinates of the starting point (starting values of the independent variables) before the first call, and the minimum at the end of the calculation.
array xold [0 : var, 1 : var]	coordinates of the vertices of the actual simplex. At the first call we need not assign any value to it.
array x [1 : var]	its elements are the independent variables of the function to be minimized. It is not necessary to assign values to its elements.
real function	the function to be minimized given as a real expression or procedure, given in terms of **array** x.
array y [0 : var]	the function values at the vertices of the simplex. Its elements will be calculated by the procedure.
real ymin	the minimum value of the function.
real eps	the maximum permitted error in ymin.
integer nconstr	number of inequalities defining the constraints. Specified by value.
real constr	expression of the elements of **array** x being negative if a given point lies outside of the permitted domain.
integer iconstr	parameter for counting the constraints.
label ERROR	this is used as an exit before the minimum is reached if count = cmax, or if there cannot be found a starting point within the permitted domain.

The listing of the procedure is

```
procedure OPTSIMP3 (var, count, cmax, step, xnew,
    xold, x, function, y, ymin, eps, nconstr, constr,
    iconstr, ERROR);
value var, cmax, eps, nconstr;
integer var, count, cmax, nconstr, iconstr;
real function, ymin, eps, constr;
array xnew, xold, x, y, step;
label ERROR;

begin integer i, j, k, s;
real w;
array a, b, r [1 : var], p, q[0 : var];
boolean B;
procedure change (A, B); real A, B;

begin real e;
e: = A;
A: = B;
B: = e
end;
```

```
procedure rearrangement (A, B, n);
value n; integer n; array A, B;

begin integer i, j, k;
for i: = 0 step 1 until n − 1 do
for j: = i + 1 step 1 until n do
if A[i] < A[j] then
begin change (A[i], A[j]);
for k: = 1 step 1 until n do
change (B[i, k], B[j, k])
end
end;

real procedure S;
begin integer i; real m; array cs[1 : nconstr];
for iconstr: = 1 step 1 until nconstr do
cs[iconstr]: = constr;
m: = 0;
for i: = 1 step 1 until nconstr do
m: = m + (if cs[i] < 0 then cs[i] else 0);
S: = m
end;
```

```
L0: if count ≠ 0 then go to L1;
for i: = 1 step 1 until var do
x[i]: = xnew[i];
if S < 0 then
begin

OPTSIMP3 (var, count, cmax, step, xnew, xold, x, — S,
y, ymin, ₁₀ —10, 0, 0, iconstr, ERROR);
if ymin > 0 then goto ERROR;
count: = 0;

goto L0
end;

L3: for i: = 2 ↑ var — 1 step — 1 until 0 do
begin p[0]: = q[0]: = 0;
for j: = 1 step 1 until var do
begin q[j]: = 2 × (q[j — 1] + p[j — 1]);
p[j]: = i : 2 ↑ (var — j) — q[j]
end;
```

```
for j: = 1 step 1 until var do
begin p[j]: = 2×p[j] – 1;
a[j]: = xnew[j] + p[j]×step[j]×sqrt((j + 1)/2/j);
b[j]: = xnew[j] + p[j]×step[j]×sqrt(1/2/j/(j + 1))
end;

B: = true;
for k: = 0 step 1 until var do
begin for j: = 1 step 1 until var do
begin
xold[k, j]: = if k < j then xnew[j] else
if k = j then a[j] else b[j];
x[j]: = xold[k, j]
end;

if S < 0 then B: = false
end;

if B then go to L4
end;
```

```
go to L5;
L4: for i: = 0 step 1 until var do
begin
for j: = 1 step 1 until var do
x[j]: = xold[i, j];
y[i]: = function
end;

L1: rearrangement (y, xold, var);
ymin: = y[var];
for i: = 1 step 1 until var do
xnew[i]: = xold[var, i];
if count > cmax then
go to ERROR;

k: = 0;
L2: for j: = 1 step 1 until var do
begin r[j]: = 0;
for i: = 0 step 1 until var do
r[j]: = xold[i, j] + r[j];
x[j]: = 2×r[j]/var − (var + 2)×xold[k, j]/var
end;
```

```
w: = function;
s: = if k < var then k + 1 else k;
if w < y[s] ∧ S = 0 then
begin for i: = 1 step 1 until var do
xold[k, i]: = x[i];
y[k]: = w;
count: = count + 1;
go to L1
end;

k: = k + 1;
if k < var + 1 then go to L2;
if abs(y[var − 1] − y[var]) ≥ eps then
L5: begin for i: = 1 step 1 until var do
step[i]: = step[i]/2;
go to L3
end
end of OPTSIMP3;
```

b. Minimizing the sum of square deviations by linearization of the model. The method has been called by many names including the Newton—Raphson method. A special case of this technique has already been shown in the example of sums of exponential functions [see Sect. 4.(C)(2)].

Starting from the sum of squares of deviations

$$S = \sum_{i=1}^{m} [y_i - f(x_{1i}, x_{2i}, ..., x_{ni}, a_1, a_2, ..., a_q)]^2$$

Expanding the function f into a Taylor series of the first order in the vicinity of the initial estimates of the constants $a_1^0, a_2^0, ..., a_q^0$, and introducing the notations

$$\frac{\partial f}{\partial a_j} = f^{(j)}$$

$$f_i = f(x_{1i}, x_{2i}, ..., x_{ni}, a_1^0, a_2^0, ..., a_q^0)$$

$$f_i^{(j)} = \frac{\partial f(x_{1i}, x_{2i}, ..., x_{ni}, a_1^0, a_2^0, ..., a_q^0)}{\partial a_j}$$

the sum of squares may be written as

$$S = \sum_{i=1}^{m} (y_i - f_i - f_i^{(1)} \, da_1 - f_i^{(2)} \, da_2 - ... - f_i^{(q)} \, da_q)^2$$

We now calculate the partial derivatives of S with respect to the constants da_j $(j = 1, 2, ..., q)$ and we put these derivatives equal to zero. After rearranging

$$[y_i f_i^{(j)}] - [f_i f_i^{(j)}] = \sum_{k=1}^{q} [f_i^{(k)} f_i^{(j)}] \, da_k \qquad (123)$$

where $[a_i b_i]$ is an abbreviation for $\sum_{i=1}^{m} a_i b_i$.

The system of linear equations (123) is to be solved for the corrections da_k of the constants. The second estimate of the constants is

$$a_k^1 = a_k^0 + da_k$$

162

Then the calculation will be repeated until the corrections become small enough. If the sum of squares S does not diminish, then the initial estimates of the constants must be replaced by another set of values.

The calculation just outlined can be carried out by the JORDAN4 procedure worked out in Algol which is named after Ch. Jordan (1871—1959). If, starting from the first estimate, the procedure is not convergent, then it searches for an appropriate set of starting values among the grid points of a rectangular lattice.

The procedure heading is

procedure JORDAN4 (var, const, m,x, F, a, meres, eps, sz, v, ERROR);
value var, const, m, sz;
integer var, const, m, sz;
real F, v, eps;
array x, a, meres;
label ERROR;

The formal parameters are

integer var	total number variables, $n + 1$,
integer const	number of constants, q
integer m	number of measurements, m
integer sz	the half number of grid points for each constant
real F	a global procedure defined below
real eps	relative stop criterion for the constants. The calculation is finished if $da_i/a_i <$ eps $(i = 1, 2, ..., q)$
real v	step factor in the systematic variation of initial estimates
array x [1 : var]	the independent variables $x(1),...,x(n)$
array a [1 : const]	the constants
array meres [1 : m, 1 : var]	the measured data
label ERROR	this is used as an exit if the iteration does not converge or if a matrix to be inverted is singular

The listing of the procedure is

```
procedure JORDAN (var, const, m, x, F, a, meres, eps,
  sz, v, ERROR);
value var, const, m, sz;
integer var, const, m, sz;
real F, v, eps;
array x, a, meres;
label ERROR;
begin
integer j, k, l;
real SO;
array b [1 : 2 * sz + 1,1: const], c, d [1 : const],
  function [0 : const, 1 : m], S [0 : const, 0 : const],
  T [1 : const, 1 : const + 1];
procedure LINEQ 1 (N, a, x, NOSOLUTION);
integer N;
array a, x;
label NOSOLUTION;
begin
integer p, i, j;
real M;
for p: = 1 step 1 until N − 1 do
begin
for i: = p + 1 step 1 until N do
begin
if a [p, p] ≠ 0 then go to L2;
if a [i, p] ≠ 0 then go to L1;
if i < N then go to L3;
go to NOSOLUTION;
·L1:
for j: = p step 1 until N + 1 do
```

```
begin
M: = a [p, j];
a [p, j]: = a [i, j];
a [i, j]: = M
end of row exchange;
go to L3;
L2:
if a[i, p] = 0 then go to L3;
M: = - a [i, p]/a [p, p];
for j: = p + 1 step 1 until N + 1 do
a [i, j]: = a [i, j] + M * a [p, j];
L3:
end for i;
end for p;
if a [N, N] = 0 then go to NOSOLUTION;
for p: = N step  1 until 1 do
begin
x[p]: = a[p, N + 1]: = a[p, N + 1]/a[p, p];
if p = 1 then go to L4;
for i: = p - 1 step - 1 until 1 do
[i, N + 1]: = a[i, N + 1] -  x[p] * a [i, p]
end for second p;
L4:
end LINEQ - 1;
for i: = 1 step 1 until 2 * sz + 1 do
for j: = 1 step 1 until const do
[i, j]: = if i = 1 then a[j] else
if i < sz + 2 then a[j] * v ↑ (i - 1) else
[j]/v ↑ (i - 1 - sz);
for i: = 1 step 1 until const do
[i]: = 1;
```

```
l: = const;
L1:
SO: = ₁₀100;
for i: = 1 step 1 until const do
a[i]: = b[c[l], i];
L2:
for i: = 0 step 1 until const do
for j: = 1 step 1 until m do
begin
for k: = 1 step 1 until var do
x[k]: = meres [j, k];
function [i, j]: = F
end:
for i: = 0 step 1 until const do
for j: = i + 1 step 1 until const do
begin
S[i, j]: = 0;
for k: = i step 1 until m do
S[i, j]: = S[i, j] + function [i, k] * function [j, k]
end;
for i: = 0 step 1 until const do
begin
S[i, i]: = 0;
for k: = 1 step 1 until m do
S[i, i]: = S[i, i] + function [i, k] ↑ 2
end;
if S[0, 0] > SO then
begin
L3: j: = c[1] + 1;
if j = 2 * sz + 1 then
```

```
begin
if l = 1 then go to ERROR;
l: = l − 1;
go to L3
end;
c[l]: = j;
if l < const then
begin
for i: = l + 1 step 1 until const do
c[i]: = 1;
l: = const
end;
go to L1
end;
S0: = S[0, 0];
for i: = 0 step 1 until const do
for j: = 0 step i until i − 1 do
S[i, j]: = S[j, i];
for i: = 1 step 1 until const do
for j: = 1 step 1 until const do
T[i, j]: = S[i, j];
for i: = 1 step 1 until const do
T[i, const + 1]: = − S[i, 0];
LINEQ1 (const, T, d, ERROR);
for i: = 1 step 1 until const do
a[i]: = a[i] + d[i];
for i: = 1 step 1 until const do
if d[i]/a[i] > eps then go to L2;
end
end of JORDAN;
```

The main program must contain a global **integer** i variable. The function F must be declared by a global procedure
real procedure F (i, x);
integer i;
array x;
F := **if** i = 0 **then** ... the equation reduced to zero
else if i = 1 **then** ... the partial derivative of f with respect to the constant a_1, ...

⋮

else 0;
If some parameters in the model are linear, then the dimensionality of the task is diminished. See Barham and Drane [43] for a discussion of such models.

References

1 E.C. Wood, in C.L. Wilson and D.W. Wilson (Eds.), Comprehensive Analytical Chemistry, Vol. IA, Elsevier, Amsterdam, 1959, pp. 76—106.
2 K. Eckschlager, Errors, Measurement and Results in Chemical Analysis, Van Nostrand-Reinhold, London, 1969.
3 N.L. Johnson and F.C. Leone, Statistics and the Physical Sciences, Vols. I and II, Wiley, New York, 1964.
4 Ch. Lipson and N.J. Sheth, Statistical Design and Analysis of Engineering Experiments, McGraw-Hill, New York, 1973.
5 J.E. Freund, Modern Elementary Statistics, Prentice Hall, New York, 3rd edn., 1967.
6 V. Graf and J.J. Henning, Formeln und Tabellen der Mathematischen Statistik, Springer, Berlin, 1953.
7 K. Doerffel, Statistik in der Analytischen Chemie, VEB Deutscher Verlag für Grundstoffindustrie, Leipzig, 1966.
8 V.G. Jenson and G.V. Jeffreys, Mathematical Methods in Chemical Engineering, Academic Press, London, New York, 1963, Chap. 10.
9 R.B. Dean and W.J. Dixon, Anal. Chem., 23 (1951) 636.
10 W.J. Dixon, Biometrics, 9 (1953) 74.
11 W.J. Dixon and F.J. Massey, Jr., Introduction to Statistical Analysis, McGraw-Hill, New York, 1951.
12 R. Sutarno, private communication.
13 P. Móritz, E. Gegus and P. Répás, Computer methods for the determination of the concentration and homogeneity of standard samples for spectrometric analysis, Euroanalysis II, 2nd European Conference on Analytical Chemistry, Budapest, 1975, Abstract of Papers, pp. 17—18.
14 Procedure for Statistical Evaluation of Analytical Data Resulting from Inter-

national Test (Fourth Draft Proposal) ISO, International Organization for Standardization, Technical Committee 102 — Iron Ores, Sub-committee 2 — Chemical Analysis, December 1977.
15 P.J. Slevin and G. Svehla, Z. Anal. Chem., 246 (1969) 5.
16 G. Svehla, Talanta, 13 (1966) 641.
17 V.V. Nalimov, The Application of Mathematical Statistics to Chemical Analysis, Pergamon Press, Oxford, 1963, pp. 36—38.
18 R.W. Frei, M.M. Frodyma and V.T. Lien, in G. Svehla (Ed.), Wilson and Wilson's Comprehensive Analytical Chemistry, Vol. IV, Elsevier, Amsterdam, 1975, Chap. 3.
19 J. Nyvlt, Chem. Prum., 9 (1959) 468.
20 D.W. Bacon and D.G. Watts, Biometrika, 58 (1971) 525.
21 M. Ahsanullah and A.K. Md. Ehsanes Saleh, Int. Stat. Rev., 40 (1972) 139.
22 S. Ergun, Ind. Eng. Chem., 48 (1956) 2063.
23 R.A. Fisher, Statistical Methods for Research Workers, 1948.
24 H.R. Schwarz, Commun. ACM, 5 (1962) 94.
25 M. Ascher and G.E. Forsythe, J. ACM, 5 (1958) 9.
26 J.H. Cadwell, Comput. J., 3 (1960) 266.
27 Ch. Jordan, Ann. Math. Stat., Ann Arbor, Michigan, 3 (1932) 257.
28 Ch. Jordan, Calculus of Finite Differences, Jordan, Budapest, 1939, Chelsea, New York, 1947, 2nd edn., pp. 426—460.
29 Ch. Jordan, Hung. Acta Math., 1 (4) (1949) 1.
30 J.W. Richards, Interpretation of Technical Data, van Nostrand, Princeton, 1967, pp. 5—14.
31 F.J. Sinibaldi, Jr., T.L. Koehler and A.H. Bobis, Chem. Eng., (1971) (May 17) 139.
32 H. Späth, Commun. ACM, 10 (2) (1967) 87.
33 G.R. Deily, Commun. ACM, 9 (1966) 85.
34 S. Oberländer, Z. Angew. Math. Mech., 48 (1963) 493.
35 E. Slaviček, Collect. Czech. Chem. Commun., 35 (1970) 2885.
36 J. Mička and O. Schmidt, Sci. Pap. Inst. Chem. Technol., Fac. Inorg. Org. Technol., Prague, (1958) 587, 620.
37 S.D. Foss, Biometrics, 25 (1969) 580.
38 A. Lemaitre and J.P. Malengé, Comput. Biomed. Res., 4 (1971) 555.
39 D.M. Himmelblau, Process Analysis by Statistical Methods, Wiley, New York, pp. 176—207.
40 W. Spendley, G.R. Hext and F.R. Himsworth, Technometrics, 4 (1962) 41.
41 C.W. Lowe, Trans. Inst. Chem. Eng., 42 (1964) T334.
42 P. Móritz, Optimization of functions by a modified sequential simplex method, 3rd Intern. Congr. Chem., Eng. Chem. Equip. Constr. Autom., Lecture Summaries D, September, 1969, Mariánské Lázně, Czechoslovakia, p. 26.
43 R.H. Barham and W. Drane, Technometrics, 14 (1972) 757.

Chapter 2

Mass spectrometry

J. GRIMSHAW

1. Introduction

Mass spectrometry employs the action of electric and magnetic fields on charged ions to separate these ions according to their ratio of mass to charge, *m/e*. It has been much used for the separation of positively charged ions and, recently, attention has also been directed to the separation of negatively charged ions. The technique was first employed, from around 1920, for the separation of isotopes, the determination of the isotopic composition of the natural elements, and the accurate determination of atomic mass. About 1945, commercial mass spectrometers were designed for use in the analysis of the positive ions obtained by bombarding petroleum hydrocarbons with electrons and this technique proved a powerful tool for the analysis of hydrocarbon mixtures. From this stage, the subject naturally developed around 1960 towards the interpretation of the cracking patterns shown by organic molecules under electron bombardment as a means of determining the molecular structure of unknown compounds. Gas chromatography and mass spectrometry have been linked in the present phase of the development of mass spectrometry as an analytical tool in organic chemistry. The first technique affords the means for separating the components of a mixture while the latter technique is a very powerful tool for the identification of a pure compound. Mass spectrometry of inorganic compounds has developed in parallel as a tool for the analysis of trace elements.

The growth of mass spectrometry can be divided into three generations of instruments. The first generation is represented by the pioneer instruments, each of which was individually designed. The second generation comprises the commercial instruments elaborated during the period 1945–1965 when emphasis was placed on the design of ion optics so as to achieve maximum sensitivity and resolving power. In the third generation of instruments, which are presently being elaborated, emphasis is placed on the design of stable solid-state electronic circuitry which functions in a highly reproducible manner; the output from the spectrometer can be readily digitated and fed to a computer for processing. This refinement very rapidly gives information which can only be obtained from a second generation instrument after much time-consuming effort on the part of the operator.

2. Instrumentation

(A) HISTORICAL MASS SPECTROMETERS

During experiments on the discharge of electricity through gases at low pressure, positively charged rays were seen emanating from the cathode in a direction away from the anode[1]. These rays were fully investigated by Thomson [2] at the Cavendish Laboratory, Cambridge, who showed that they were due to positively charged particles. In his mass spectrometer (Fig. 1), a stream of these particles was deflected by parallel magnetic and electric fields simultaneously applied and the deflected beam recorded by its impact on a photographic plate. The original narrow beam of particles was

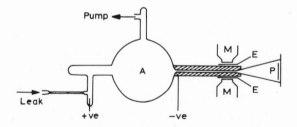

Fig. 1. Thomson's parabola mass spectrograph. The gaseous specimen is subjected to a silent electric discharge in bulb A. Positive ions pass through a fine bore tube, magnet M, and electrostatic plates E and fall on a photographic plate P.

172

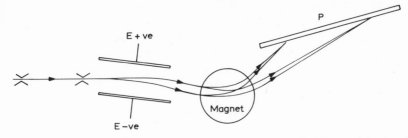

Fig. 2. Schematic diagram of Aston's mass spectrograph. The positive ion beam passes through slits, electrostatic deflector plates E and a magnetic field and falls on a photographic plate P.

deflected to form a series of parabolae from which the ratio of mass to charge, m/e, for the particles in the beam could be calculated. Particles having a given value of m/e are deflected along one parabola and this parabolic distribution arises because the method by which the ions are generated results in particles with a wide distribution of velocities.

A mass spectrometer which can focus ions having a range of velocities, and therefore energies, was constructed by Aston [3] at the Cavendish Laboratory. The ion beam was deflected first by an electric field and then by a magnetic field arranged perpendicular to the first. This arrangement (Fig. 2) focused the beam as a series of images of the entrance slit in a plane where they could be recorded on a photographic plate. The value of m/e for the ions forming a particular image could be calculated.

If positive ions can be generated in such a way that they all possess the same kinetic energy, a magnetic field alone is sufficient to deflect the particles to a focus according to their value of m/e. A method of analysis based on this principle was devised by Dempster [4] (Fig. 3) at the Ryerson Physical Laboratory, Chicago, at the same time as Aston's work. Positive ions of low energy were generated by heating metallic salts on platinum strips or by bombarding the salts with electrons. These ions were then accelerated through a potential field, V, and a narrow beam of ions with a small spread of kinetic energy was emitted through a slit into a homogeneous magnetic field, H, where they travelled in a circular path, the radius of which was varied by altering the acceleration potential V. At appropriate values of V, the ions passed through a second slit into the detector chamber. The

References pp. 268—273

173

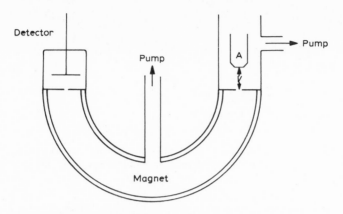

Fig. 3. Schematic diagram of Dempster's mass spectrometer. Ions are produced in A by heating a metallic salt, accelerated by potential V and pass into a homogeneous magnetic field to be focused on the detector slit.

peaks on the graph of ionic current against potential V correspond to particles of definite values of m/e and the heights of these peaks to the relative amounts of the particles present in the beam. This is the principle of operation of many modern mass spectrometers, although it is now usual to scan the mass spectrum by varying H rather than V.

Using these complementary methods, Aston and Dempster were able to demonstrate the isotopic composition of a number of elements and later workers refined their techniques.

(B) SAMPLE FLOW IN THE MASS SPECTROMETER

A block diagram of the essential parts of a mass spectrometer is shown in Fig. 4. The operation of the spectrometer requires a col-

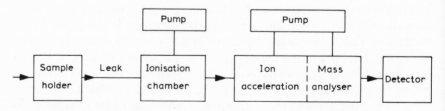

Fig. 4. Block diagram of the functional parts of a mass spectrometer.

174

lision-free flight path for the ions and to achieve this, the pressure in the spectrometer should be less than 10^{-6} Torr. Volatile and gaseous samples are introduced through a leak into the ionisation chamber from a sample holder of 1—2 l capacity, where they are held at 0.1 Torr. Less volatile samples can be introduced by means of a probe directly into the ionisation chamber and then heated to give the required vapour pressure. The sample flow rate is very small; 2 l of gas at 100°C and 0.1 Torr contain 10^{-5} g mole and this sample is sufficient for several hours of operation under ordinary conditions. In the ionisation chamber, the sample is given sufficient energy to form ions which are then accelerated and passed to the mass analyser. Finally, the ions are collected at a focus according to their value of m/e. Several variations of each of the components of a mass spectrometer are available commercially. The design of individual components is discussed and the characteristics of a number of commercial mass spectrometers are summarised in Table 1 (p. 198).

The flow rate of the gaseous sample under the low pressure conditions which exist in a mass spectrometer must be considered. The flow rate of a gas through orifices and tubes of small diameter depends on whether the mean free path of the gas molecules is smaller or larger than the diameter of the tube. The mean free path of a particle depends upon its collision diameter and concentration but is independent of velocity. From kinetic theory, the mean free path, L, of a pure gas is given by

$$L = \frac{1}{n\sigma^2\sqrt{2\pi}}$$

where n is the number of molecules cm^{-3} and σ is the collision diameter. The collision diameters of most gases do not differ sufficiently to affect approximate calculations so that we can write

$$L = \frac{5 \times 10^3}{P} \text{ cm}$$

where P is the pressure in Torr. Viscous flow occurs when the mean free path is smaller than the tube diameter and molecular flow when it is greater than the tube diameter.

Viscous flow. In this condition, the gas flow through a tube of radius r and length l (both in cm) is governed by the Poiseuille equation for normal fluid flow. With a pressure difference ΔP (dyne

cm^{-2}) across the tube, the flow rate Q (1 s^{-1}) is given by

$$Q = \frac{\Delta P \pi r^4}{8l\eta}$$

where η (Poise) is the viscosity of the gas.

Molecular flow. Under these conditions, the molecules move completely independently of each other and most collisions occur with the walls of the tube. The flow rate is now given by the expression

$$Q = \sqrt{\frac{\pi RT}{M}} \cdot \frac{d^3}{4l}\Delta P$$

where M is the molecular weight of the gas and d (cm) the diameter of the tube.

Molecular flow exists in the space between the ionisation chamber and the vacuum pumps and at the low pressure end of the leak. Under normal operating conditions, viscous flow occurs at the high pressure end of the leak. When the sample is a mixture of molecules of different masses, molecular flow will cause fractionation according to mass since Q is inversely proportional to $M^{1/2}$. Fractionation of a mixture does not occur under conditions of viscous flow. However, the viscosity of a gas mixture may vary with its composition and this causes a variation in Q for a given value of ΔP. Thus, the sample pressure in the ionisation chamber and therefore the total ion current may vary with the viscosity. These problems are connected with the analysis of gas mixtures by mass spectrometry and they will be discussed further on p. 244. They are irrelevant to the mass spectrometry of pure substances.

(C) SAMPLE INLET SYSTEMS

All the ion sources described in the next section, with the exception of the spark and the thermal ionisation sources, require the sample to be introduced as vapour at about 10^{-6} Torr. This is achieved by volatilising the sample into a heated container of up to 2 l capacity at a pressure of around 10^{-1} Torr which is monitored by a pressure gauge. The sample is then passed into the ionisation chamber through a suitable leak. The pressure inside the ionisation chamber can be monitored by a suitable pressure gauge and by the total ion current produced.

The sample container, Fig. 5, is usually fitted with several sample

176

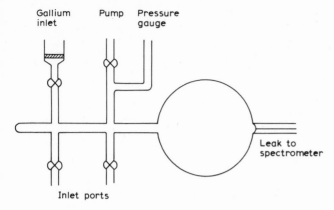

Fig. 5. Inlet system for admitting volatile samples to the mass spectrometer.

inlets to facilitate the introduction of samples for analysis and also standards as required. One inlet allows the introduction of gases or volatile liquids and solids from small containers. A second type of inlet consists of a glass frit sealed from the atmosphere by a layer of molten gallium. Gallium is chosen as the sealing liquid because of its low vapour pressure and low melting point. Volatile liquids can be introduced through this gallium inlet from a lambda pipette by touching the tip of the pipette on the glass frit. The sample is then drawn into the evacuated inlet system.

Solids with a very low vapour pressure can be introduced directly into an entrance to the ionisation chamber on a silica or platinum probe and then volatilised by gently heating the probe. This is termed the direct insertion method and is very useful for the qualitative analysis of organic compounds of low volatility. The probe is first placed in an antechamber and evacuated to the pressure of the spectrometer. A connecting valve is then opened and the probe inserted near the ionisation chamber by means of a magnetically operated device.

(D) ION SOURCES

(1) Electron bombardment sources

Positive ions are formed by passing a beam of electrons through the sample at a pressure of 10^{-4} to 10^{-6} Torr. Most such sources are

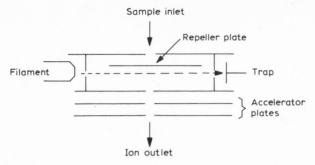

Fig. 6. Schematic diagram of an electron bombardment ion source.

based on the designs of Bleakney [5] and Nier [6]. A schematic diagram is given in Fig. 6. By careful design of the entrance and exit slits, the sample pressure inside the ionisation chamber can be made higher than that in the remainder of the spectrometer. This gives a higher ratio of the sample to the residual gas pressure and a higher ion current for a given sample flow rate.

The electron beam is produced from electrons emitted by a heated filament, accelerated through an aperture into the ionisation chamber and finally collected in the trap. This trap current is usually stabilised to within 1 part in 10^4 at some value between 10 and 100 μA. Excessive scattering of the electron beam is prevented by application of a weak parallel magnetic field. The positive ion current begins when the electron beam has an energy greater than the ionisation potential of the molecules (ca. 10 eV). It increases to a maximum with increasing electron acceleration voltage and then falls slowly. For analytical purposes, the most reproducible results are obtained with an electron energy close to that which gives the maximum in this curve, and a value around 70 eV is most commonly used. The energy spread of the electron beam, which is about 3 eV, is not important at an acceleration potential of 70 V.

It is often useful to operate with an electron beam of much lower energy as the resulting mass spectrum may then show fewer fragment ions. This simplified fragmentation pattern can often be more readily interpreted in terms of molecular structure, particularly when a complex molecule is being examined. However, the energy spread of the electron beam does not allow an accurate determination of

178

the appearance potential of an ion. To make possible such measurements, Clarke [7] used an electrostatic filter to provide an electron beam with a narrow energy spread and Fox et al. [8] devised a type of electron gun which creates a monoenergetic electron beam.

Energy is transmitted to molecules from the bombarding electron beam by the repulsive effect between the molecular electronic cloud and the electron beam. An electron is ejected from the molecule and the transferred energy gives rise to a highly vibrationally excited molecular ion. Fragmentation of the molecular ion occurs and the resulting daughter ions may have kinetic energy as a result of thermal motion and from the kinetic energy of separation of the fragments. Thus, the resulting ion beam has a small spread of energy of about 0.05 eV.

This ion source is the one most commonly used to obtain mass spectra from volatile organic compounds. It can be used with both single- and double-focusing mass analysers.

An incandescent tungsten or rhenium filament is most commonly used as the electron source. The ionisation chamber is designed so as to keep the flow of gas between the ionisation region and the filament as small as possible since, at the temperature of the filament, thermal decomposition of many materials is to be expected. Some samples may decompose in contact with the walls of the ionisation chamber which are at a temperature of around 200°C. Most hydrocarbons react with incandescent tungsten to form a layer of tungsten carbide [9] and this reaction cannot be totally avoided. The effect of this is to change the electrical resistance of the filament and some compensation must be made in order to maintain a constant electron beam current. For stable operation, the surface layers of the filament should consist of W_2C. The carbon content of a filament can be controlled by the introduction of appropriate compounds, oxygen for reducing the carbon content and acetylene or but-2-ene for carbonizing.

Rhenium is an alternative filament material which has the advantage that it does not form carbides and so the filament is more stable in use. The disadvantage of rhenium is that its evaporation rate is greater than that of tungsten for a given emission. Other filament materials which have been suggested include carbon, tantalum, thoria—iridium and thoria—tungsten. Thoria—tungsten has the advantage of increased lifetime under extreme conditions and so is sometimes used in small mass spectrometers which function as leak detectors.

(2) Chemical ionisation sources

These sources use an electron bombardment technique to form the positive ions but the mechanism of ionisation is quite different from that which operates in the simple electron bombardment source [10—12]. The ionisation chamber is fed with a mixture of a reactant gas, usually methane or ethane, at a pressure of 0.3—3 Torr and the sample at a pressure of 10^{-6} Torr or lower. The ionisation chamber must be relatively more gas-tight than in a simple electron bombardment source. Because of the low abundance of the sample, virtually all the primary ionisation caused by the electron bombardment occurs to the reactant gas. In subsequent ion—molecule reactions, CH_5^+ is generated from methane and $C_2H_7^+$ from ethane. These ions can protonate the sample molecule to generate a quasi-molecular ion $M^+ + H$ which may, in some cases, readily lose a hydrogen molecule to leave $M^+ - H$. Such processes for ionisation of the sample molecule occur with a much lower transfer of energy than in the case for direct electron bombardment. In consequence, fragmentation of the quasi-molecular ion is greatly reduced so that it is often the most prominent ion in the spectrum. $M^+ + 1$ ions are prominent from molecules with a Brønsted base centre such as an oxygen or a nitrogen atom while $M^+ - 1$ ions are prominent from hydrocarbons.

(3) Field ionisation sources

Molecules and atoms produce positive ions when subjected to intense electric fields of about 10^8 V cm^{-1}. In commercial field ionisation sources, the ionisation chamber is fed with the sample at a pressure of around 10^{-6} Torr and the electric field is obtained in the vicinity of a fine wire or a sharp metal edge held at a positive potential of 5—10 KV. A strongly divergent positive ion beam is formed and this is focused onto the mass analyser entrance slit by allowing it to pass through apertures in a series of negatively charged discs. Gomer and Inghram [13] were the first to show that the mass spectra of organic molecules obtained using field emission sources are much simpler than those obtained by electron impact. This is because the energy supplied to the molecules is only a little above the ionisation potential. The resulting mass spectra show a strong molecular ion and only a few fragment ions. An ion $M^+ + 1$ is often intense and is formed by the reaction between a molecule of the

180

sample and water absorbed on the high potential surface [14,15]. At a given sample pressure, the total ion current is much less from field ionisation than from electron bombardment sources. This means that, even though the spectrum obtained by field ionisation is simpler, it may require more sample to make a satisfactory record. A further disadvantage is that the emitter surface may be unstable and severely affected by its past history. Commercial sources are available which combine the field ionisation and electron bombardment modes of operation.

(4) Photoionisation sources

The majority of gaseous organic molecules have an ionisation potential in the range 8—11 eV and they can be ionised by photon bombardment using light of wavelength lower than 150 nm. Ion sources which operate on this principle are designed similar to electron impact sources except that the electron beam is replaced by a beam of light [16]. Light is produced in a discharge tube containing hydrogen or krypton at pressures of 10^{-1}—10 Torr and passed into the ionisation chamber through a window capable of transmitting far-ultraviolet light. Most window materials will not transmit below 100 nm and this limits the energy of the electron beam to less than 11.8 eV.

Photoionisation mass spectra are very similar to the mass spectra obtained by electron bombardment with an electron beam energy of 10—12 eV. The molecular ion is usually predominating in these mass spectra and only a few fragment ions are formed. For a given sample pressure, the electron bombardment source operated at this low energy has the advantage of greater sensitivity. However, the fragmentation pattern obtained by low-energy electron bombardment is subject to large variations in both absolute and relative ion intensities due to changes in other nearby potential fields. By contrast, variations in the spectral composition of the light in a photoionisation source do not lead to such serious changes in ion intensities and the light beam is unaffected by nearby potential fields. Low-energy electron beam bombardment is, however, more generally used for obtaining simplified mass spectra because the same source can be used without modification for conventional work at 70 eV.

(5) Spark ionisation sources

The ion sources which have been described so far require a specimen to be in the vapour state. Spark ionisation sources can be used with a solid sample and are particularly suitable for the elemental analysis of inorganic materials. The material to be analysed is formed into a electrode. If the material is non-conducting, it can be mixed with electrically conducting material such as graphite or packed into a thin-walled conducting tube. A radio frequency voltage of the order of 5×10^4 to 10×10^4 V is supplied between this and a secondary electrode to maintain a spark discharge which vaporises material from the primary electrode. Some material is vaporised in the form of positive ions which are then accelerated and limited in angular spread by means of slits. The ions have a large energy spread and a double-focusing mass analyser must be used to obtain a well-resolved mass spectrum [17]. Also, the intensity of the ion beam fluctuates rapidly so that photographic recording of the whole spectrum must be used to obtain meaningful intensity measurements on individual ions. This source is really restricted to use with a mass analyser of Mattauch—Herzog design. Recent developments of the spark ionisation technique which aim to control the ion beam intensity are discussed in Sect. 5.

The first spark ionisation source was constructed in 1951 for the analysis of steel samples [18]. The steps of calibration followed by analysis of unknown samples used in this technique very much resemble the steps used during spectrographic analysis. Mass spectrometry has the advantage over spectrographic analysis of inorganic samples in that it gives a sensitivity which is approximately equal for all the elements, varying, generally, by not more than a factor of three [19]. This makes the technique very suitable for trace element analysis.

(6) Thermal ionisation sources

Positive ions can be produced directly from some solids when these are heated on a metal filament. This method cannot be used for the quantitative chemical analysis of mixtures because of the very large variations in the ionisation efficiency between the different elements. Its main application is to the measurement of isotopic abundances because it yields a beam of ions with a very narrow energy

182

spread which can be analysed by a single-focusing mass analyser. A small quantity of the sample is heated on a tungsten or platinum filament and the resulting ions are accelerated through a potential field and then collimated [20]. The ion beam is almost free of ions generated by ionisation of the background gas in the spectrometer.

(E) MASS ANALYSERS

(1) Single-focusing mass spectrometers with magnetic sector field

Positive ions generated in the ionisation chamber pass through a slit into the acceleration chamber where they fall through an electric field, V, commonly 3—6 KV, and are then allowed to pass as a narrowly divergent beam through a slit into a magnetic field, H. The ion beam is deflected by the magnetic field and ions with the correct relationship between m/e and the quantities V and H are focused onto the collector slit. This arrangement is depicted in Fig. 7.

The spread of translation energies in the positive ion beam at its source should be as small as possible so that the kinetic energy of the ions entering the magnetic field is derived from the potential energy acquired in the accelerating field. Thus, the ion velocity, v, in the magnetic field is given by the expression

$$Ve = \tfrac{1}{2}mv^2$$

As the ions enter the magnetic field, they experience a force orthog-

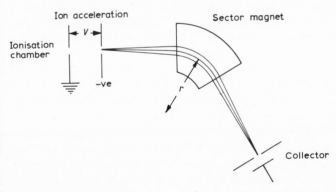

Fig. 7. Schematic diagram of a 60° magnetic sector single-focusing mass spectrometer.

onal both to the magnetic lines of force and to their line of flight. In consequence, the ions travel in a circular path whose radius r is given by

$$Hev = \frac{mv^2}{r}$$

or

$$He = \frac{mv}{r}$$

For singly charged ions, the radius of curvature of the flight path is dependent on the magnetic field strength and the ion momentum, mv. Elimination of the velocity term in the above equations gives the relationship

$$\frac{m}{e} = \frac{H^2 r^2}{2V}$$

Thus at a fixed radius, r, which is determined by the geometry of the instrument, ions of different mass-to-charge ratio can be deflected to the collector by varying either the magnetic field or the acceleration voltage and, in this fashion, the mass spectrum is scanned.

Ions emerge from the accelerator as a divergent beam. When certain conditions of the magnetic field sector angle and the distances between the slits and the magnetic field are met, directional focusing of the ion beam onto the detector slit is achieved. Commercial instruments employ 60° (Neir design [21]), 90° (Hipple design [22]) or 180° (Dempster design [4]) sectors. The radius r determines the mass dispersion, i.e. the distance by which two mass peaks are separated. The resolution (defined on p. 193) is directly proportional to the sector radius, r, and inversely proportional to the sum of the ion source and collector slit widths. Decreasing the slit width naturally lowers the sensitivity of the instrument and magnetic mass spectrometers are equipped with variable slits to provide a choice of the optimum resolution and sensitivity.

Commercial mass spectrometers of this design are manufactured with various values of the sector radius, r, to suit particular applications. The instruments of low resolving power can be quite small and rugged. For most applications, it is preferable to maintain a constant acceleration voltage and to scan the mass spectrum by varying the

184

magnetic field. This is because the efficiency with which ions pass through the slits in the acceleration region depends on the acceleration voltage and such a variation in efficiency is unsatisfactory for quantitative analysis.

(2) Double-focusing mass spectrometers with magnetic sector field

In presenting the theory of operation of a magnetic sector field mass spectrometer, it was assumed that the kinetic energy of the ions entering the magnetic field is determined only by the acceleration voltage. In practice, this is only approximately true as, for example, when the ions enter the acceleration field they already possess a range of kinetic energies. Single-focusing mass spectrometers are directional focusing but not velocity focusing for ions of a given mass so their resolving power is limited because of the spread in kinetic energy of the ions leaving the acceleration chamber. A considerable improvement in resolving power is achieved by passing the ions through an electric field, E, prior to their deflection by the magnetic field. This electric field is placed between two curved parallel plates of mean radius r_e so that the field is orthogonal to the ion flight path. The ions experience an accelerating force Ee, and if the value of E is correctly chosen such that

$$Ee = \frac{mv^2}{r_e}$$

and

$$Ve = \tfrac{1}{2}mv^2$$

i.e.

$$r_e = \frac{2V}{E}$$

the ions will travel in a circular path of radius r_e. All ions having the same kinetic energy will be brought to a common focus, regardless of mass.

Commercial double-focusing mass spectrometers use a radial electrostatic field and a homogeneous magnetic sector field to bring an ion beam which is divergent both in direction and velocity to a focus dependent on the ion mass. A slit placed between the electrostatic and magnetic analysers can be used to improve the resolving power of the instrument, at the expense of sensitivity, by restricting the range of energy of the ions which are allowed to pass on.

a. Nier—Johnson design. Some commercial mass spectrometers use the design pioneered by Johnson and Nier [23] shown in Fig. 8. The dimensions of the instrument are optimised for both directional and velocity focusing. Ions of only one value of m/e will be in focus at one time on the collector slit and the mass spectrum is scanned by varying the magnetic field, H. For some applications, particularly the accurate measurement of ion mass, the mass spectrum is scanned by varying the accelerator voltage, V, at constant H. In the latter case, the relationship.

$$r_e = \frac{2V}{E}$$

must be maintained by varying the electrostatic field, E, in unison with V.

Reversed Nier—Johnson geometry is used in some spectrometers where the ion beam undergoes magnetic focusing first and electrostatic focusing second.

b. Mattauch—Herzog design. Other commercial mass spectrometers use the design of Mattauch and Herzog [24] which is illustrated in

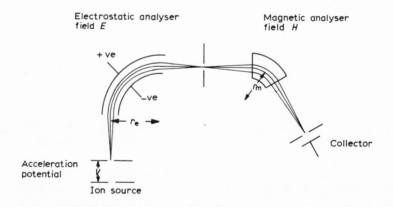

Fig. 8. Schematic diagram of the Nier—Johnson double-focusing mass spectrometer.

186

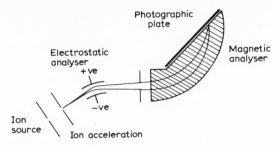

Photographic plate

Electrostatic analyser +ve

Magnetic analyser

−ve

Ion source

Ion acceleration

Fig. 9. Schematic diagram of the Mattauch—Herzog double-focusing mass spectrograph.

Fig. 9. The electrostatic field is designed to generate an ion beam dispersed according to energy and focused at infinity. This beam is then focused by a 90° magnetic sector. Ions of all values of m/e are simultaneously in focus as parallel images of the source slit in one plane within the magnetic field. Consequently, a photographic plate can be used for ion detection.

Since the photographic method of recording integrates the ion signal, it will give a truer record of relative ion intensities in situations where the total ion intensity fluctuates rapidly. Such conditions arise during the analysis of metal ions produced by spark discharge between two electrodes and may also occur during GLC—mass spectrum analysis. Electrical detection of the ion beam is also possible with this spectrometer design. For this, the photographic plate is replaced by a fixed collector slit and detector system. A mass spectrum is then generated by varying H so that each ion in turn is focused on the collector slit.

(3) Omegatron mass spectrometer

Sommer et al. [25a] first described this compact mass spectrometer, which is very suitable for use in vacuum technology, for the analysis of residual gases. Ions are generated by electron bombardment of the gas within the spectrometer, Fig. 10, and subjected simultaneously to a constant magnetic field, H, and a sinusoidal radio frequency field of frequency f. When ions are generated in a magnetic field, they adopt a circular orbit of radius r which depends on the velocity v so that the force they experience as a result of the

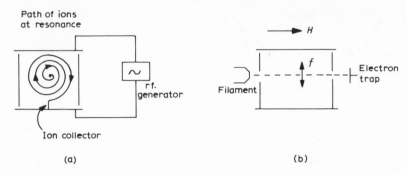

rf.
generator

Filament

Electron
trap

Ion collector

(a)   (b)

Fig. 10. Schematic diagram of the omegatron mass spectrometer. The instrument is placed in a uniform magnetic field H and supplied with an rf. signal f. (a) View along the magnetic lines of force; (b) view normal to the magnetic field.

magnetic field is balanced by the orbital angular momentum.

$$Hev = \frac{mv^2}{r}$$

The orbital angular velocity, Ω radians s^{-1}, which is termed the cyclotron velocity of the particle is given by

$$\Omega = \frac{v}{r} = \frac{He}{m}$$

When the electric field frequency is such that

$$2\pi f = \Omega$$

the ion will adopt a spiral path and eventually strike the ion collector. Thus the condition for ion collection is

$$\frac{m}{e} = \frac{H}{2\pi f}$$

Ions with a slightly different mass will beat with the radio frequency field and pass through maximum and minimum radii. Other ions strike the walls of the chamber. The resolution is limited by this beating effect and is given by the expression

$$Resolution = \frac{r_m H^2 e}{V_0 m}$$

where r_m is the distance from the collector to the point of origin of

188

the ions and V_0 is the amplitude of the radiofrequency field. Resolution varies inversely with m. Commonly, r_m is about 1 cm and the instrument is useful up to m/e values of about 50.

The omegatron mass spectrometer has the advantage of high sensitivity since virtually all the ions produced by the electron beam are collected. Absolute partial pressures can be calculated from the geometry of the electron beam and the known ionisation cross-section of the gas. The mass spectrometer has the disadvantage of a long ion flight path which increases the probability of ion—molecule collisions and a consequent loss in resolving power.

(4) Ion cyclotron resonance mass spectrometer

This instrument resembles the omegatron mass spectrometer. Ions are generated in an electron bombardment source and allowed to drift through a magnetic field, H, towards the negatively charged ion collector. In the magnetic field, the ions adopt circular trajectories with orbital angular velocity Ω, as described in the previous section. When an alternating electric field of frequency ω is applied normal to H, the ions will absorb energy when

$$2\pi\omega = \Omega$$

Such absorption can be detected. A mass spectrum is therefore obtained by measuring the absorption at a fixed frequency while scanning the magnetic field [25b]. A block diagram of the apparatus is given in Fig. 11.

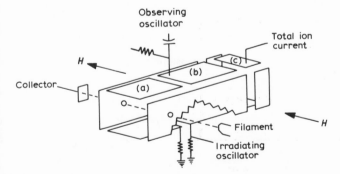

Fig. 11. View of the cyclotron resonance cell. The cell is placed in a uniform magnetic field H. (a) Ion source region; (b) ion analyser region; (c) ion collector.

Ions drift slowly from the source to the ion collector plates and have residence times inside the spectrometer of the order of milliseconds. This gives a high probability for ion—molecule reactions even at low pressures and the ion cyclotron resonance spectrometer is particularly used to study such reactions.

(5) Cycloidal mass spectrometer

An ion beam is accelerated by a potential field V and passed through a slit into homogeneous and orthogonal magnetic and electrostatic fields, H and E. The entrance slit is parallel to the magnetic field. Ions describe a circular path under the influence of the magnetic field alone and a linear path under the influence of the electric field alone. Under the influence of the combined fields, the ions execute a cycloidial path (Fig. 12) and with the appropriate conditions they can be brought to a combined directional and velocity focus. If E, H and V are held constant, all values of m/e are brought to focus as images of the entrance slit in a common plane and the mass scale is linear. The instrument can be used in this mode with photographic detection. Alternatively, an exit slit leading to an electrical detector can be placed in the focal plane and the mass spectrum scanned either by keeping E and V constant while H is varied, or E and V are varied while keeping H and the ratio E/V constant.

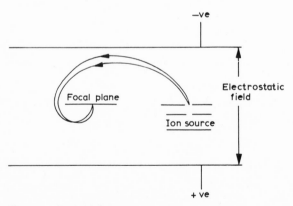

Fig. 12. Cycloidal-focusing mass spectrometer. View of the ion path looking along the magnetic lines of force.

The resolving power of a cycloidial mass spectrometer is limited by the difficulties in attaining sufficiently homogeneous electrical and magnetic fields. This type of instrument is usually designed for gas analysis in the range 12—150 a.m.u.

(6) Quadrupole and duodecapole mass spectrometers

This type of mass spectrometer, first designed by Paul and Stein-wedel [26] does not require a magnetic field. In the quadrupole version, mass analysis is achieved using a radio frequency electric field established between four rod electrodes which may be either circular or hyperbolic in cross-section, see Fig. 13. Opposite pairs of electrodes are electrically connected. The duodecapole mass analyser has two trimmer electrodes in each quadrant to permit easier adjustment of the equipotential field. A voltage consisting of a constant component U and a sinusoidal radio frequency component of frequency f and amplitude V_0 is applied to this quadrupole array. Ions are allowed to pass into the array through a circular aperture and a range of ion energies can be tolerated. Each ion continues in the z direction along the axis of the array with constant velocity but suffers perturbations in the x and y directions. When U, V_0, f and the quadrupole array dimension r_0 bear a special relationship to the m/e value, ions will oscillate in the x and y directions and eventually pass through the quadrupole to be collected by a large aperture detector. Other ions not in stable oscillation are captured by the quadrupole array.

The theory of the quadrupole mass analyser has been developed

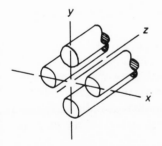

Fig. 13. The quadrupole mass filter. Rf. and d.c. voltages are applied to four precision-ground metal rods to produce a time-varying electric field which permits mass analysis of ions travelling in the z direction.

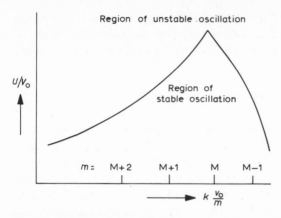

Fig. 14. Stability diagram for the quadrupole mass filter.

[26,27] and this can be summarised as a diagram of the region of oscillatory stability given in Fig. 14. The region is defined by values of U/V_0 and KV_0/m where K is a function containing f and r_0. Mass scanning is accomplished either by varying U and V_0 while keeping the ratio U/V_0 constant, or by varying f. The mass resolution of the array is dependent on the value of U/V_0. Ideally, this value should be close to the apex of the stability region but a practical limit is set because of departures from ideal geometry of the constructed array.

(7) Time-of-flight mass spectrometer

This type of mass spectrometer also does not require a magnetic field. A short duration pulse of ions is accelerated through a potential field into a field-free region towards the detector. In an ideal situation, all ions in the pulse will enter the field-free region with the same kinetic energy. Ions of different mass will have different velocities and will arrive at the detector at different times. The detector output will then constitute a mass spectrum. It is usually displayed on an oscilloscope which can be triggered by the ionisation pulse generator.

In practice, there is a contribution to the ion kinetic energy from thermal motions and the energy released during ion fragmentation reactions. This, together with the duration of the ion pulse, limit the resolving power of time-of-flight mass analysers. The main advantage of this type of mass analyser is that, because the ionisation pulse is of very short duration, 100—300 ns, the analysis of substances taking part in fast gas phase reactions can be achieved.

(F) RESOLVING POWER

The most important parameter of a mass analyser is the mass resolution. Recorded ion peaks are usually Gaussian in shape and two peaks of equal height h separated by a mass difference Δm are said to be resolved when the height, Δh, of the valley between them is 10% or less of the peak height, i.e. $\Delta h/h \leqslant 0.1$. The resolving power, R, is then given by the value

$$R = \frac{m}{\Delta m}$$

where m is the mass of the lower peak. This definition of resolving power is loosely termed "10% valley resolution". Other definitions of resolving power based on $\Delta h/h = 0.01$ or 0.5 are sometimes used.

The overlap of Gaussian-shaped peaks where $\Delta h/h = 0.1$ is illustrated in Fig. 15. This degree of overlap is close to the limit which can be tolerated without significant distortion of the peak heights. It is obvious that a much greater degree of overlap could be tolerated before the peaks could be considered as just on the limit of distinct resolution.

It is unlikely that the resolving power of an instrument can be determined by direct measurement of the overlap of two adjacent peaks of equal height. However, R, as defined above, can be easily calculated from the shape of any isolated peak which is Gaussian in shape. The value of Δm required to calculate the resolving power is

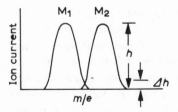

Fig. 15. Ion peaks illustrating 10% valley resolution, $\Delta h/h = 0.1$.

given by the width of this peak at 5% height. Also

$$\Delta m = 2.08\Gamma_{1/2}$$

where $\Gamma_{1/2}$ is the width of the peak in a.m.u. at half-height. With sufficient accuracy

$$R = \frac{m}{2\Gamma_{1/2}}$$

The importance of resolving power as a description of mass resolution can be illustrated in two ways using Fig. 15. If, in the example, m_1 is 650 and m_2 651, then the resolution would be 650. An instrument of this resolving power is suitable for recording most mass spectra but it cannot be used to check the composition of individual ion peaks by accurate mass measurement as discussed on p. 210. In order to show the order of magnitude of the resolving power which is necessary to make accurate mass measurements, consider the example where both ions have the same nominal mass 176 and m_1 is $C_{12}H_{16}O$ with 176.1201 a.m.u. while m_2 is $C_{11}H_{12}O_2$ with 176.0837 a.m.u., then

$$R = \frac{176}{0.0364} = 4835$$

(G) DETECTORS AND RECORDERS

(1) Graphical and magnetic tape recording

In these systems of recording, the mass spectrum is scanned over a fixed detector slit by varying either the ion accelerator potential or the magnetic deflector field and the detector output is graphed using a suitable recorder. Modern amplifier and recorder systems can detect extremely low currents in the region of 10^{-18} to 10^{-19} A which allows the mass spectrometer to be operated at fast scan rates. The limiting speed at which a spectrum can be scanned is determined by the response time of the recording galvanometer and also by the need to have a statistically significant number of ions in the peak which is recorded. The current corresponding to one single charged ion per second is 1.6×10^{-19} A. If six ions can be collected, this is sufficient to give an estimate of the intensity of a given ion peak but a reasonable statistical sample requires 20—25 ions. The number of ions which are collected depends on the rate at which the spectrometer scans the peak. This scan rate is expressed as the time to scan from mass m to mass $10m$ and the fastest scan rates employed are 2—6 s/decade.

If we take a typical scan over the range 300—30 a.m.u. at a scan rate S with a resolving power R, the peak width at half height, $\Gamma_{1/2}$, is given by

$$\Gamma_{1/2} = \frac{m}{2R}$$

and the time to scan one peak, approximated to the time to scan over $\Gamma_{1/2}$ is given with sufficient accuracy by $S/2R$. If n ions are collected in this sweep time, the peak current is $(2nR/S) \times 1.6 \times 10^{-19}$ A. If this is the average peak current, then the total ion current integrated over 300 a.m.u. is $(600nR/S) \times 1.6 \times 10^{-19}$ A. This expression can be used to judge the total ion current which will be necessary to give a meaningful mass spectrum. Thus, if n is 6, R is 1000 and S is 6 s/decade, the total ion current is approximately 9×10^{-14} A and the individual peak current is 3×10^{-16} A.

Currents in the region of 10^{-18} to 10^{-19} A can be amplified with an electron multiplier and recorded. The conditions outlined in the previous paragraph will easily give a recordable spectrum even when only six ions are collected in a given ion peak. There is a danger, therefore, that with fast recording an insufficient number of ions will be collected to give significant intensity values for individual ion peaks. For quantitative work, it is necessary to use slower scan rates and higher total ion currents to be sure that each peak is due to a large sample of ions.

a. Faraday cup collector and electrometer amplifier. The electrometer valve is the preferred method of current amplification when high precision and accuracy of measurement of the ion beam intensity are more important than speed and sensitivity. The ion beam is collected in a Faraday cup and the resulting current applied to the grid of an electrometer valve from where it is leaked to earth through a resistance R_i and a condenser of capacity C_i. Electrometer valves are designed so that R_i is as high as possible and C_i as low as practicable. In a system designed to amplify an ion signal of 10^{-14} A, R_i can be as high as 10^{11} Ω while C_i is around 2 pF. The time constant of the circuit is given by $R_i \times C_i$ and is 0.2 s. Individual mass peaks will have to be separated by at least this time interval in order to be recorded individually and scan rates of 60 s, or longer, per decade can be tolerated.

An independent amplifier of this type is usually used to monitor the total ion beam current and, for this purpose, a sample of the total ion beam is collected from close to the mid point of the beam path.

b. Electron multiplier collector and electrometer amplifier. The electron multiplier collector is used in qualitative and less accurate quantitative mass spectrometry. It produces a current amplification of 10^3–10^8. The ions impinge on a surface and produce electrons. These electrons impinge on a second surface and so on for a number of times until the resulting cascade effect gives the required signal amplification. The final current is fed to an electrometer valve but, since the current is higher, the input resistance can be considerably lower than in the previous example for a given amplification factor. The time constant for the system is thus lowered and faster scan rates than those possible with the Faraday cup detector can be used.

The electron multiplier is virtually noiseless and gives no background current but the gain varies depending on the previous operating conditions. Also, overloading and saturation effects can occur. The sensitivity depends on the mass of the impinging ion and also, in some cases, on its structure and atomic composition. These are disadvantages for quantitative analysis. The advantage of the electron multiplier is that it enables fast scan rates to be used and this is particularly important for the GLC—mass spectrometry application.

The Daly detector is useful for the detection of metastable ions as well as the ordinary mass spectrum and is discussed on p. 234. It allows the ions to impinge on a plate generating secondary electrons which are detected by means of a scintillator and photomultiplier tube.

c. Potentiometric and galvanometric recorders. The type of recorder used to register mass spectra depends on the work being carried out. Spectra for the qualitative analysis or fingerprint identification of organic compounds are most easily recorded using a galvanometric recorder. This recorder is capable of very rapid response but it does not give great accuracy in the measurement of peak heights. The ratio of maximum to minimum peak height in a spectrum is likely to

196

be of the order of 10^2 and, to cover this range, either the recorder response is arranged to be logarithmic or three galvanometric recorders are arranged to write on the same chart strip and are connected to amplifiers set with their gains in the ratio of 100 : 10 : 1.

When quantitative analysis by peak height measurement is to be carried out, then the spectrum must be recorded with greater accuracy. Potentiometric recorders are used and trip switches are usually fitted so that the amplifier gain is automatically adjusted to enable peaks of widely differing heights to be recorded on the same chart. The use of such a system necessitates a slow scan rate which is also compatible with the use of a high precision electrometer amplifier.

d. Magnetic tape recording. The output from an amplifier can be passed to an analogue to digital converter and then recorded on magnetic tape. This system is particularly used for recording data from a GLC—mass spectrometry system and a number of commercial designs are available. The data can subsequently be processed by computer to give the mass and intensity of each peak in digital form or to present the mass spectrum in graphical form.

e. Calibration of the mass scale. The second generation of mass spectrometers, developed during the period 1945—1965, are not generally fitted with very high precision electronic circuitry. It was not found practicable to incorporate mass calibration circuits into these instruments and so each mass spectrum has to be individually calibrated. A mass spectrum usually shows a peak at each atomic mass unit, particularly when run at high amplification. Thus it is possible to calibrate the mass scale of the spectrum by counting peaks from the low molecular weight peaks due to H_2O, N_2 and O_2 present as residual gases on the spectrometer. These peaks form an easily recognised pattern. This method is tedious but is often necessary since a non-linear scale results from using a magnetic deflection mass spectrometer. Standard substances which produce a known and easily recognised pattern of peaks at high mass can be added to the sample to facilitate calibration of the mass scale and this technique is discussed on p. 211.

The third generation of mass spectrometers, developed after 1970, have very high precision solid state electronic circuitry. Some commercial mass spectrometers incorporate a Hall probe to monitor the intensity of the magnetic deflector field. This probe then activates

TABLE 1

Commercial mass spectrometers

(a) Single-focus magnetic sector type

Manufacturer [a]	Model	Magnetic sector radius (mm)	Mass range (a.m.u.)	Resolving power	Accessories etc.
DuPont	DP-1				GL, GLC, data processing
Edwards Instruments Ltd.	E-60		Up to 500	ca. 300	Pye-Unicam GLC
LKB Produkter AB	LKB9000	200; 60°	12—1000	1200—5000	Mass marker, peak matcher, GLC data processing
VG Micromass Ltd.	MM1	10; 180°	2— 240	22	Leak detector
	MM2	10; 180°	2— 240	44	Leak detector
	MM6	60; 90°	1— 500	Up to 400	Quantitative analysis, isotopic ratio determination
	MM12	120; 60°	1— 900	1500	CI, FD, mass marker, GLC, data processing, isotopic ratio determination
	MM16	160; 60°	1—1400	1800	CI, FD, mass marker, GLC, data processing
	MM30F	305; —	Up to 3750	2500—6000	CI, FD, mass marker, GLC, data processing
Niagra Instruments	PNS10	—; 90°	1—1000	1000	CI
Varian Associates Ltd.	EM600	—; 60°			
	CH7A				Isotopic ratio determination with solid samples
	TH5				Isotopic ratio determination with gaseous samples
	MAT230				

[a] AEI Scientific Apparatus Ltd., Urmston, Manchester M31 2CD, Gt. Britain.

Hewlett-Packard, Palo Alto, California 94304, U.S.A.
JEOL Ltd., 1418 Nakagami, Akishima, Tokyo 196, Japan.
LKB Produkter AB, S-161 25, Bromma 1, Sweden.
VG Micromass Ltd., Nat Lane, Winsford, Cheshire, Gt. Britain.
Niagra Instruments, 246 Lake Shore Road, St. Catherins, Ontario, Canada.
Nuclide Spectra, 642 East College Avenue, State College, Pennsylvania 16801, U.S.A.
Varian Associates Ltd., Palo Alto, California 94304, U.S.A.

(b) Double-focus electrostatic and magnetic sector type. Manufacturers provide an extensive range of accessories, e.g. CI, FD, FI, GLC, data processing, peak matching

Manufacturer [a]	Model	Design type	Electrostatic sector radius (mm)	Magnetic sector radius (mm)	Mass range (a.m.u.)	Resolving power	Other accessories
AEI Scientific Apparatus Ltd.	MS902	Nier–Johnson	380; —	300; —	8–7200	40 000	Metastable scanning, appearance potential determination
	MS1073	Nier–Johnson	132;90°	102;90°	1–1200	1 000	
	MS3076	Nier–Johnson	216;90°	165;90°	4–2400	20 000	
	MS5074	Nier–Johnson	380;90°	305;90°	8–8000	70 000	Metastable scanning, appearance potential determinations
	MS702	Mattauch–Herzog			16– 580	3 000	Spark source, photographic recording, ion probe surface analysis
JEOL Ltd.	JMS D100	Nier–Johnson	150;90°	150; 60°	Up to 2400	10 000	Mass marker
	JMS 01	Mattauch–Herzog			1–4000	20 000	Spark source, photographic recording, ion probe surface analysis, mass marker

Table 1b (continued)

(b) Double-focus electrostatic and magnetic sector type. Manufacturers provide an extensive range of accessories, e.g. CI, FD, FI, GLC, data processing, peak matching

Manufacturer[a]	Model	Design type	Electrostatic sector radius (mm)	Magnetic sector radius (mm)	Mass Range (a.m.u.)	Resolving power	Other accessories
VG Micromass Ltd.	MM ZAB-2F	Reversed Nier—Johnson	380; —	300; —	1–1200	40 000	Ion probe surface analysis
	MM70	Nier—Johnson		127; —	1–2600	>20 000	
Nuclide Spectra	12·90-G GRAF-3	Mattauch—Herzog					
Varian Associates Ltd.	MAT311	Reversed Nier—Johnson	—; 90°	—; 90°	2–3600	>25 000	CI, FD, FI, GLC, data processing
	MAT711	Mattauch—Herzog				80 000	FD, FI, mass marker, GLC, data processing, metastable scanning, photographic recording

a For manufacturers' addresses, see Table 1(a).

(c) Quadrupole and duodecapole type

Manufacturer[a]	Model	Mass range (a.m.u.)	Resolving power	Accessories etc.
Finnigan Instruments	400	1— 400		
	750	1— 750	500	GLC
	1015	1— 750	500	Mass marker, GLC, data processing, multiple ion detector

Company	Model	Range		Notes
Hewlett-Packard Ltd.	5992	10— 800		CI, mass marker, GLC—MS combination with data processing
Varian Associates Ltd.,	MAT44	2—1200		CI, GLC, data processing
VG Micromass Ltd.	Q4	2— 60		Leak detector
	Q7	1— 120		Leak detector
	Q8	1— 300	300	
	Q9	1— 400	300	
	Q50	1— 800	1500	Mass marker

electronic circuits which record a scale on the mass spectrum as it is being scanned. Quadrupole mass spectrometers show a linear variation of mass in focus with the quadrupole field voltage so that it is particularly easy to fit such a spectrometer with an automatic mass calibration unit. Other modern mass spectrometers are equipped with a dedicated computer and magnetic tape recording. The mass scale remains stable for some time so that it can be calibrated once using standards and the system will then plot calibrated spectra.

(2) Photographic recording

A spectrometer of the Mattauch—Herzog design allows simultaneous focusing of ions in a plane and so the mass spectrum can be recorded on a photographic plate. Spectra can be recorded in this way with a resolution equal to the maximum attainable by virtue of the mass analyser design. The disadvantage of photographic recording is that the sensitivity varies with the mass of the ion recorded and the calibration necessary for quantitative work is somewhat tedious. The advantage is that photographic recording is an integration method so that it can be used with a fluctuating signal to give a meaningful result, as in the spark ionisation of inorganic specimens. In terms of the quantity of sample used, photographic recording is as sensitive as any other method of recording. For GLC—mass spectrometry studies at high resolution where the recording time has to match the gas chromatographic peak width, photographic recording may give more satisfactory results. This is because, in the fastest scanning modes, resolution is limited by the time constant of the amplifier—recorder system rather than by the design of the mass analyser.

(H) COMMERCIAL MASS SPECTROMETERS

Mass spectrometers have a modular construction and consist of an ion source, a mass analyser and a detector system. Manufacturers offer a range of modules to suit the demands of a variety of users. The mass analyser systems available from some manufacturers are listed in Table 1. These are supplied with electron impact ion sources and some means of recording the mass spectrum in graphical form. Other ion sources available are listed. The larger instruments can also be equipped for connection to a GLC apparatus and a combined GLC—mass spectrometry system is often available from the manufacrurer. Computer assisted data processing is also commonly available. Some manufacturers produce mass spectrometers suitable for accurate isotopic ratio determination by the simultaneous collection of two or three ion peak currents.

3. Mass spectrometry of pure organic compounds

(A) THE MASS SPECTRUM AND THEORIES OF FRAGMENTATION

The majority of covalent organic compounds can be made to give sufficient vapour pressure so that a mass spectrum can be recorded. Volatile samples are stored as vapour in a heated globe and slowly introduced into the ionisation chamber. Less volatile samples are introduced on a probe directly into the ionisation chamber and then heated to produce sufficient vapour pressure.

Mass spectra are routinely obtained using an electron bombardment ion source with an electron beam energy of 70 eV. Energy is transferred to the molecule and this results in the removal of one electron to give the molecular ion $[M]^{\cdot +}$. The positively charged and radical character of this species is indicated by the symbol $\cdot +$. Decomposition of the molecular ion occurs to give daughter ions, neutral radicals and neutral molecules.

The mass spectrometer usually employed has a mass analyser of the magnetic sector field type and records the steady state concentration of these ions within the ion source, less those ions which decompose along the ion flight path and are lost. The resulting mass spectrum is a finger print for the identification of an organic compound. Changes in the physical dimensions of the spectrometer and small changes in the electron beam energy can cause small changes in the relative intensities of peaks in the mass spectrum. This effect is relatively unimportant for purposes of qualitative identification but it must be considered if the quantitative analysis of mixtures is required. On the other hand, large changes in the electron beam energy can cause substantial changes in the mass spectrum, as can a change in the type of ion source which is used.

The quasi-equilibrium theory [28] of mass spectrometric fragmentation assumes that the internal energy acquired by the molecular ion at the time of its formation is distributed over all the accessible energy states by radiationless transitions. A certain fraction of these energy states corresponds to activation complexes which decompose. According to this theory, the ion—radical character of the molecular ion is distributed throughout the bonds of the molecule. The principle fragment ion peaks correspond to the most stable ion products of the most favourable reaction pathways. Under the usual conditions of mass spectrometry, there are no molecular collisions so there can be no equilibration of energy among individual molecular ions. The decomposition rates here are not equivalent to the rates found in solution chemistry where there is a statistical distribution of internal energy among millions of molecules.

An alternative theory [29] of mass spectrometric fragmentation has been applied by Djerassi and his coworkers. They suppose that the charge in a molecular ion may be considered as localised in some particular bond, lone pair, or conjugated π-system in the molecule. For an aliphatic ether or amine, the site of charge localisation is the heteroatom because one of the lone pair electrons would be easiest to remove. The charge site is then considered to trigger the fragmentation.

To illustrate the symbolism used in each of these theories, we can consider two important fragmentation processes of the molecular ions of aliphatic ketones, the McLafferty rearrangement and the α-cleavage process. These are symbolised in the quasi-equilibrium theory as decomposition reactions of a delocalised molecular ion each yielding a neutral species and a more favoured ion which in the case of the McLafferty rearrangement is also a molecular ion. The symbol $\overset{z}{\underset{z}{\big\}}$ indicates which bond is broken in the decomposition process. The normal arrow symbol $\nearrow$ is reserved to indicate movement of an electron pair.

In the localised charge theory, these are symbolised as radical reactions and the symbol $\nearrow$ indicates movement of one electron.

204

The localised charge theory is useful in that it focuses attention on the most probable decomposition processes. However, it is an over-simplification to consider that ion—radicals possess a localised charge and this theory is now considered to be a useful mnemonic with little predicative value [30,31].

The quasi-equilibrium theory can be used in a predicative sense but this requires a calculation of the properties of the molecular ion using molecular orbital theory. This approach is too complex to be of value in the broad field of organic mass spectrometry and so some general criterion is needed to define which bond in a molecule is most likely to be broken. When bond cleavage occurs in a molecular ion or a fragment ion, the charge distribution in the transition state for this process will approach that in the products. Bond cleavage reactions are endothermic processes and so the low-energy transition states for decomposition of a given ion will be those leading to products which are stabilised by inductive or resonance effects. The criterion used here is that the most favoured pathway for bond breaking is the one giving the most stable products.

(B) DETERMINATION OF A MOLECULAR FORMULA

(1) The molecular ion

The first step in examining a mass spectrum is usually to decide which peak corresponds to the molecular ion. The spectrum is counted to increasing m/e values using the techniques described on p. 207 until no more ions are observed and this is then the molecular ion region. The abundance of a molecular ion varies greatly with the nature of the molecule. In general, aromatic compounds and others containing an extended π-system show a strong molecular ion while aliphatic compounds show a weak one. Identification of a molecular ion can be checked in several ways.

(i) The resulting mass number must be compatible with a stable molecule and not a radical. For example, compounds of C, H, O and S containing an even number of N atoms have an even mass number while molecules containing an odd number of N atoms have an odd mass number.

Often, loss of a hydrogen atom from the molecular ion is an important fragmentation process and the M − 1 peak is intense. Thus the most abundant ion of the cluster around the molecular ion is not necessarily the molecular ion itself. Ions that arise via ion molecule collisions may appear at masses higher than the molecular mass of the compound but these have very low abundance under the conditions normally used to obtain a mass spectrum. Exceptions to this rule are sometimes found with the M + 1 peak for molecules which possess a strongly basic oxygen or nitrogen centre where the corresponding protonated form is a stable entity.

(ii) The intensity of the molecular ion is proportional to the sample pressure and the intensity decreases of the molecular ion and related fragments as a function of pressure are equal. This is useful to check for ions thought to arise from material remaining in the spectrometer from a previous run. Also, large M + 1 peaks due to ion—molecule reactions decrease sharply with pressure.

(iii) The fragmentation pattern in the region of the molecular ion should be consistent with the loss of small radicals or molecules such as $H^{\cdot}$, $CH_3^{\cdot}$, H_2O, CO, etc.

(iv) Final identification of the molecular ion must be accompanied by a partial interpretation of the most abundant peaks in the mass spectrum.

(2) Isotope clusters

The molecular ion is never represented by the peak at highest mass number. This is because natural carbon contains a significant percentage (1.1%) of the ^{13}C isotope and molecules containing one atom of this isotope give rise to a molecular ion at M + 1. By convention, mass spectrometrists calculate the molecular weight of a compound on the basis of the most abundant isotope of each element present. The isotopic composition of some elements commonly encountered in organic mass spectrometry is listed in Table 2. Further data on accurate mass numbers of isotopes [32,34] and the isotopic composition of elements [33,34] is to be found in the references cited.

For compounds containing C, H and O, an approximate idea of

TABLE 2

Relative atomic mass and abundance of natural isotopes of selected elements

Element	Isotope	Relative atomic mass (C^{12} = 12.000000)	Natural abundance (%)
Hydrogen	^{1}H	1.007825	99.985
	^{2}H	2.014102	0.015
Boron	^{10}B	10.012943	18.7
	^{11}B	11.009309	81.3
Carbon	^{12}C	12.000000	98.9
	^{13}C	13.003354	1.1
Nitrogen	^{14}N	14.003074	99.64
	^{15}N	15.000108	0.36
Oxygen	^{16}O	15.994915	99.8
	^{17}O	16.999133	0.04
	^{18}O	17.999160	0.2
Fluorine	^{19}F	18.998405	100
Silicon	^{28}Si	27.976927	92.2
	^{29}Si	28.976491	4.70
	^{30}Si	29.973761	3.1
Phosphorus	^{31}P	30.973763	100
Sulphur	^{32}S	31.972074	95.0
	^{33}S	32.971461	0.76
	^{34}S	33.967865	4.2
Chlorine	^{35}Cl	34.968855	75.8
	^{37}Cl	36.965896	24.2
Bromine	^{79}Br	78.918348	50.5
	^{81}Br	80.916344	49.5
Iodine	^{127}I	126.904352	100

the number of carbon atoms in the molecule can be gained from the relative heights, h and $h + 1$, of the M and M + 1 ions, respectively.

$$\text{No. of carbon atoms in molecular ion} \simeq \frac{100(h + 1)}{1.1h}$$

This relationship is not very accurate unless corrections are made for the isotopic contribution from lower mass peaks to the molecular ion peak but it may serve to identify the M − 1, M and M + 1 peaks in some circumstances. Simple application of the rule leads to erroneous results when the compound contains an element such as N or S with a relatively abundant higher nucleide.

The relative isotopic abundances for the possible combinations of C, H, O and N have been tabulated [35] whilst another tabulation includes combinations with S and Cl [36]. These tables can be used as a basis for calculating the isotopic abundances of any molecular ion. To illustrate the principle of such calculations, consider a compound of the elements A and B each of which is a mixture of isotopes separated by one mass unit, A_1 A_2 A_3 and B_1 B_2 B_3. First normalize the isotope composition, a_1, b_1 etc., by putting $a_1 = b_1 = 1$. Then set up a matrix with the isotope composition of element A as the first row and the isotope composition of element B as the first column. Each entry in the first row is multiplied by the first column entry and the result entered underneath; each entry in the first row is multiplied by the second column entry and the result entered one column to the right; the process is repeated for each column entry. Finally the sum of each column gives the intensity of a particular mass peak.

	$a_1 = 1$	a_2	a_3		
$b_1 = 1$	1	$1 \times a_2$	$1 \times a_3$		
b_2		$b_2 \times 1$	$b_2 \times a_2$	$b_2 \times a_3$	
b_3			$b_3 \times 1$	$b_3 \times a_2$	$b_3 \times a_3$
m/e	$M = M_{A_1} + M_{B_1}$	$M+1$	$M+2$	$M+3$	$M+4$
Relative abundance	1	$a_2 + b_2$	$a_3 + b_3 + a_2 b_2$	$a_2 b_3 + a_3 b_2$	$a_3 b_3$

This process can be repeated to obtain the isotopic composition of an ion ABC where C is a new element by using the result for AB as the first line of a new matrix with the isotopic composition of C as the first column. When the isotopic composition of an organic compound which is not in the tables is required, the nearest composition found in the tables can be taken as A in the above matrix while B becomes the residual elements.

Example. To find the isotopic abundances of the molecular ion of ferrocene, $C_{10}H_{10}Fe$.
From Mass and Abundance Tables, the isotopic composition of $C_{10}H_{10}$ is

Mass no.	130	131	132
Normalized abundance	1.000	0.110	0.005

Iron has the isotopic composition

Mass no.	54	56	57	58
% Abundance	5.9	91.6	2.2	0.33
Normalized abundance	0.064	1.000	0.024	0.004

We can now set up the matrix to find the isotopic abundances of $C_{10}H_{10}Fe$.

m/e	$C_{10}H_{10}$	130	131	132			
Fe		1.000	0.110	0.005			
54	0.064	0.064	0.007	0.000			
55	0.000		0.000	0.000	0.000		
56	1.000		1.000	0.110	0.005		
57	0.024			0.024	0.003	0.000	
58	0.004				0.004	0.000	
$C_{10}H_{10}Fe$ abundance		0.064	0.007	1.000	0.134	0.012	0.000
m/e		184	185	186	187	188	189

In general, the isotope peak intensities due to a complex ion are given by simple probability. For n atoms of an element with isotopes of abundance a_1, a_2, a_3 ... in order of atomic mass increasing by one mass unit, expansion of the expression $(a_1 + a_2 + a_3 + ...)^n$ gives the proportions of the isotopic ions due to that element. When n atoms of element A and m atoms of element B are present, the proportions of isotopic ions are given by the expansion of $(a_1 + a_2 + ...)^n$ $(b_1 + b_2 + ...)^m$ where a_1, a_2 ... are the relative amounts of isotopes of element A and b_1, b_2 ... are the relative amounts of isotopes of element B.

The isotope peak intensities due to two atoms of an element with isotope abundances a_1, a_2 and a_3 are given by

Expansion of
$(a_1 + a_2 + a_3)^2$: a_1a_1 $2a_1a_2$ $(2a_1a_3 + a_2a_2)$ $2a_2a_3$ a_3a_3
Corresponding
mass number: M   M + 1   M + 2   M + 3   M + 4

The isotope peak intensities due to one atom of an element with isotope abundances a_1 and a_2, and one atom of an element with isotope abundances b_1 and b_2 are given by

Expansion of $(a_1 + a_2)(b_1 + b_2)$: a_1b_1 $(a_1b_2 + a_2b_1)$ a_2b_2
Corresponding mass number: M   M + 1   M + 2

Tables of the isotope clusters resulting from combinations of other elements commonly encountered in organic mass spectroscopy are given in Tables 3—6. Commonly encountered elements which are

References pp. 268—273

TABLE 3

Isotope clusters in the spectra of S_n^+ ions
M is the mass of the ion containing only ^{32}S atoms.

Atoms present	Relative probability of occurrence of ions of masses				
	M	$M+1$	$M+2$	$M+3$	$M+4$
S_1	1.000	0.008	0.044		
S_2	1.000	0.016	0.088	0.001	0.002
S_3	1.000	0.024	0.133	0.002	0.006
S_4	1.000	0.032	0.177	0.004	0.012
S_5	1.000	0.040	0.222	0.007	0.020

monoisotopic include ^1H (for most practical purposes), ^{19}F, ^{31}P and ^{127}I.

(3) Determination of the molecular formula

The results of an elemental analysis by the combustion or other methods together with the molecular weight of the substance determined from the mass number of the molecular ion are usually suffi-

TABLE 4

Isotope clusters in the spectra of B_n^+ ions.
M is the mass of the ion containing only ^{11}B atoms. The ratio $^{11}B/^{10}B$ is taken as 4.00.

Atoms present	Relative probability of occurrence of ions at masses						
	$M-6$	$M-5$	$M-4$	$M-3$	$M-2$	$M-1$	M
B_1						0.250	1.000
B_2					0.063	0.500	1.000
B_3				0.016	0.188	0.750	1.000
B_4			0.004	0.063	0.375	1.000	1.000
B_5		0.001	0.016	0.125	0.500	1.000	0.800
B_6		0.004	0.039	0.208	0.625	1.000	0.667
B_7	0.001	0.011	0.078	0.313	0.750	1.000	0.572
B_8	0.004	0.026	0.137	0.438	0.875	1.000	0.500
B_9	0.009	0.051	0.219	0.583	1.000	1.000	0.444
B_{10}	0.019	0.092	0.328	0.750	1.000	1.000	0.400

cient data from which to calculate the molecular formula of an organic compound. The elemental analysis also gives a check on the purity of the compound under investigation by mass spectrometry. It is sometimes overlooked that a sample may contain substances of lower molecular weight which give rise to molecular ion peaks that are mistaken for fragment ions of the molecular ion of highest mass number. Also, the sample may contain involatile material which will not give rise to a mass spectrum.

The molecular formula of any ion can also be determined from a knowledge of the mass of this ion with an accuracy of 1 part in 10^5. Mass measurements to this accuracy can be made with a double-focusing mass spectrometer having a resolving power of the order of 10^4. The unknown signal is compared with a peak of accurately known mass due to an added standard. Peaks due to $C_n F_m$ fragments from added perfluorokerosene [37] or heptacosfluorotributylamine are used for comparison [38]. Perfluorokerosene has usable peaks up to m/e 800 and heptacosfluorotributylamine has a molecular ion at m/e 614. Perfluoroalkyltriazines [39] have been proposed as standards for higher mass numbers up to m/e 1600.

With a spectrometer of the Nier—Johnson design, the procedure [40] is to first adjust the magnetic field, H, so as to focus the peak of unknown mass m on the exit slit of the mass analyser. Next, the magnetic field is maintained constant while the accelerator potential is switched from its value V_1 to a second value V_2 such that an ion of known mass m_2 is focused on the exit slit of the mass analyser. Switching is performed automatically while V_2 is varied so that in the final adjustment first one ion then the second ion is in focus. The electrostatic deflector field, E, must also be varied so that the ratio V/E remains constant. From the formulae on pp. 184 and 186

$$m_1 V_1 = m_2 V_2 = \frac{H^2 r^2 e}{2}$$

H and r are fixed so m_1 can be determined from

$$m_1 = m_2 \frac{V_2}{V_1}$$

With a spectrometer of the Mattauch—Herzog design, the peaks due to fragments of the added perfluoro compound are used to calibrate the mass scale. Assuming a linear scale between adjacent peaks,

TABLE 5

Isotopic clusters in the spectra of $Cl_n Br_m^+$ ions.
M is the mass of the ion containing only ^{35}Cl and ^{79}Br ions.

Atoms present	Relative probabilities of occurrence of ions at masses								
	M	$M+2$	$M+4$	$M+6$	$M+8$	$M+10$	$M+12$	$M+14$	$M+16$
Br_1	1.000	0.977							
Br_2	0.512	1.000	0.489						
Br_3	0.341	1.000	0.978	0.319					
Br_4	0.174	0.682	1.000	0.652	0.159				
Br_5	0.105	0.511	1.000	0.998	0.478	0.093			
Br_6	0.053	0.314	0.767	1.000	0.733	0.286	0.047		
Br_7	0.031	0.209	0.614	1.000	0.977	0.573	0.186	0.026	
Br_8	0.016	0.123	0.419	0.819	1.000	0.782	0.382	0.107	0.013
Cl_1	1.000	0.326							
Cl_2	1.000	0.653	0.106						
Cl_3	1.000	0.979	0.319	0.035					

Cl_7	0.438	1.000	0.979	0.532	0.174	0.034	0.004	0.000
Cl_8	0.336	0.876	1.000	0.653	0.266	0.069	0.011	0.001
$BrCl$	0.767	1.000	0.245					
$BrCl_2$	0.614	1.000	0.457	0.064	0.017			
$BrCl_3$	0.511	1.000	0.652	0.177	0.064	0.005		
$BrCl_4$	0.438	1.000	0.839	0.334	0.022	0.001		
$BrCl_5$	0.376	0.981	0.522	0.149				
Br_2Cl	0.438	1.000	0.698	0.137				
Br_2Cl_2	0.384	1.000	0.896	0.319	0.039	0.010		
Br_2Cl_3	0.314	0.920	1.000	0.500	0.117	0.037	0.003	
Br_2Cl_4	0.241	0.786	1.000	0.635	0.215			
Br_3Cl	0.262	0.852	1.000	0.489	0.080	0.020		
Br_3Cl_2	0.205	0.734	1.000	0.638	0.187	0.066	0.005	
Br_3Cl_3	0.165	0.646	1.000	0.778	0.319			
Br_4Cl	0.143	0.604	1.000	0.799	0.304	0.043		
Br_4Cl_2	0.119	0.544	1.000	0.940	0.472	0.118	0.012	
Br_5Cl	0.080	0.419	0.895	1.000	0.611	0.191	0.024	

TABLE 6

Isotope clusters in the spectra of Si_n^+ ions
M is the mass of the ion containing only ^{28}Si atoms.

Atoms present	Relative probability of occurrence of ions at masses						
	M	$M + 1$	$M + 2$	$M + 3$	$M + 4$	$M + 5$	$M + 6$
Si_1	1.000	0.051	0.034				
Si_2	1.000	0.102	0.070	0.003	0.001		
Si_3	1.000	0.153	0.109	0.011	0.004		
Si_4	1.000	0.204	0.151	0.021	0.008	0.001	
Si_5	1.000	0.256	0.195	0.036	0.014	0.002	
Si_6	1.000	0.307	0.242	0.055	0.023	0.004	0.001
Si_7	1.000	0.358	0.292	0.077	0.034	0.007	0.002
Si_8	1.000	0.409	0.344	0.104	0.047	0.011	0.003

the mass of an unknown peak can be determined by micrometer measurements on the mass spectrum.

Some mass spectrometers are fitted with twin ion paths. In these double-beam spectrometers, each ion beam can be fed with a separate sample but each beam experiences identical electrical and magnetic fields. With such a spectrometer, the reference ion beam can be maintained and recorded or not alongside the sample mass spectrum.

Computer techniques are available for reducing the raw mass spectral data to give a printed list of each ion and its molecular formula.

Individual isotopes have masses which are not exactly whole numbers but which are known, relative to $^{12}C = 12.000000$, with high accuracy. Because of this, the various possible formulae giving a particular mass number can be differentiated provided that the mass of the molecular ion is measured with an accuracy of 1 part in 10^5. Table 2 lists the accurate masses of some common isotopes and further data are to be found in the reference cited. Tables of accurate masses are available for various combinations of C, H, O, and N [35]. As an illustration of the use of accurate mass measurements, the masses corresponding to molecular formulae containing C, H, O, and N of molecular weight 74 are listed in Table 7. A circular slide rule has been devised for calculating formulae from accurate mass data and vice versa [41].

TABLE 7

Molecular formulae corresponding to a molecular weight of 74

Molecular formulae	Possible structural formula	Mass
$C_2H_6N_2O$	$\begin{array}{c} O^- \\ \| \ _+ \\ CH_3N{=}NCH_3 \end{array}$	74.048010
$C_3H_{10}N_2$	$H_2N{-}CH_2CH_2CH_2{-}NH_2$	74.084394
$C_4H_{10}O$	C_4H_9OH	74.073161
C_6H_2	$H{-}(C{=}C)_3{-}H$	74.015649

Perfluoro compounds are convenient standards for accurate mass measurement because fluorine has a relatively large mass defect from a whole number value. Because of this, peaks due to C_nF_m fragments do not overlap with peaks of the same nominal mass due to C-, H-, O-, and N-containing fragments. Perfluorocompounds are also inert and relatively volatile for compounds of high molecular weight so they can be pumped out quickly from the mass spectrometer.

(C) INTERPRETATION OF FRAGMENTATION PATTERNS

Interpretation of the fragmentation reactions leading to the most abundant fragments in a mass spectrum can give information on the structure of the parent compound. In addition to these fragments of high abundance, the mass spectrum contains many minor fragments which arise by indirect pathways and which yield no useful structural information. In the space available, it is only possible to discuss the most usually observed fragmentation reactions and their application to structural analysis. Further information can be obtained from the many monographs on the mass spectrometry of organic compounds. When faced with a specific problem, it is useful to read a monograph on the fragmentation patterns encountered in the particular class of compound to which the sample belongs.

The fragmentation pattern is characteristic of a particular compound so a comparison of mass spectra can be used as a method of confirming the identity of two compounds. Libraries of mass spectra taken with an electron bombardment ionisation source operated at 70 eV are available for computer-assisted comparison with an unknown compound. These have been listed in the bibliography section. Particularly where an attempt has been made to include

most of the known commercially important compounds, such computer-assisted identification can be of great value. The library lists
the molecular ion and a number (5—10) of the most important fragmentation peaks in the mass spectrum. After comparing the
unknown spectrum with this library, a computer is programmed to
print out the most probable identifications together with correlation
coefficients.

(1) Carbon—carbon bond cleavage to give a carbonium ion

   a. Illustrations from the mass spectra of hydrocarbons. Fragmentations of this type involve cleavage of a carbon—carbon single bond
with retention of the positive charge at a carbon atom as a carbonium ion. Fragmentation of a hydrocarbon molecular ion gives a carbonium ion and a radical and the carbonium ion can suffer further
fragmentation with loss of an alkene.

$$C_nH_{2n+2}\rceil^{\bullet+} \longrightarrow R-\overset{|}{\underset{|}{C}}-\overset{|}{\underset{|}{C}}-\overset{|}{\underset{|}{C^+}} + R^{\bullet}$$

$$R-\overset{|}{\underset{|}{C}}-\overset{|}{\underset{|}{C}}-\overset{|}{\underset{|}{C}}{}^{\bullet} \longrightarrow R-\overset{|}{\underset{|}{C^+}} + {}^{\backslash}C=C^{/}$$

The increase in stabilisation of a carbonium ion by alkyl substituents leads to the preferred fragmentation of hydrocarbon chains at
branching points. The effect of this stabilisation can be demonstrated
by examining the mass spectra of isomeric hydrocarbons C_9H_{20} illustrated in Fig. 16.

   The mass spectrum of a straight-chain hydrocarbon has a typical
appearance with important peaks at masses corresponding to the ions
$C_nH_{2n+1}^+$ which show a maximum intensity when $n = 3$, 4 or 5 and
then fall off in intensity at higher mass numbers. Typically, the
molecular ion is of low abundance. This is because the larger fragments contain many bonds and have a high probability of further
fragmentation whereas the smaller fragments arrived at by multiple
fragmentations lack the activation energy to decompose further.
Such a sequence of ion intensities is shown in the mass spectrum of
nonane [42] [Fig. 16(a)]. The peak due to M — CH_3 is almost absent
because loss of the methyl radical is less favoured than loss of ethyl
or a higher homologue where the radical centre is stabilised by alkyl
substitution.

216

The position of the chain branch in the spectra of 3-ethylheptane [43] and 4-ethylheptane [44] (Fig. 16) is indicated by the ion peaks which do not exhibit this smooth variation of abundance with mass number. In the case of 3-ethylheptane, these diagnostic fragments are

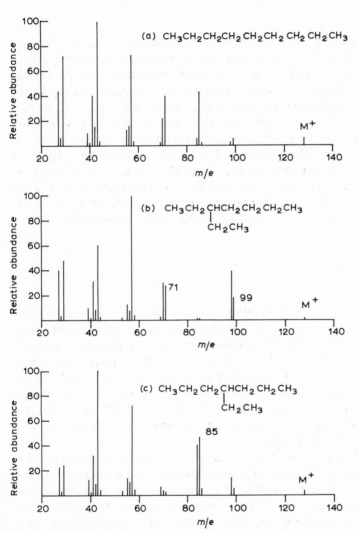

Fig. 16. Mass spectra of some alkanes, C_9H_{20}. (a) Nonane; (b) 3-ethylheptane; (c) 4-ethylheptane.

at m/e 71 and 99 corresponding to the fragmentation

$$CH_3CH_2\overset{|}{C}HCH_2CH_2CH_2CH_3 \Big]^{\cdot+}$$
$$\overset{|}{C}H_2CH_3$$
$$m/e\ 128$$

$$CH_3CH_2\overset{+}{C}H(CH_2)_3CH_3 + C_2H_5^{\cdot}$$
$$m/e\ 99$$

$$(CH_3CH_2)_2\overset{+}{C}H + C_4H_9^{\cdot}$$
$$m/e\ 71$$

The corresponding fragments for 4-ethylheptane are at m/e 99 and 85 due to loss of an ethyl radical and a propyl radical, respectively, from the branch point. From symmetry considerations, the probability of loss of a propyl radical from 4-ethylheptane is twice that for loss of an ethyl radical and this is one reason why the peak at m/e 85 is at much higher abundance. The peak at m/e 99 could easily be overlooked as being due to fragmentation at a chain branch because of its low abundance, but it exhibits the characteristic of fragmentation at such centres that the peak one mass unit lower is of equal, or higher, abundance. Such peaks one mass unit lower arise by fragmentation of two bonds and have a greater probability of occurring at a branch point than with a straight-chain hydrocarbon.

$$CH_3CH_2CH_2-\overset{\overset{H}{|}}{\underset{\underset{H_3CH_2C}{|}}{C}}-\overset{\overset{H}{|}}{\underset{\underset{H}{|}}{C}}-CH_2CH_3 \Big]^{\cdot+} \longrightarrow CH_3CH_2CH_2CH=CHCH_2CH_3 \Big]^{\cdot+} + C_2H_6$$
$$m/e\ 98$$

Provided only a small number of chain branches are present, the mass spectrum of an alkane will indicate the nature of this branching. Use has been made of this in some structural studies on alkanes bearing functional groups [45]. The mass spectrum of the parent alkane, obtained by chemical manipulation to remove the functional groups, has a mass spectrum dominated by fragmentation at the branch points while the mass spectrum of the substance itself is dominated by the influence of the functional groups. This gives information useful for structure determination.

In the mass spectra of alkenes, fragmentation of an allylic bond is favoured because of resonance stabilisation of the resulting allylic carbonium ion. In spite of this, the mass spectra of alicyclic olefins are not suitable for determining the position of the olefin group because this group shows a greater probability for migration along the carbon chain. Thus the mass spectra of isomeric straight chain olefins greatly resemble each other. The mass spectra of 1-octene [46]

218

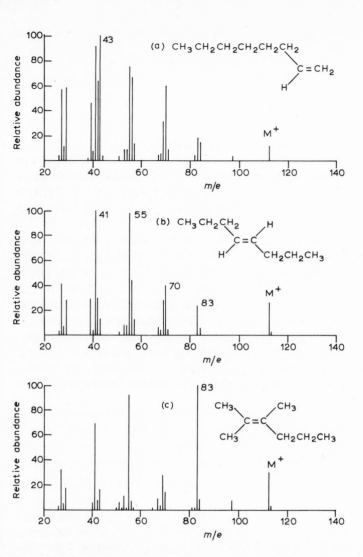

Fig. 17. Mass spectra of some alkenes, C_8H_{16}. (a) Oct-1-ene; (b) oct-4-ene; (c) 2,3-dimethylhex-2-ene.

and 4-octene [47] are compared in Fig. 17 and show only minor differences. In general, the mass spectra of alicyclic alkenes are characterised by peaks at mass numbers C_nH_{2n-1} due to allylic cleavage of the olefins formed by isomerisation as just discussed and peaks at

C_nH_{2n} due to the McLafferty rearrangement and fragmentation of these olefins.

The McLafferty rearrangement is discussed on p. 231.

Only tetrasubstituted alicyclic olefins show a sufficiently low probability of olefin migration that ions due to allylic fragmentation of the parent olefin become dominant in the mass spectrum. Thus, in the spectrum of 2,3-dimethyl-2-hexane [48] [Fig. 17(c)] there is an abundant fragment at m/e 83 due to the allylic ion $C_6H_{11}^+$.

Less heavily substituted alicyclic olefins must be transformed chemically into a derivative and the latter examined by mass spectrometry in order to demonstrate the position of the double bond. This approach to the determination of olefin structure will be discussed on p. 224 as one example of the fragmentation patterns observed in the mass spectra of alkanes bearing functional groups.

Alkyl-substituted benzenes show a strong fragmentation to give the $C_7H_7^+$ ion with m/e 91. This ion has been shown to have the tropylium structure rather than the benzyl ion structure which might be expected. The mass spectra of alkyl-substituted aromatic compounds

m/e 132   *m/e* 91

pounds are thus quite distinctive and show only weak fragments at mass numbers lower than 91. As an example, the mass spectrum of butylbenzene [49] is illustrated in Fig. 18. Di- and higher-substituted benzenes usually give an intense substituted tropylium ion fragment by the benzylic cleavage. 1-Methyl-4 isopropylbenzene [50], for example, gives an intense ion at m/e 119. Due to this type of fragmentation.

$(CH_3)_2C_7H_5^+$ + $CH_3^{\bullet}$
m/e 119
Relative abundance 100%

220

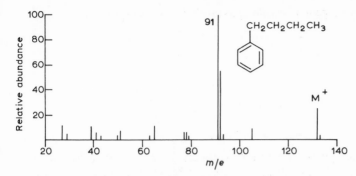

Fig. 18. Mass spectrum of butylbenzene.

This cleavage also serves to demonstrate that the stability of the carbonium ion fragment can compensate for the simultaneous formation of a relatively high-energy methyl radical. In the cases of alkane fragmentation, it was noted previously that loss of a methyl radical had low probability.

An exception to the high probability of benzylic cleavage is shown by the isomeric xylenes. These give an abundant fragment at m/e 91 due to formation of the tropylium ion $C_7H_7^+$ with loss of a methyl radical and no ion due to simple benzylic cleavage with loss of a hydrogen atom to give the substituted tropylium ion CH_3—$C_7H_6^+$ which is a process requiring substantially higher energy.

Alkylbenzenes with a sufficiently long alkyl chain also undergo a McLafferty rearrangement (see p. 231) with fragmentation of two bonds. The ion at m/e 92 in the mass spectrum of butylbenzene Fig. 18) is due to this rearrangement.

$b.$ $Illustrations$ $from$ the $mass$ $spectra$ of $alcohol$ $ethers$ and $amines.$ Oxygen and nitrogen have a lone pair of electrons which can be donated towards a positive centre. Thus, in the mass spectra of aliphatic alcohols, ethers, and amines, there is a high probability of fragmentation to give a resonance-stabilised carbonium ion.

$$\left[\begin{array}{c} R^1 \\ R^2 \end{array}\!\!\!\diagdown\!\!\!\text{CHOH}\right]^{\cdot+} \longrightarrow \begin{array}{l} R^1CH\!=\!\overset{+}{O}H \ + \ R^{2\,\cdot} \\[6pt] R^2CH\!=\!\overset{+}{O}H \ + \ R^{1\,\cdot} \end{array}$$

$$\left[\begin{array}{c} R^1 \\ R^2 \end{array}\!\!\!\diagdown\!\!\!\text{CHNH}_2\right]^{\cdot+} \longrightarrow \begin{array}{l} R^1CH\!=\!\overset{+}{N}H_2 \ + \ R^{2\,\cdot} \\[6pt] R^2CH\!=\!\overset{+}{N}H_2 \ + \ R^{1\,\cdot} \end{array}$$

This fragmentation provides a means for the location of the hydroxyl group in an aliphatic alcohol when the alcohol group is secondary or tertiary. The mass spectrum of 2-octanol [15] [Fig. 19(b)], for example, shows an abundant ion at m/e 45 due to the ion $CH_3 CH\!=\!\overset{+}{O}H$. The mass spectra of a number of alcohols have been compared [52] and there is usually greater probability for loss of the larger carbon chain as a radical and very low probability for loss of

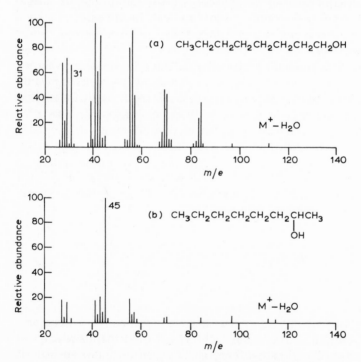

Fig. 19. Mass spectra of two alkanols, $C_8H_{18}O$. (a) 1-Octanol; (b) 2-octanol.

222

a methyl radical where the radical centre is not stabilised by alkyl substituents.

The molecular ion is very weak in the mass spectra of alcohols and usually the peak at highest mass corresponds to $M - H_2O$. In the mass spectrum of octanol, this occurs at m/e 112 and in the spectrum of 2-octanol there is also a peak at m/e 115 due to $M - CH_3$. Primary alcohols show a high tendency for dehydration in the mass spectrometer and the spectra of 1-alkanols strongly resemble the mass spectra of the olefins formed by loss of water [compare the spectrum of 1-octanol [53], Fig. 19(a), with that of 1-octene, Fig. 17(a)]. The fragment $CH_2=CHOH$, m/e 31, in the mass spectra of 1-octanols is of little diagnostic value and also becomes a less abundant fragment.

Dehydration may occur in the heated inlet system of the mass spectrometer by a thermally induced 1,2-elimination but the molecular ion can also undergo elimination. Deuterium-labelling experiments have shown that dehydration of the molecular ion occurs predominantly by a 1,4-elimination [54].

Low molecular weight aliphatic ethers show abundant fragments due to such resonance-stabilised oxonium ions but ethers containing four-membered and longer carbon chains exhibit mass spectra which are dominated by hydrocarbon fragments [55]. Thus, ethyl isobutyl ether [56] shows an abundant fragment due to the oxonium ion.

$$CH_3CH_2OCH_2CH(CH_3)_2 \rightarrow CH_3CH_2\overset{+}{O}=CH_2 + C_3H_7 \cdot$$
m/e 59
relative abundance 100%

while di-n-butyl ether [57] shows abundant hydrocarbon fragments and only 15% abundance of the oxonium ion $C_4H_9\overset{+}{O}=CH_2$ at m/e 87.

Nitrogen has a greater electron-releasing power than oxygen and so the mass spectra of amines show a high abundance of fragments due to nitrogen-stabilised carbonium ions. In the spectrum of tri-n-butylamine [58] (Fig. 20), the peak at m/e 142 is due to this fragmentation.

$$(CH_3CH_2CH_2CH_2)_3N \rightarrow (CH_3CH_2CH_2CH_2)_2\overset{+}{N}=CH_2 + C_3H_7 \cdot$$
m/e 142

Further fragmentation of this ion, as discussed on p. 231, leads to the ion at m/e 100.

This ability of nitrogen to locate the fragmentation of long carbon chains has been utilised in a method for determining the position of

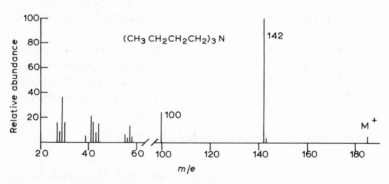

the olefinic group in an alkane [59]. In this process, the alkene is converted to its epoxide by reaction with a peracid and the epoxide is reacted with dimethylamine to give a mixture of two aminoalcohols. Electron bombardment of β-aminoalcohol in the mass spectrometer leads to fragmentation to the nitrogen-stabilised carbonium ion together with a radical stabilised by an adjacent oxygen function. This fragmentation is highly favoured. Examination of the mixture of aminoalcohols by mass spectrometry leads to the location of the position of the double bond in the carbon chain of the original olefin when the fragments due to $RCH=\overset{+}{N}Me_2$ have been identified.

The fragmentation of ethers to give an oxonium ion has been used in another process for locating the double bond in an olefin [60]. The alkene is oxidised with osmium tetroxide to give a glycol which

Fig. 20. Mass spectrum of tri-n-butylamine.

224

is converted to the cyclic acetal with acetone. Fragmentation in the

$$R^1CH=CHR^2 \longrightarrow R^1\overset{|}{\underset{|}{C}}H-\overset{|}{\underset{|}{C}}HR^2 \longrightarrow R^1\overset{|}{\underset{|}{C}}H-\overset{|}{\underset{|}{C}}HR^2$$

mass spectrometer of the molecular ion from this acetal leads to ions formed by loss of the $R_1^{\cdot}$ and $R_2^{\cdot}$ radicals, respectively, and so the location of the original double bond can be deduced. As would be expected from the previous discussion of the fragmentation of ethers, these characteristic ions are of low abundance. One means of aiding their identification is to enrich the osmium tetroxide and the acetone used in the preparation of the acetal with ^{18}O. The peaks being sought will then show a characteristic isotope pattern.

Alternatively, acetaldehyde can be used in place of acetone to make a second acetal which is examined in the mass spectrometer. The peaks due to the structurally important fragment will now be at 14 mass units lower and by comparison of both mass spectra it should be possible to identify them.

c. The acylium ion in the mass spectra of ketones. The acylium ion is a second type of carbonium ion which owes its stability to mesomeric electron donation from an oxygen atom. Ketone mass spectra [61] show abundant fragments due to such ions provided that the carbonyl group is not present in a ring system. Acylium ions show a high probability for loss of the stable molecule carbon monoxide to form a carbonium ion. The abundance of acylium ions

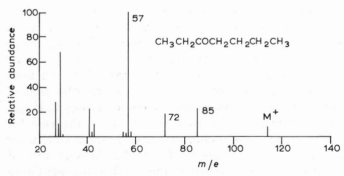

Fig. 21. Mass spectrum of heptan-3-one.

in the mass spectra of ketones is a function of these two factors. The mass spectrum of heptan-3-one (Fig. 21) shows peaks at m/e 57 and 85 due to the two acylium ions $C_2H_5\overset{+}{C}O$ and $C_4H_9\overset{+}{C}O$, respectively. Loss of carbon monoxide from these affords the carbonium ions with m/e 29 and 37. In general, cleavage of the ketone molecular ion leads to greater probability for loss of the longer carbon chain because this gives the more stable radical.

A problem arises in the interpretation of the mass spectra of ketones because the acylium ion has the same nominal mass as a carbonium ion with one extra carbon atom; for example, C_2H_5CO has the same nominal mass as C_4H_9, so the two ion peaks would be superimposed on the mass spectrum as usually presented. Confirmation of the formula of an acylium ion can be obtained in several ways. If the mass spectrum is run at high resolution, it is possible to record the previously superimposed ions as two separate peaks and to confirm their formulae by determining their accurate masses. Alternatively, the ketone can first be labelled with ^{18}O by equilibrating it with water enriched with this isotope [62]. Oxygen-containing ions in the mass spectrum of this ^{18}O-enriched ketone will then give rise to isotope peaks and so can be easily identified.

The McLafferty rearrangement, which is a second important process in the fragmentation of the molecular ion from an alicyclic ketone, is discussed on p. 231.

(2) Carbon—heteroatom bond cleavage

Fragmentation of the C—X bond in a molecular ion is a process which competes with the fragmentations discussed in Sect. 3. (C)(1). It is an important process in the mass spectral fragmentation of alkyl halides but is of very low probability when X is oxygen or nitrogen. Cleavage of the carbon-to-halogen bond in a molecular ion occurs to give a carbonium ion and the halogen radical as would be expected because of the electronegativity of halogen atoms. Thus, the mass spectrum of an alkyl halide would be expected to resemble that of the corresponding alkane with abundant peaks at masses C_nH_{2n+1} and a relatively abundant M — halogen peak. This is the case for bromides and iodides [63]. Alkyl chlorides and fluorides [60] show a high probability for fragmentation of two bonds with loss of hydrogen chloride or hydrogen fluoride and further fragmentation of the initially formed species leads to abundant peaks at masses

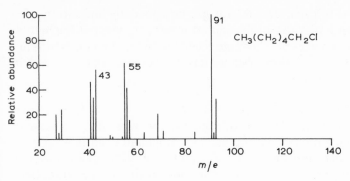

Fig. 22. Mass spectrum of 1-chlorohexane.

C_nH_{2n-1}. A 1,3-elimination has been shown to be the preferred process for alkyl chlorides [64]. In the mass spectra of chlorides and fluorides, abundant hydrocarbon peaks at C_nH_{2n-1} and C_nH_{2n+1} are seen. In addition, primary alkyl chlorides and bromides give rise to an ion $C_4H_8X^+$ (X = Cl or Br) which is given a cyclic structure. In the mass spectrum of 1-chlorohexane [63] (Fig. 22), this latter process gives rise to the ions with m/e 91 and 93 containing the ^{35}Cl and ^{37}Cl isotopes of chlorine, respectively. The molecular ion is of

negligible abundance in this mass spectrum as is usually the case for alkyl halides.

Aromatic compounds [65] with fluorine, chlorine or bromine substituents on the ring show similar probabilities for fragmentation of the molecular ion either by cleavage of the carbon—halogen bond or, when a suitable substituent is present, by the benzylic side chain cleavage typical of alkylbenzenes. Iodobenzenes show preferred fragmentation of the carbon—iodine bond in the mass spectrometer.

(3) Simultaneous cleavage of two bonds

The simultaneous cleavage of two bonds is a process which occurs frequently when the energy required is partly offset by the formation of new bonds to give two molecules of low energy content.

Processes of this type, which are accompanied by migration of a hydrogen atom along the molecular structure, are termed fragmentations with rearrangement. Cleavage reactions involving two bonds are in competition with other fragmentation processes encountered in the two previous sections.

a. Diels—Alder type cleavage of cyclohexenes. The Diels—Alder cleavage is a common feature of the mass spectra of cyclohexenes yielding a butadiene and an alkene. This fragmentation is illustrated by the mass spectrum (Fig. 23) of 1-methyl-4-isopropylcyclohex-3-ene [66]. The fragment at *m/e* 96 is due to the ionised butadiene which results from Diels—Alder cleavage of the molecular ion. Further cleavage of this fragment occurs with loss of either H˙ or CH₃ giving an ion which, in each case, can be drawn as an allylic carbonium ion.

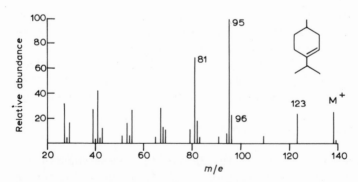

Either fragment from a Diels—Alder cleavage can carry the charge originally present on the molecular ion. In the fragmentation of 1-methyl-4-isopropylcyclohex-3-ene, there is a probability that the charge will remain on the monoene product. This is represented by

Fig. 23. Mass spectrum of 1-methyl-4-isopropylcyclohex-3-ene.

228

the $C_3H_6^+$ ion at m/e 42 and its further fragmentation to the allyl ion at m/e 41. The relative probabilities of the charge residing on either the butadiene or the alkene product depend on the ionisation energies of these products in the ground state [67]. More heavily substituted alkenes have lower ionisation energies which can be lower than the ionisation energy of the butadiene fragment.

The double bond which directs this type of fragmentation can also be part of an aromatic system. Thus, the principle cleavage of 2-methyltetralin [68] yields the Diels—Alder fragment $C_8H_8^+$ while a minor fragmentation pathway of the molecular ion is loss of a methyl radical. In contrast, loss of the methyl radical is the principal

m/e 131 (19%) + $CH_3^{\cdot}$ ←—— [2-methyltetralin]$^{\cdot+}$ ——→ [o-xylylene type]$^{\cdot+}$ + C_3H_6
m/e 104 (100%)

m/e 131 (100%) + $CH_3^{\cdot}$ ←—— [1-methyltetralin]$^{\cdot+}$ ——→ [styrene type CHCH$_3$]$^{\cdot+}$ + C_2H_4
m/e 118 (18%)

fragmentation pathway of 1-methyltetralin [69] because this leads to the low-energy benzyl carbonium ion while the Diels—Alder cleavage reaction is a minor pathway.

b. Fragmentation with rearrangement

Site-unspecific rearrangements. Some functional groups undergo elimination from the molecular ion with simultaneous abstraction of a hydrogen atom. The elimination can be represented formally as

$$\left[\begin{array}{c} C-X \\ C_n \\ C \end{array} \right]^{\cdot+} \longrightarrow \left[\begin{array}{c} C \\ C_n \\ C \end{array} \right]^{\cdot+} + HX$$

It is in competition with other fragmentation reactions of the molecular ion. The alternative α-carbon bond cleavage reaction which leads to the $C=X^+$ ion is of high probability when X is nitrogen. Fragmentation of the C—X bond is an alternative process of high probability when X is fluorine, chlorine or oxygen. Where elimination of HX occurs, the ionic product undergoes further fragmentation in the same manner as the corresponding olefin.

Experiments with deuterium-labelled samples have indicated that the hydrogen atom eliminated in these reactions does not come from a specific site. Where X = OH, the elimination of water from higher alcohols occurs predominately by 1,4-elimination ($n = 2$) via a six-membered transition state but smaller transition states are also possible [70,71]. In contrast, elimination of acetic acid from alkyl acetates (X = OCOCH$_3$) occurs by 1,2-elimination to the extent of 55% and by 1,3-elimination to the extent of 45% of the reaction [70]. Elimination of hydrogen chloride (X = Cl) occurs predominantly by 1,3-elimination ($n = 1$) [64].

A related rearrangement occurs during the further fragmentation of the stabilised carbonium ions which result from cleavage of ethers, thioethers and amines.

Here, when X is oxygen [72] or nitrogen [73], hydrogen transfer occurs by 4-, 5- or 6-membered transition states. When X is sulphur [74], reaction mainly occurs through a 4-membered transition state.

Site-specific rearrangements. The aromatic ortho-effect. The fragmentation patterns of isomeric benzene derivatives are often very similar but some *ortho*-disubstituted benzenes fragment in quite a different way to their *meta* and *para* isomers. This is often called the *ortho*-effect and it is most pronounced for derivatives of salicylic [75] anthranilic [76] and *ortho*-methylbenzoic [77] acids. Loss of methanol from the molecular ion of methyl salicylate occurs readily whereas the corresponding *meta* and *para* isomers show loss of MeO˙ only.

Salicylic acid shows a similar fragmentation with loss of water. Long-chain esters of salicylic acid first fragment to give the alkene and salicylic acid which then loses water.

Esters of anthranilic acid fragment to the ion C$_7$H$_5$NO$_2^+$ and esters of *ortho*-methylbenzoic acid give the ion C$_8$H$_6$O$^+$ by analogous

230

$$+ \quad C_4H_8$$

processes. The elimination of a hydroxyl radical from *ortho*-nitro-toluene is another example of *ortho*-effect [78].

The McLafferty rearrangement. As a result of his study of the mass spectra of aliphatic ketones, McLafferty [79] showed that the β-cleavage of long-chain ketones with migration of hydrogen from carbon to oxygen is a high probability process. It has been established from the mass spectra of various deuterated ketones that this is a site-specific rearrangement, only the γ-hydrogen atom being transferred in the fragmentation.

The enolic product does not revert to the keto form before under-going further decomposition. An identical reaction occurs with esters (X = OR) and acids (X = OH). The reaction may be represented as a concerted or a stepwise process [80]. Many investigations have established the site-specific nature of this reaction for ketones and esters. Suprisingly, the corresponding rearrangement of long-chain aliphatic aldehydes to yield the ion $CH_2=CH-OH^{\neg\cdot+}$ is not site-specific [81].

The McLafferty rearrangement peak is often very prominent in the mass spectra of carbonyl compounds. In the mass spectrum of heptan-3-one (Fig. 21), the ion at m/e 72 arises by a McLafferty rearrangement.

$$m/e \ 72$$

A carbon—nitrogen bond or a carbon double bond can also initiate a McLafferty rearrangement although the resulting fragment peaks are not usually so prominent.

Olefins undergo the McLafferty rearrangement and this is a fragmentation useful for structive elucidation for highly substituted olefins which show less tendency to isomerisation of the double bond before fragmentation (see p. 218). The olefin bond which takes part in this reaction can be part of an aromatic system. The fragmentation of long-chain alkyl benzenes (p. 221) is an example of this general reaction.

(4) Metastable ions

Fragmentation of ions occurs both within the ionisation chamber and along the whole ion flight path. The normal ion peaks which are recorded in the mass spectrum come from ions which are generated in the ionisation chamber and survive the flight path from being given their initial kinetic energy in the ion accelerator to a point beyond the magnetic mass analyser. When an ion fragments along the flight path, its kinetic energy is divided between the daughter ion and the neutral fragment so that the daughter ion has less kinetic energy than the same ion when formed in the ionisation chamber. Such ions are termed metastable ions. Metastable ions formed in certain field-

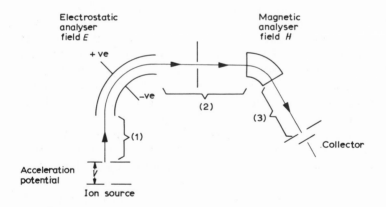

Fig. 24. Diagram of a double-focusing mass spectrometer showing the three field-free regions, (1), (2) and (3). Ion fragmentations which occur in these regions can be detected by appropriate scanning methods.

free regions of the flight path can be focused and recorded as part of the mass spectrum. Metastable ions formed in the magnetic or electrical field of a spectrometer follow random paths and are lost. Readily observable metastable ion peaks are found in the spectra from single- and double-focusing mass spectrometers with magnetic sector field mass analysers. The three field-free regions encountered in such spectrometers are classified in Fig. 24 with reference to a double-focusing instrument. In single-focusing instruments, there is no first field-free region and the second field-free region extends from the ion accelerator to the magnetic analyser.

a. Second field-free region. The metastable peaks encountered on a mass spectrum, as recorded in the conventional manner, are generated in this region. Parent ions of mass m_1 are accelerated through a potential V and enter this region with a velocity v given (see p. 183) by the expression

$$eV = \tfrac{1}{2} m_1 v^2$$

When the parent fragments to a daughter ion of mass m_2, the velocity of the particles will remain unchanged provided there is no energy change during the fragmentation process. The daughter ion continues along the flight path and is deflected by the magnetic field H along a path of radius r given (see p. 184) by

$$\frac{m_2 v^2}{r} = Hev$$

Elimination of v between the equations gives the expression

$$\frac{m_2^2}{m_1 e} = \frac{H^2 r^2}{2V}$$

Thus, the daughter ion, m_2, travels along the focal radius, r, as if it has an apparant mass m^* given by [83]

$$m^* = \frac{m_2^2}{m_1}$$

Such metastable ion peaks can be used to determine the precursor and daughter ions from which they arise. A metastable peak can be expected only if there are peaks of considerable intensity for both the parent and daughter ions. If a metastable peak at mass m^* can be related to the masses of two other peaks m_1 and m_2 by the above equation, it can be assumed with reasonable certainty that the fragment of mass m_2 arises by decomposition of the species of mass m_1. The part which is ejected may be lost as one particle or as two particles in two steps following in rapid succession [84]. Fragmentation requires less than 10^{-11} s and the ions spend a considerably longer time than this in the field free region, so multiple fragmentations are possible. Consideration of the molecular formula of the ejected portion usually indicates if this was lost as one or two species. The absence of a metastable peak does not exclude the formation of one ion from another because the intensity of the expected metastable peak may be low for various reasons.

Metastable peaks are usually broad with a Gaussian shape and some appear with a distinctly flat top. The factors which influence their shape have been discussed by Beynon and Fontaine [85]. Fragmentation of the parent ion is not an energetically neutral process as was supposed in the simple theory of metastable peaks. Energy is released in this process and since the laws of conservation of energy and momentum apply, this additional energy appears as kinetic energy of recoil between the two particles formed. Recoil may take place at an angle to the line of flight. It results in a spread of velocity for the daughter ions and, in consequence, the metastable peaks are broadened.

b. Third field-free region. Because the third field-free region lies after all focusing has occurred, the parent and daughter metastable ions formed in this region are brought to the same focus at the collector with the m/e value of the parent ion. The metastable ions possess less translational energy than any normal ions formed prior to the ion acceleration chamber and this difference can be used as a means of separating normal from metastable ions. In the form of detector due to Daly et al. [87a] (Fig. 25), a glass scintillator, spluttered with aluminium so as to give an electrically conducting surface, is held in a metal disc B in front of a photomultiplier tube. The aluminium coating is sufficiently thick to stop ions but not electrons. The potential of B is maintained at some value V_B while V is the ion beam accelerator potential. When $V_B > V$, all the ions are repelled and fall onto a collector plate, emitting secondary electrons. These electrons are accelerated by V_B and cause the scintillator to emit light which is detected by the photomultiplier tube. In this mode of operation with $V_B > V$, the detector records a normal mass spec-

234

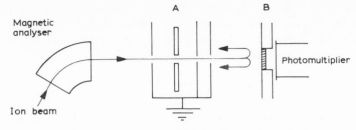

A B

Magnetic
analyser

Photomultiplier

Ion beam

Fig. 25. Schematic diagram of the Daly detector for metastable ions. A and B are repeller plates and B holds a glass scintillator disc.

trum. When $V_B < V$, normal ions (mass m_1) are not recorded but daughter ions (mass m_2) can be detected provided that $V_B > Vm_2/m_1$. Daughter ions formed in the third field-free region are detected at an apparant mass of m_1. Those formed in the second field-free region are detected at an apparent mass of m_2^2/m_1.

The various modes of decomposition of a parent ion, m_1, can be found by first adjusting the spectrometer to pass this ion. V_B is then made less than V so that the parent ion is not recorded but daughter ions formed in the third field-free region will still be recorded. The voltage, V_A, on retarding grid A is then increased slowly and the photomultiplier output monitored. Daughter ions, m_2, will be removed by grid A when $V_A > Vm_2/m_1$ so that steps are formed in the plot of photomultiplier output against V_A which allow identification of the masses of the daughter ions. Ions which pass through the box containing A will have their original kinetic energy restored since both the entrance and exit slits are earthed.

c. First field-free region. The first field-free region lies between the ion accelerator and the electrostatic analyser of a double-focusing instrument. It does not exist in a single-focusing instrument. Metastable ions formed in this region are rejected by the electrostatic analyser when this is operated in the conventional manner since they have a lower energy than normal ions. By use of what is termed a *defocusing technique*, it is possible to eliminate the normal mass spectrum and record only metastable ions formed in the first field-free region.

In one defocusing technique [86a], the accelerator voltage V, the electrostatic deflector voltage E and the magnetic field H are adjusted to focus the daughter ion m_2 of interest. Now, the accelerator voltage is increase to a value $(m_1/m_2)V$ when metastable ions of mass m_2 produced in the first field-free region by fragmentation of ions of mass m_1 will be recorded. The parent ions, m_1, leave the accelerator with a velocity v given by the expression

$$\frac{em_1 V}{m_2} = \tfrac{1}{2}m_1 v^2$$

which rearranges to

$$eV = \tfrac{1}{2}m_2 v^2$$

Thus, after fragmentation, the metastable daughter ion travels on with the same velocity and it has the correct energy and momentum to be focused by both the electrostatic and magnetic deflectors. In practice, V is continuously varied so that a spectrum of metastable transitions which give rise to m_2 can be recorded. This method for recording metastable transitions is slow if all the spectrum is to be covered, but it uniquely relates daughter and parent ions.

In a second defocusing technique [86b], a detector is placed in the ion beam at the focus of the electrostatic analyser. When V and E are set for normal operation, this detector records the total ion current. Metastable ions formed in the first field-free region by fragmentation of an ion of mass m_1 have the velocity, v, of the parent ion which was defined by the equation

$$eV = \tfrac{1}{2}m_1 v^2$$

but a mass of m_2. When E is reduced to a value of $(m_2/m_1)E$, the metastable ion will travel along a path of radius r_e given by the equation (see p. 185).

$$\frac{m_2}{m_1} Ee = \frac{m_2 v^2}{r_e}$$

which rearranges to

$$Ee = \frac{m_1 v^2}{r_e}$$

Elimination of v between these equations gives

$$r_e = \frac{2V}{E}$$

which is the focal radius of the electrostatic analyser under normal

236

operation. Thus the metastable ion m_2 has been brought into focus. When E is continuously varied from its normal value to zero, the recorder registers all the metastable transitions generated in the first field-free region. Only the value of m_2/m_1 for each transition can be obtained, i.e. daughter and parent ions are not uniquely related, but this is a rapid method for recording all metastable transitions.

More recently, methods which involve the simultaneous scanning of both the acceleration voltage and the electrostatic analyser potential have been used to characterise fragmentation processes. If the instrument is first set for normal operation focused on a particular parent ion, then scanning V and E in such a way as to keep the ratio E^2/V constant gives a spectrum of all the daughter ions from the chosen parent ion which have been formed in the first field-free region [87b]. This same result can also be obtained by scanning the magnetic and electrostatic fields so that the ratio H/E is constant [87c]. The second method of scanning gives the more satisfactory results but it is limited to spectrometers which are fitted with a Hall probe and the necessary circuitry for accurate control of magnetic field.

If the spectrometer is first set for normal operation and focused on a particular daughter ion, then scanning H and E in such a way that the ratio H^2/E is constant gives a mass spectrum showing peaks due to the fragmentation of various parent ions to this particular daughter ion [87c].

(5) Alternative ionisation techniques

Mass spectra of organic molecules are most easily obtained using electron bombardment ionisation at 70 eV. The resulting spectra are characteristic of the organic compound, only minor variations in ion intensities are encountered between one instrument and another, and, in general, the spectra are relatively insensitive to small changes in the operating conditions. A large body of fragmentation data is available for purposes of comparison.

The disadvantage of these mass spectra from the point of view of structural investigations is the wealth of detail they contain which may obscure the major fragmentation pathways. Ionisation techniques which result in simplified spectra have been developed and the more important of these are discussed here.

a. Low-energy electron bombardment ionisation. Reduction of the accelerator voltage of the ionisation electron beam from 70 eV to about 10—15 eV results in much less energy being transferred to the molecular ion in the ionisation process. In this situation, the low-energy fragmentation processes gain greater prominence and the resulting mass spectrum is simplified. This is at the expense of a decrease in the total ion current for a given sample concentration. These low-energy fragmentations are usually the fragmentations most readily interpretable in terms of the molecular structure. The simplification in the spectrum which results can be seen (Fig. 26) in the mass spectrum of genipin [88] taken at 70 eV and 10 eV.

b. Chemical ionisation. In this ionisation technique, a mixture of the substrate with a great excess of either helium or methane is subjected to bombardment with electrons of energy often in excess of

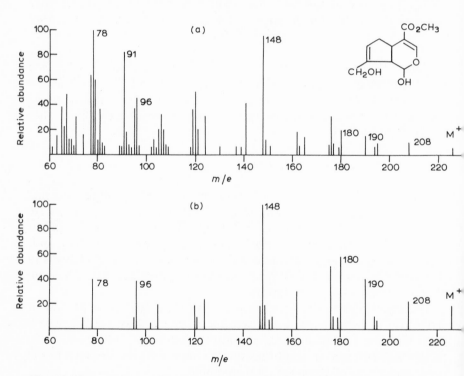

Fig. 26. Mass spectrum of genipin, $C_{11}H_{14}O_5$. (a) at 70 eV; (b) at ca. 10 eV.

238

100 eV. All the initial ionisation occurs on the carrier gas molecules and any ions which are formed rapidly loose their excess energy by collision with carrier gas molecules. When helium [89] is the carrier gas, the ionisation process on substrate molecules is one of charge transfer to helium ions which results in the formation of a substrate molecular ion of low internal energy. The fragmentation which follows is mostly due to low-energy processes and so a simplified mass spectrum results which is similar to the spectrum obtained by low-energy electron bombardment ionisation.

When methane [89] is used as the carrier gas, the principle ion formed in the plasma is CH_5^+, while $C_2H_5^+$ and $C_3H_5^+$ are also present. These ions interact with the substrate. The CH_5^+ ion protonates the substrate molecule to give an $(M + 1)^+$ ion and this can lose hydrogen to give and $(M - 1)^+$ ion. These two ions derived from the substrate molecule are termed *pseudo molecular ions* and usually they give rise

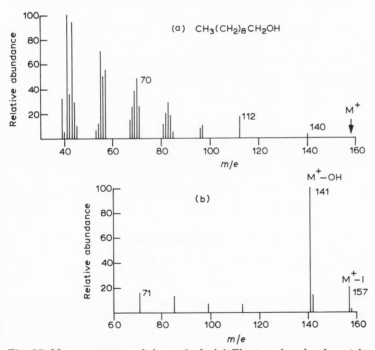

Fig. 27. Mass spectrum of decan-1-ol. (a) Electron bombardment ion source at 70 eV; (b) chemical ionisation using isobutane. An ion peak occurs at m/e 215 (20%) due to the associated ion. $C_{10}H_{21}\overset{+}{O}\overset{\diagup H}{\diagdown C_4H_9}$

References pp. 268—273

to the two most intense peaks in the mass spectrum. The ionisation process results in the formation of ions of low internal energy so that fragmentation reactions are limited. Fragmentation of pseudo molecular ions proceeds according to the influence of bond energy and ion stabilisation considerations as for fragmentation under electron bombardment. The probability for fragmentation of a bond is increased if this gives rise to a stabilised ion or a stable molecule. Since this chemical ionisation process is intrinsically different from electron bombardment ionisation, the two types of mass spectra, derived from the same molecule, can have entirely different fragmentation patterns. These differences are illustrated in the mass spectra of decan-1-ol [97] (Fig. 27).

Ions at $(M + 29)^+$ and $(M + 41)^+$ can be formed, usually in low abundance, when methane is used as the carrier gas. These result by addition of $C_2H_5^+$ and $C_3H_5^+$, respectively, to the substrate molecule. Other low molecular weight alkanes, ethane and isobutane [89], have been used in place of methane as the carrier gas. These also lead to protonation of the substrate molecule as the primary ionisation route. In these cases, the species responsible for protonation has a lower energy than CH_5^+ from methane so that the pseudo molecular ion is formed with a lower internal energy. This results in an even more simplified fragmentation pattern.

Hydrocarbons. Alkanes are protonated on a σ-bond by attack at the point of highest electron density along the bond to give an $(M + 1)^+$ ion with a five co-ordinate carbon atom [90]. The principle mode of decomposition of this ion is by loss of hydrogen to give a carbonium ion $(M - 1)^+$ and the latter is usually the strongest ion peak in the mass spectrum [91]. Fragment ions of the general formula $C_nH_{2n-1}^+$ comprise the remainder of the spectrum. The preferential cleavage of electron bombardment mass spectra, is not so marked. Cycloparaffins [92] give rise to strong ions at $(M + 1)^+$ and $(M - 1)^+$ and further fragments of general formulae $C_nH_{2n+1}^+$ and $C_nH_{2n-1}^+$ are formed.

Alkenes [93] are protonated at one of two sites, either on the π-bond to give a strong $(M + 1)^+$ ion which on fragmentation yields a series of daughter ions of formula $C_nH_{2n+1}^+$, or on the allylic σ-bond. Protonation at the latter site followed by loss of hydrogen gives the $(M - 1)^+$ ion which is also usually strong. This ion is the allyl carbonium ion and further fragmentation yields a series of daughter ions of formula $C_nH_{2n-1}^+$.

240

The π-system in aromatic molecules provides a relatively strongly basic site for protonation so an intense $(M + 1)^+$ ion peak is observed in their mass spectra [94]. Aromatic rings are too stable to undergo fragmentation on chemical ionisation. Elimination of a side chain can, however, occur if the resulting carbonium ion is stabilised and a common example of this is the loss of a *tert*-alkyl side chain such as *tert*-butyl from *tert*-butylbenzene.

$$\underset{m/e\ 135}{\overset{H^+}{\text{C}_6\text{H}_5\text{C(CH}_3)_3}} \longrightarrow \underset{}{\text{C}_6\text{H}_6} + \underset{m/e\ 57}{(CH_3)_3C^+}$$

Alkylbenzenes show little tendency to protonate on the side chain, which is a weakly basic site. Protonation here would be a principal route leading to formation of the benzyl ion $C_7H_7^+$. Thus, this $C_7H_7^+$ ion does not dominate the chemical ionisation mass spectra of alkylbenzenes as it does the electron bombardment mass spectra.

Functional groups with an electron lone pair. Oxygen-, sulphur- and nitrogen-containing functional groups and also the halogens are readily protonated on the hetero atom possessing a lone pair of electrons. Molecules containing these functional groups give a strong $(M + 1)^+$ ion peak and a peak due to the fragment ion formed by loss of the protonated functional group. The probability of this fragmentation is dependent on the stability of the resulting carbonium ion and the strength of the bond which is broken.

Halogen-containing alkyl compounds [95] readily lose the hydrogen halide to form an alkyl carbonium ion which is usually the dominant peak in the mass spectrum. This is also the case for fluorocarbons [96] and perfluorohexane, for example, shows an abundant fragment at m/e 319.

$$n\text{-}C_6F_{14} + CH_5^+ \rightarrow \underset{m/e\ 319}{C_6F_{13}^+} + HF + CH_4$$

Alcohols [95,97] as well as their ethers [98] and acetates [99] show a similar loss of the protonated functional group as a neutral molecule to leave a carbonium ion. There is a much lower probability for loss of the protonated amino group [98].

The protonated ester group can undergo cleavage by two routes. One reaction yields a carbonium ion from the alcohol part of the molecule and the second yields an acylcarbonium ion.

$$R^1-\underset{\underset{H}{|}}{\overset{\overset{O}{\|}}{C}}-\overset{+}{\underset{}{O}}-R^2 \Bigg\langle \begin{array}{l} R^1-CO_2H \ + \ [R^2]^+ \\[2em] R^1-CO^+ \ + \ R^2OH \end{array}$$

Protonation of an amide leads to cleavage to give, principally, the acylcarbonium ion.

$$R^1-\underset{\underset{H}{|}}{\overset{\overset{O}{\|}}{C}}-\underset{\underset{H}{|}}{\overset{\overset{H}{|}}{N}}-R^2 \quad\longrightarrow\quad R^1-CO^+ \ + \ R^2NH_2$$

Two bond fragmentation reactions. Some of the reactions found in chemical ionisation mass spectrometry are analogous to the McLafferty rearrangement in electron bombardment mass spectrometry. This type of fragmentation is found particularly in the chemical ionisation mass spectra of esters [99] where the first ionisation step is protonation of the alcohol oxygen. Fragmentation then occurs according to

$$R-\overset{H}{\underset{\underset{H}{|}}{\overset{|}{C}}}\overset{CHR}{\underset{CH_2}{\cdots}} \quad\longrightarrow\quad R-C\overset{OH}{\underset{\overset{+}{O}H}{\diagdown}} \ + \ RCH_2=CHR$$

Ketones do not exhibit the McLafferty rearrangement in chemical ionisation mass spectrometry. This is because the carbonyl function is protonated in the methane plasma giving an abundant $(M + 1)^+$ peak and this deactivates the carbonyl lone pair as a trigger for the McLafferty rearrangement.

c. Field ionisation and field desorption. Field ionisation [14,100] occurs when an atom or molecule is acted upon by an electric field of the order of 10^7-10^8 V cm^{-1}. Such fields occur in the vicinity of sharp tips and fine wires held at potentials of the order of 10^4 V. Ionisation is confined to the tips of microneedles which cover the surface of the emitter and prior activation of the emitter with benzonitrile [101] leads to a large increase in the active emitter surface.

The removal of an electron from the molecule occurs by the quantum mechanical tunnelling effect and ions are produced with a low internal energy content. The resulting mass spectra show an intense

242

molecular ion or $(M + 1)^+$ ion peak and relatively few fragments ions. Favoured fragmentation processes are those which involve the cleavage of weak bonds and the formation of stabilised carbonium ions.

The substrate may be present in the gas phase at low pressure but it is also possible to introduce the substrate as a coat on the emitting surface by prior application of a solution of the sample. When the coated emitter is raised to a high potential and gently heated, the absorbed molecules tend to migrate towards the tips of any microneedles where they are ionised and desorbed. This *field desorption* technique is particularly applicable to involatile samples. The part of the internal energy of the resulting ions which arises from the temperature of the source is reduced by this technique so that strong molecular ions can be obtained from unpromising substrates. A comparison of the field ionisation and electron bombardment ionisation

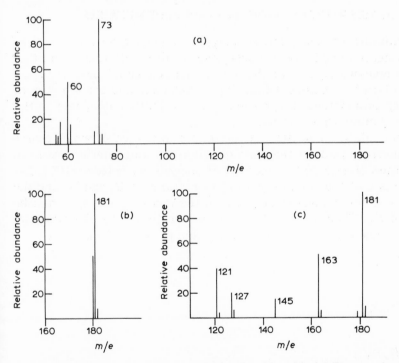

Fig. 28. Mass spectrum of glucose, $C_6H_{12}O_6$, $M = 180$. (a) Electron bombardment ion source at 70 eV; (b) field desorption technique; (c) field ionisation technique showing ions $M^+ + 1$-$n(H_2O)$ where $n = 0, 1, 2$ and 3.

References pp. 268—273   243

mass spectra of glucose [102] (Fig. 28) illustrates this point. Field ionisation and field desorption can generate the molecular ion and large fragment ions from high molecular weight compounds of biological importance such as peptides [103] and oligosaccharides [104].

Ionisation by the field desorption technique usually gives an abundant $(M + 1)^+$ ion peak. This ion results from protonation of a basic functional group in the molecule either by water which is also absorbed on the emitter, or by a second molecule of the substrate. Even when the sample is originally introduced in the gas phase, field ionisation can give rise to a strong $(M + 1)^+$ ion if the molecule contains a basic group.

4. Analysis of mixtures of gaseous and volatile compounds

(A) ANALYSIS WITHOUT PRIOR SEPARATION OF MIXTURES

A mixture of gases can be analysed by mass spectrometry if the components exhibit spectra significantly different from each other in the cracking pattern or the relative abundance of ions. The spectrum of each component must be reproducible and the ion current directly proportional to pressure over the limits anticipated in the mixtures under examination.

If the mixture consists of two components, each of which gives a characteristic peak in the mass spectrum without interference from the other, then a simple method of analysis is available [105]. Let the masses of the components in the mixtures be M_a and M_b and the heights of the characteristic peaks in the mixture mass spectrum be H_a and H_b. Then, if S_a and S_b are the sensitivities for the two pure compounds in units of recorder deflection per unit of mass

$$H_a = M_a S_a$$

and

$$H_b = M_b S_b$$

Thus

$$\frac{M_a}{M_b} = \frac{H_a}{H_b} \frac{S_b}{S_a}$$

$$= K \frac{H_a}{H_b}$$

244

where K is a constant the value of which can be determined by calibration with one or more mixtures of known composition. This method is relatively simple and does not depend on the pressure measurement which is necessary to deal with mixtures having overlapping spectra of components. It is particularly applicable to spectra obtained by chemical and field ionisation methods which show few fragment peaks so that the possibility of overlapping peaks is reduced. A known amount of an internal standard can be added to assist in the analysis of multicomponent mixtures provided no overlapping of peaks occurs. Then, the ratio of the mass of each component relative to the internal standard can be determined.

If one of the peak heights, say H_b, in the mixture mass spectrum contains a small component due to a fragment from compound a, this can be allowed for in the calculation. The ratio, R, of peak heights H_b and H_a in the mass spectrum of the pure compound a will be known. When H_a has been measured for the mixture mass spectrum,

TABLE 8

Partial mass spectra of some cyclohexane derivatives

m/e	Relative abundance		
	Cyclohexane	Cyclohexene	1,3-Cyclohexadiene
85	4.87		
84	74.9		
83	4.61	2.45	
82	0.22	38.0	
81	0.47	10.8	3.66
80	0.08	0.93	57.0
79	0.53	6.13	100
78	0.19	1.35	18.6
77	0.79	4.92	37.8
69	22.4	0.12	0.01
68	1.84	5.38	0.05
67	3.26	100	0.92
57	4.91	0.04	0.00
56	100	0.45	0.00
55	35.0	6.31	0.20
54	7.07	80.0	5.62
Symbol for compound	A	B	C
Sensitivity for base peak divisions/μm	$S_a = 41.9$	$S_b = 44.6$	$S_c = 41.2$

RH_a can be subtracted from the value of H_b for the mixture leaving the contribution to this peak from compound b alone which is then inserted into the calculation of mixture composition.

The more rigorous method for analysis of complex mixtures is best illustrated with a specific example. This method is used in petroleum analysis [106], for the analysis of many other mixtures of organic compounds [107], and for the analysis of gaseous mixtures such as those from fractionation of liquid air [108] and some polluted atmospheres. It is very suitable for the routine analysis of mixtures of similar composition where the calculations are best performed by a small on-line computer. Calibration spectra of the individual components of the mixture are recorded first and then the mass spectrum of the mixture is recorded under the same conditions.

The analysis of a mixture of cyclohexane, cyclohexene and 1,3-cyclohexadiene is taken as an example. The spectra of these hydrocarbons are given in Table 8 where the most abundant ion in each spectrum is given the intensity of 100. Spectrometer sensitivities in chart divisions per micron of sample pressure are given at the bottom of the table. This data was taken from reference tables but for the most accurate work it is best to prepare calibration spectra since mass spectra are dependent to some degree on individual spectrometer design. The spectrometer sensitivity is usually calibrated with respect to the sensitivity to the peak at m/e 43 in the spectrum of butane so that compensation can be made for differences in sensitivity between individual spectrometers.

The minimum of information necessary to determine the composition of a mixture of three components is the heights of three peaks in the mixture spectrum. The most suitable peaks are at m/e 79, 67 and 56. Simultaneous equations can be set up with the meaning of the symbols defined in Table 8.

Height of peak m/e 79 = $0.0053AS_a + 0.0613BS_b + 1.00CS_c$
Height of peak m/e 67 = $0.0326AS_a + 1.00BS_b + 0.0092CS_c$
Height of peak m/e 56 = $1.00AS_a + 0.0045BS_b$

where A, B and C are the partial pressures of compounds A, B and C. Solution of these equations then gives the value of AS_a, BS_b and CS_c. More than three simultaneous equations can be set up and solved by one of the standard computer methods so as to eliminate random errors. The final values of AS_a, BS_b and CS_c are then divided by the appropriate sensitivity factors given in the table to give the partial pressures of A, B and C in the mixture. Normalising to give a total pressure of 100 results in the mole-percentage composition of the mixture.

The method depends upon the mass spectrum of a component being highly reproducible both for cracking pattern and sensitivity over a long period. A common cause of poor reproducibility is the condition of the filament which serves as electron source in the ionisation chamber. If the filament is made of tungsten, then it has to be suitably carbonised as discussed on p. 179. Rhenium which is not subject to carbonisation is frequently used as the filament material.

The sample vapour is usually held in a vessel of up to 2 l capacity from which it flows through a leak into the spectrometer. Sample flow through the leak is partly viscous and partly molecular. Over the region of viscous flow, the flow rate and hence the spectrometer sensitivity at a given pressure difference is dependent on the viscosity of the mixture. A change in viscosity should affect the sensitivity factors for the various components of the mixture to the same relative degrees. However, for the most reliable results, it is usual to determine the sensitivity factors by analysis of a prepared mixture of known composition. Molecular flow along the low pressure end of the leak causes a fractionation of material according to mass. The composition of the sample must change during the time it is leaked into the spectrometer and a relatively large volume of sample is used so that this change becomes negligible.

(B) ANALYSIS WITH GLC–MS COMBINATION

Mass spectrometry and gas—liquid chromatography are separately both very powerful tools for the analysis of mixtures of volatile compounds. Various approaches to the combination of these two techniques have been made since about 1960. Gas—liquid chromatography is used to separate the components of a mixture and, even when capilliary columns are used, the amount of sample available is still more than adequate to provide a mass spectrum. The separated specimens are transferred in the gas stream to the spectrometer.

This combination of two instruments affords a choice of ways in which the gas—liquid chromatogram can be recorded. The chromatograph effluent can be split and part passed to a normal gas—liquid chromatograph detector. Alternatively, a record of the chromatograph can be made from the response of the mass spectrometer. The total ion current can be recorded and this will rise in proportion to the amount of material which is eluted from the chromatography column. This method of recording will detect all compounds eluted from the column. With either of these modes of detection, a mass spectrum can be obtained for every compound as it is eluted from the column by activating the mass scanning system. Clearly, the scanning time should be commensurate with the half-width of the chromatography peak under investigation and this may necessitate rapid scanning. Mass spectra obtained in this way are used for qualitative analysis of the component of the chromatography peak.

In a third method of recording a chromatogram, the mass spectrometer can be focused on a particular mass number and the intensity of this ion beam recorded as the gas—liquid chromatogram proceeds. If the mass spectra of the components of an overlapping pair of chromatography peaks have been recorded previously, it is usually possible to choose an ion which is present only in the mass spectrum of the one component of interest. In this way, the selectivity of both instruments is utilised to record only one pair of overlapping chromatography peaks.

In choosing a stationary phase for the chromatography column, some consideration has to be given to the consequences which result from bleeding of this phase. Bleeding gives rise to a background mass spectrum. This can be subtracted from the observed spectra but it may obscure important details if excessive bleeding occurs.

Whatever interface between the two instruments is chosen from the more important types discussed here, the mixture entering the mass spectrometer will contain a proportion of the carrier gas. A carrier gas with a high ionisation potential must be chosen so that the acceleration voltage for the electron bombardment ion source can be adjusted to a level where organic molecules will be ionised in preference to the carrier gas. Commonly used carrier gases have the ionisation potentials helium 24.46 eV, hydrogen 15.6 eV, nitrogen 15.5 eV, while the ionisation potentials of most organic molecules are in the region of 7—11 eV. The most satisfactory carrier gas is helium and this is used in conjunction with an electron bombardment ion source with an acceleration potential of 20—25 V.

Specimen flow into the mass spectrometer is unlikely to exceed $0.1—0.3$ ml min^{-1} at atmospheric pressure. This volume flow rate becomes faster at lower pressures. It is important that the take-off from the column to the spectrometer is operated at low pressure so that the dead space in this region is swept rapidly. Otherwise, mixing of closely spaced chromatography fractions will occur in this dead space and there could also be a delay of many seconds between a substance leaving the column and entering the spectrometer.

(1) Interfacing without a molecular separator

The outflow from a gas chromatograph may be connected directly to the mass spectrometer. However, spectrometers fitted with only a single stage pumping system to maintain a high vacuum in the ion flight path can seldom accept a higher rate of gas flow than 0.1—0.3

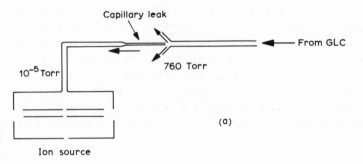

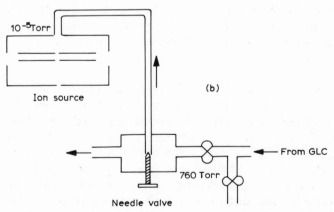

Fig. 29. Simple connectors between gas—liquid chromatography and mass spectrometer systems. (a) With capillary leak; (b) with needle valve.

References pp. 268—273   249

ml min^{-1} at atmospheric pressure. Some spectrometers are fitted with a differential pumping system where one pump evacuates the ionisation chamber and a second pump evacuates the ion flight path. Such instruments can accept a flow rate of 1—2 ml min^{-1} of gas at atmospheric pressure.

One group of workers [109] has applied the total outflow from a capillary column gas chromatograph to the mass spectrometer through a long length of fine capillary tubing to maintain a pressure difference between the spectrometer and the operating gas chromatograph.

Fortunately, the mass spectrometer is a very sensitive instrument so that only a part of the outflow from a gas chromatograph need be applied to the spectrometer. When a flow divider which gives only 1% transfer of this outflow to the spectrometer is used, injection of 10^{-3} g of compound onto the chromatography column will result in delivery of 10^{-5} g to the spectrometer and this is sufficient to give a spectrum from a component present to the extent of only 1% of the total injection. Very simple flow dividers have been built [110] for the transfer of small amounts to the spectrometer and these are illustrated in Fig. 29. They are all designed so that the connection to the spectrometer operates at reduced pressure when the volume flow rate is larger.

(2) Interfacing with a molecular separator

Several types of molecular separator, using a variety of principles, have been used to fractionate the chromatograph exudate so as to increase the concentration of organic material in the gas stream supplied to the mass spectrometer.

a. Fractionation of gases in an expanding jet. The differential rates of diffusion of gases in an expanding jet can be used to accomplish fractionation of gaseous mixtures. The commonly used, two-stage Ryhage separator [111] is shown diagrammatically in Fig. 30. It is usually constructed in stainless steel and a similar separator has been constructed in glass [112]. Effluent gas from the chromatograph passes through an orifice and expands into the evacuated area between this and a second orifice. For each component of the gas stream, diffusion flow in this region of expansion is a function of the molecular weight and is proportional to the diffusion coefficient. A

250

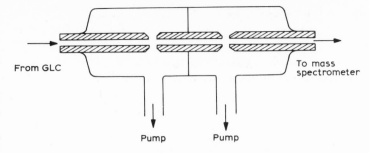

Fig. 30. The Ryhage jet molecular separator.

separation effect occurs according to molecular weight and the core of the gas stream is enriched in the heavier component. The concentrated effluent passes through a second separation and on into the spectrometer.

This type of separator gives excellent enrichment with effluent flow rates of 30 ml min^{-1} and spectrometer flow rates of 0.2 ml min^{-1} under atmospheric pressure. The average pressure in the expanding jet must be high enough to ensure a viscous flow condition in the jet core. If this condition is lost, the separation factor drops rapidly [113]. The separator design naturally lends itself to a fast throughput of effluent and hold-up by dead volume is absent. In general, use of this separator does not lead to chromatography peak broadening. Particular care must be taken to see that the jets do not become clogged by fine particles.

b. Enrichment by effusion through fine pores or slits. The principle of differential effusion through a frit has been used in several designs of separator. Watson and Biemann introduced the glass frit separator (Fig. 31) in 1964 [114]. This consists of an ultrafine porosity sintered glass tube enclosed in a vacuum envelope with glass capillaries at the entrance and exit. When the pressure of gas in the sintered glass is sufficiently reduced for molecular flow to occur, the rate at which a gas effuses to the vacuum will be inversely proportional to the square root of the molecular weight and directly proportional to the partial pressure. This results in a faster rate of effusion for the carrier gas so that the flow entering the spectrometer is enriched in the higher molecular weight components.

Several types of metal and ceramic porous separators have been

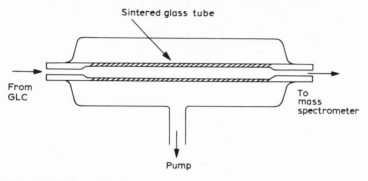

Fig. 31. Watson—Biemann effusion separator.

proposed which employ the principle of the Watson—Biemann separator. Krueger and McCloskey [115] have constructed a separator with a replaceable stainless steel fritted tube, and silver frits [116] have also been used. These metallic separators are less fragile than the original glass design.

Separators of the Watson—Biemann type can be designed to give maximum efficiency at only one flow rate. To give more flexibility, Brunnée et al. [117] devised an effusion-type separator with a variable conductance effusion path. The apparatus (Fig. 32) consists of a flat cover plate, mounted on a micrometer screw, which can be closed over two annular rings, about 2 cm diameter, raised on a base

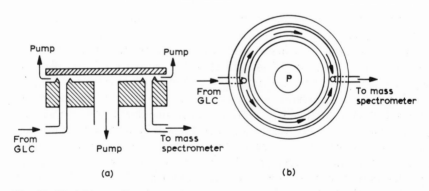

Fig. 32. Variable conductance separator of Brunnée. (a) Sectional view; (b) top view.

252

plate. The whole is mounted in a vacuum jacket. When the cover plate is in a closed position, the chromatograph effluent passes round the annulus into the mass spectrometer. Raising the base plate forms a narrow slit through which molecular flow can occur and so the helium carrier gas effuses preferentially. The slit width can be varied to suit the effluent flow rate from various types of chromatography column.

c. *Preferential diffusion through a semipermeable membrane.* In one commonly used separator of this type, use is made of the preferential solubility of organic compounds in silicone rubber which is impervious to helium. The chromatograph effluent flows past a silicone rubber diaphragm which is supported by a coarse metal frit and the far side of the diaphragm is connected to the spectrometer inlet system. The diaphragm must be held at a sufficiently high temperature so that the organic compound diffuses rapidly through it but the temperature must not be so high as to make negligible the solubility of the organic compound in the silicone rubber. Several studies of the optimum conditions have been made. The optimum temperature for the diaphragm is usually about 70° below the boiling point of the material to be conducted. Too low an operating temperature leads to chromatography peak broadening and tailing in the effluent which reaches the spectrometer due to slow diffusion of the organic material through the diaphragm.

The advantage of the silicone rubber separator is cheapness and simplicity. As well as the disadvantage discussed above, the membrane cannot be operated above 250°C. Rupture of the membrane causing a surge of gas to the spectrometer is always a possibility.

At temperatures above 250°C, Teflon is permeable to helium. This has been utilised [118] in a design of separator where the effluent gas is passed through a long thin-walled Teflon capillary held in an evacuated container and then to the mass spectrometer. Although this Teflon separator is easily constructed, it has not been widely used. Peak distortion and time delays have been reported and the Teflon tube fractures relatively easily at the high temperatures necessary.

Palladium—silver alloys are highly permeable to hydrogen and virtually impervious to other gases. This suggested the use of palladium in a semipermeable membrane separator [119] when hydrogen is used as the carrier gas. Such separators operate at temperatures in the range 200—250°C and can be constructed to have a hydrogen flux in excess of 60 ml min⁻¹ at atmospheric pressure. This separator suffers one serious objection to general use in that catalytic reduction of the unknown may occur during its passage along the membrane surface and complete reduction of easily reduced organic molecules has indeed been observed [120]. High concentrations of sulphur and iodine compounds completely inhibit the palladium.

5. Elemental analysis of involatile inorganic samples

(A) SPARK IONISATION

(1) Introduction

Sparking of electrically conducting samples in a vacuum is of general application for the production of positive ions from all elements in the periodic table. Many sparking devices have been employed but the high voltage ratio frequency spark is the one most generally used. It produces ions with a very wide energy range centred on the applied spark potential which is normally around 19—20 kV. The width of the energy distribution curve at 10% of the maximum ion energy is of the order of 2000 eV. Such an ion beam can only be analysed using a double-focusing mass spectrometer and, to achieve good resolution, it is necessary to limit the energy range of the ions which pass from the electrostatic to the magnetic analyser. The spark process produces an ion beam of widely fluctuating intensity and often the spark becomes extinguished and requires manual relighting. A spectrometer of the Mattauch—Herzog design can accommodate these proporties of the spark source and such spectrometers were employed in the period from 1960 when the technique was developed [121]. It has the facility for photographic recording of the entire mass spectrum and thus automatically gives a spectrum which is integrated over a period of exposure.

From about 1970, more stable sparking systems have been developed which will run continuously for several hours [18]. Systems for electrical recording have also be developed where the ratio of the individual ion current to the total ion current is continuously monitored and recorded [18,122]. This overcomes the problem of fluctuating ion currents provided that the spectrum is recorded at

254

relatively slow speeds of the order of 10 min or slower for the full mass range from 7 to 238 a.m.u. Spark ionisation with electrical recording requires a double-focusing mass spectrometer but this need not be of the Mattauch—Herzog design. Indeed, another design of mass spectrometer was used by Hipple and his coworkers [123] in 1951 when they first pointed out the advantages of electrical recording.

Manufacturers of instruments for the mass spectrographic analysis of inorganic solids also provide the means for analysis of selected areas on the surface of the sample. The sample is viewed through a microscope and a high-energy ion beam can be focused onto an interesting area. The material at the beam focus is vaporised and ionised and the resulting ions are conveyed to the mass spectrometer.

The spark ionisation, mass spectrometric method of analysis has two very great advantages. Firstly, the relative sensitivity for all the elements in the periodic table differ from each other by a factor of 3—5 at the most [19]. Secondly, the electrical detector response is linear over a range of several orders of magnitude of concentration. The results from photographic recording need to be processed in one extra step to give a result which is proportional to concentration, again over several orders of magnitude [124]. From the mass spectrum, it is possible to obtain an order of magnitude estimation of the concentration of the elements present. More accurate analytical figures can be obtained after comparison with the spectrum from a single standard sample.

(2) Sample preparation and ionisation procedure

The sample must be electrically conducting and fabricated into two small rods which are clamped in the spark source. The source is evacuated and a radio frequency spark is struck between the two electrodes with an applied potential in the range of 10—40 kV.

a. Metallic samples. Massive samples can be machined to rods of a suitable size. Samples of metallic powders can be compressed into rods, compressed onto the tip of a graphite rod and then used as one electrode, or mixed with graphite and compressed into rods as discussed for non-conducting samples. An internal standard in the form of one metallic element can be added if required and the alloy melted to give a homogeneous mixture. This step is sometimes used to aid trace analysis.

Liquid metals, gallium and mercury, have been supported for analysis on copper strips at liquid nitrogen temperatures [125].

b. Non-conducting solids. If the sample can be reduced to a fine powder, then the powdered material is mixed with graphite or silver powder in an approximately equal weight ratio and compressed into rods [126]. Particular care has to be taken to prevent contamination in the grinding process. Very hard and inert samples can be powdered in a hard steel pestle and mortar. Finally, this powdered sample is washed with high-purity acid to dissolve any steel particles and then rinsed with high-purity water.

The sampling of biological materials for heavy metal contaminants is relatively easily accomplished by burning the sample and then mixing the ash with graphite or silver powder and compressing the resulting mixture into rods. Graphite and silver are chosen as matrix materials because they can be obtained very pure and free from trace elements.

Because of the very real danger of contamination during a grinding process, techniques have been developed for the spark analysis of chips of non-conducting materials. These chips have been pressed into the end of a graphite, silver, or aluminium electrode. Some workers have used silver with epoxy resin as a cement but the resulting mass spectrum contained hydrocarbon fragments derived from the cement. Fortunately, the mass spectrometer has sufficient resolving power to separate metal ions from hydrocarbon fragments of the same nominal mass. Small single crystals of non-conducting material have been mounted in gold foil by pressing and then sparked with a gold counter electrode.

c. Liquids. Liquid samples are absorbed onto pure graphite powder, dried and then compressed into rods. A known amount of an internal standard, such as a solution containing the silver ion, can be added to the liquid sample, if desired, before it is absorbed onto the graphite.

(3) Qualitative analysis by mass spectrometry

The types of ion generated by sparking and the qualitative analysis of spectra are considered here. The methods which are available for recording the spectrum are considered in relation to quantitative analysis.

Spark ionisation produces monoatomic ions from all the elements of the periodic table. Because of the high-energy spark which is used, singly and multiply charged ions result. The mass analyser focuses ions according to their m/e value so multiply charged ions are likely to appear at fractions of an atomic mass unit. In general, multiply charged ions are of lower intensity than the singly charged.

Some elements give rise to polyatomic ions but these, with one exception, are of weak abundance. The exception is carbon which, in the form of graphite, is frequently used as a matrix material. Graphite gives a series of ions of the general formula C_n^+ where n is an integer. Each of these ions has an isotope peak one mass unit higher.

Many types of molecules containing two or more elements are seen in spark spectra. These are oxides, hydroxides, dioxides, carbides, dicarbides and hydrides of the major constituent elements in the specimen. Generally, these molecular ions are of low abundance. Their abundance can be further substantially reduced by cooling the specimen with liquid nitrogen [127] and devices are available commercially for this purpose. Cooling also substantially reduces the formation of hydrocarbon ions from samples which contain organic residues.

A technique for the interpretation of the major lines in a spectrum due to mono-charged atomic ions can be considered. Each element gives rise to its characteristic isotope distribution and the isotope peaks overlap between neighbouring elements. Fortunately, with one exception, each element has an isotope at a characteristic mass number where no other element is found. In a qualitative analysis, the spectrum is first scanned to look for these characteristic isotope peaks. Next, the identification of an element is confirmed by its isotope distribution pattern. Finally, suitable ion peaks are chosen for quantitative analysis. Tables of isotope masses [32], relative abundances [34] and mass spectrum lines [128—130] have been published to assist in the spectrum analysis. A list of the mono-charged atomic ions of the natural elements is given in Table 9. During interpretation of the spectrum, it is important to beware of the occurrence of doubly charged ions from isotopes of even mass number. These ions are focused at a position $m/2$ which may be mistaken for the line due to an isotope of another element.

TABLE 9

Mono-atomic singly charged ions of the natural elements

m/e	Element and relative isotope abundance	m/e	Element and relative isotope abundance
1	$_1$H (0.99985)	41	$_{19}$K (0.0688)
2	$_1$H (1.5×10^{-4})	42	$_{20}$Ca (6.4×10^{-3})
3	$_2$He (1.3×10^{-6})	43	$_{20}$Ca (1.4×10^{-3})
4	$_2$He (1.00)	44	$_{20}$Ca (0.021)
6	$_3$Li (0.074)	45	$_{21}$Sc (1.00)
7	$_3$Li (0.926)	46	$_{22}$To (0.080)
9	$_4$Be (1.00)	47	$_{22}$Ti (0.073)
10	$_5$B (0.196)	48	$_{22}$Ti (0.738)
11	$_5$B (0.804)	49	$_{22}$Ti (0.055)
12	$_6$C (0.989)	50	$_{22}$Ti (0.054);
13	$_6$C (1.1×10^{-2})		$_{23}$V (2.3×10^{-3});
14	$_7$N (0.996)		$_{24}$Cr (0.044)
15	$_7$N (3.65×10^{-3})	51	$_{23}$V (0.998)
16	$_8$O (0.9976)	52	$_{24}$Cr (0.838)
17	$_8$O (3.7×10^{-4})	53	$_{24}$Cr (0.095)
18	$_8$O (2.0×10^{-3})	54	$_{26}$Fe (0.059);
19	$_9$F (1.00)		$_{24}$Cr (0.024)
20	$_{10}$Ne (0.909)	55	$_{25}$Mn (1.00)
21	$_{10}$Ne (3.0×10^{-3})	56	$_{26}$Fe (0.916)
22	$_{10}$Ne (0.088)	57	$_{26}$Fe (0.022)
23	$_{11}$Na (1.00)	58	$_{28}$Ni (0.679)
24	$_{12}$Mg (0.787)		$_{26}$Fe (3.3×10^{-3})
25	$_{12}$Mg (0.101)	59	$_{27}$Co (1.00)
26	$_{12}$Mg (0.112)	60	$_{28}$Ni (0.262)
27	$_{13}$Al (1.00)	61	$_{28}$Ni (0.012)
28	$_{14}$Si (0.922)	62	$_{28}$Ni (0.037)
29	$_{14}$Si (0.047)	63	$_{29}$Cu (0.691)
30	$_{14}$Si (0.031)	64	$_{30}$Zn (0.489)
31	$_{15}$P (1.00)	65	$_{29}$Cu (0.309)
32	$_{16}$S (0.950)	66	$_{30}$Zn (0.278)
33	$_{16}$S (7.4×10^{-3})	67	$_{30}$Zn (0.041)
34	$_{16}$S (0.0424)	68	$_{30}$Zn (0.186)
35	$_{17}$Cl (0.755)	69	$_{31}$Ga (0.604)
36	$_{16}$S (1.7×10^{-4});	70	$_{32}$Ge (0.205);
	$_{18}$Ar (3.4×10^{-3})		$_{30}$Zn (6.2×10^{-3})
37	$_{17}$Cl (0.245)	71	$_{31}$Ga (0.396)
38	$_{18}$Ar (6.0×10^{-4})	72	$_{32}$Ge (0.274)
39	$_{19}$K (0.931)	73	$_{32}$Ge (0.078)
40	$_{18}$Ar (0.996);	74	$_{32}$Ge (0.365);
	$_{20}$Ca (0.969);		$_{34}$Se (8.7×10^{-3})
	$_{19}$K (1.2×10^{-4})	75	$_{33}$As (1.00)

258

TABLE 9 (continued)

m/e	Element and relative isotope abundance	m/e	Element and relative isotope abundance
76	$_{34}$Se (0.090); $_{32}$Ge (0.077)	107	$_{47}$Ag (0.518)
77	$_{34}$Se (0.076)	108	$_{46}$Pd (0.267); $_{48}$Cd (8.9 × 10^{-3})
78	$_{34}$Se (0.235); $_{36}$Kr (3.5 × 10^{-3})	109	$_{47}$Ag (0.482)
79	$_{35}$Br (0.505)	110	$_{48}$Cd (0.124); $_{46}$Pd (0.118)
80	$_{34}$Se (0.498); $_{36}$Kr (0.023)	111	$_{48}$Cd (0.128)
81	$_{35}$Br (0.495)	112	$_{48}$Cd (0.241); $_{50}$Sn (9.6 × 10^{-3})
82	$_{36}$Kr (0.116); $_{34}$Se (0.092)	113	$_{48}$Cd (0.127); $_{49}$In (0.042)
83	$_{36}$Kr (0.116)	114	$_{48}$Cd (0.289); $_{50}$Sn (6.5 × 10^{-3})
84	$_{36}$Kr (0.569); $_{38}$Sr (5.6 × 10^{-3})	115	$_{49}$In (0.957); $_{50}$Sn (3.5 × 10^{-3})
85	$_{37}$Rb (0.722)	116	$_{50}$Sn (0.143)
86	$_{36}$Kr (0.174); $_{38}$Sr (0.099)	117	$_{50}$Sn (0.076)
87	$_{37}$Rb (0.278); $_{38}$Sr (0.070)	118	$_{50}$Sn (0.240)
88	$_{38}$Sr (0.826)	119	$_{50}$Sn (0.086)
89	$_{39}$Y (1.00)	120	$_{50}$Sn (0.329); $_{52}$Te (9 × 10^{-4})
90	$_{40}$Zr (0.515)	121	$_{51}$Sb (0.573)
91	$_{40}$Zr (0.112)	122	$_{50}$Sn (0.049); $_{52}$Te (0.025)
92	$_{40}$Zr (0.171); $_{42}$Mo (0.158)	123	$_{51}$Sb (0.427); $_{52}$Te (8.7 × 10^{-3})
93	$_{41}$Nb (1.00)	124	$_{52}$Te (0.046); $_{54}$Xe (9.5 × 10^{-4})
94	$_{42}$Mo (0.090)	125	$_{52}$Te (0.070)
95	$_{42}$Mo (0.157)	126	$_{52}$Te (0.187); $_{54}$Xe (8.8 × 10^{-4})
96	$_{42}$Mo (0.165); $_{44}$Ru (0.055)	127	$_{53}$I (1.000)
97	$_{42}$Mo (0.095)	128	$_{52}$Te (0.318); $_{54}$Xe (0.019)
98	$_{42}$Mo (0.238); $_{44}$Ru (0.019)	129	$_{54}$Xe (0.262)
99	$_{44}$Ru (0.127)	130	$_{52}$Te (0.345); $_{54}$Xe (0.041); $_{56}$Ba (1.03 × 10^{-3})
100	$_{44}$Ru (0.126)	131	$_{54}$Xe (0.212)
101	$_{44}$Ru (0.171)	132	$_{54}$Xe (0.269); $_{56}$Ba (9.7 × 10^{-4})
102	$_{44}$Ru (0.316); $_{46}$Pd (9.6 × 10^{-3})	133	$_{55}$Cs (1.00)
103	$_{45}$Rh (1.00)	134	$_{54}$Xe (0.104); $_{56}$Ba (0.024)
104	$_{44}$Ru (0.186); $_{46}$Pd (0.110)		
105	$_{46}$Pd (0.222)		
106	$_{46}$Pd (0.273); $_{48}$Cd (0.012)		

TABLE 9 (continued)

m/e	Element and relative isotope abundance	m/e	Element and relative isotope abundance
135	$_{56}$Ba (0.066)	169	$_{69}$Tm (1.00)
136	$_{56}$Ba (0.078);	170	$_{70}$Yb (0.030)
	$_{58}$Ce (2.5×10^{-3})	171	$_{70}$Yb (0.143)
137	$_{56}$Ba (0.113)	172	$_{70}$Yb (0.218)
138	$_{56}$Ba (0.717);	173	$_{70}$Yb (0.162)
	$_{57}$La (8.9×10^{-4});	174	$_{70}$Yb (0.318);
	$_{58}$Ce (2.6×10^{-3})		$_{72}$Hf (1.8×10^{-3})
139	$_{57}$La (0.999)	175	$_{71}$Lu (0.974)
140	$_{58}$Ce (0.885)	176	$_{70}$Yb (0.126);
141	$_{59}$Pr (1.00)		$_{72}$Hf (0.052);
142	$_{60}$Nd (0.271);		$_{71}$Lu (0.026)
	$_{58}$Ce (0.111)	177	$_{72}$Hf (0.184)
143	$_{60}$Nd (0.122)	178	$_{72}$Hf (0.271)
144	$_{60}$Nd (0.239); $_{62}$Sm (0.031)	179	$_{72}$Hf (0.138)
145	$_{60}$Nd (0.083)	180	$_{72}$Hf (0.352);
146	$_{60}$Nd (0.172)		$_{74}$W (1.3×10^{-3});
147	$_{62}$Sm (0.150)		$_{71}$Ta (1.2×10^{-4})
148	$_{62}$Sm (0.112)	181	$_{73}$Ta (1.00)
149	$_{62}$Sm (0.138)	182	$_{74}$W (0.264)
150	$_{62}$Sm (0.074)	183	$_{74}$W (0.144)
151	$_{63}$Eu (0.478)	184	$_{74}$W (0.306);
152	$_{62}$Sm (0.266);		$_{76}$Os (1.8×10^{-4})
	$_{64}$Gd (2×10^{-3})	185	$_{75}$Re (0.371)
153	$_{63}$Eu (0.522)	186	$_{74}$W (0.284);
154	$_{64}$Gd (0.022)		$_{76}$Os (0.016)
155	$_{64}$Gd (0.147)	187	$_{75}$Re (0.629);
156	$_{64}$Gd (0.205);		$_{76}$Os (0.016)
	$_{66}$Dy (5.2×10^{-4})	188	$_{76}$Os (0.133)
157	$_{64}$Gd (0.157)	189	$_{76}$Os (0.161)
158	$_{64}$Gd (0.249);	190	$_{76}$Os (0.264);
	$_{66}$Dy (9×10^{-4})		$_{78}$Pt (1.2×10^{-4})
159	$_{65}$Tb (1.00)	191	$_{77}$Ir (0.373)
160	$_{66}$Dy (0.023)	192	$_{76}$Os (0.410);
161	$_{66}$Dy (0.189)		$_{78}$Pt (7.8×10^{-3})
162	$_{66}$Dy (0.255);	193	$_{77}$Ir (0.627)
	$_{68}$Er (1.4×10^{-3})	194	$_{78}$Pt (0.329)
163	$_{66}$Dy (0.250)	195	$_{78}$Pt (0.338)
164	$_{66}$Dy (0.282)	196	$_{78}$Pt (0.253);
165	$_{67}$Ho (1.00)		$_{80}$Hg (1.6×10^{-3})
166	$_{68}$Er (0.334)	197	$_{79}$Au (1.00)
167	$_{68}$Er (0.229)	198	$_{80}$Hg (0.100);
168	$_{68}$Er (0.271);		$_{78}$Pt (0.072)
	$_{70}$Yb (1.4×10^{-3})	199	$_{80}$Hg (0.168)

Table 9 (continued)

m/e	Element and relative isotope abundance	m/e	Element and relative isotope abundance
200	$_{80}$Hg (0.231)	207	$_{82}$Pb (0.226)
201	$_{80}$Hg (0.132)	208	$_{82}$Pb (0.523)
	$_{80}$Hg (0.298)	209	$_{83}$Bi (1.00)
203	$_{81}$Tl (0.295)	232	$_{90}$Th (1.00)
204	$_{80}$Hg (0.069);	234	$_{92}$U (5.6×10^{-5})
	$_{82}$Pb (0.015)	235	$_{92}$U (7.2×10^{-3})
205	$_{81}$Tl (0.705)	236	$_{92}$U (0.993)
206	$_{82}$Pb (0.236)		

(4) Quantitative analysis. Recording and processing of the spectrum. A knowledge of the relative sensitivity of the recording device to individual elements is essential when carrying out quantitative analysis. Relative sensitivity is usually determined by analysis of a standard sample. This can vary markedly with the ion acceleration potential [131] so it is necessary to standardise on a setting of the acceleration potential to avoid errors from this effect. Variation of the spark voltage, repetition rate or pulse length has a much lower effect on sensitivity but again these variables should be maintained constant during an analysis [124].

During an analysis, it is necessary to vary the strength of the ion beam over several orders of magnitude. This is best accomplished by using an ion beam chopper to reject a portion of the beam while maintaining the spark parameters constant. The ion beam intensity can be changed by altering the spark parameters but this is bad practice when attempting to carry out precise determinations.

a. Electrical recording. Detection of the ion beam is made with an electron multiplier situated behind the exit slit in the focal plane of the instrument. The exit slit is usually set slightly wider than a single line width. Where a general survey analysis is required, the spectrum is scanned by varying the magnetic field so that each ion beam is swept successively across the detector slit. The detector signal and a signal from a total ion beam monitor are fed to separate logarithmic amplifiers, the amplified signals are subtracted, and the resultant is fed to a recorder. The recorder trace represents the proportion of the total ion beam corresponding to the m/e value in focus on the exit slit and this is plotted on a logarithmic scale. By taking three scans at different electron multiplier gains, the full concentration range of nearly all elements in a sample can be examined down to the 0.1 ppm range.

In a preliminary survey, it is usually assumed that the ion source produces ions from all elements with the same efficiency. This approach is particularly useful in the analysis for trace elements at the parts per million level where an order of magnitude estimation of concentration is required. With this assumption, the intensity of an ion peak becomes proportional to the concentration of that isotope in the specimen. The ratio of the intensity of an impurity isotope peak, I_2, to the ion peak, I_1, of an isotope of the matrix element is multiplied by the ratio of the relative isotope abundances, A_1 and A_2, to give the concentration of the impurity element. The relative concentration, C_2, of the impurity element is given by the expression

$$C_2 = K \frac{I_2}{I_1} \frac{A_1}{A_2}$$

where K is a constant whose value has been assumed to be unity. In a more refined analysis, a standard sample is used to determine the value of K.

In the most precise measuring technique, each isotope beam in turn is held stationary on the detector and integrated along with the total ion current for a controlled reference total ion beam charge. Fast switching from one ion beam to another is achieved by changing the accelerator voltage while maintaining a constant magnetic field. The integrated ion beam intensity is proportional to the concentration of the isotope under examination and analysis of a standard sample is used to determine the proportionality factor. This method is suitable for the analysis of both trace elements and major constituents with a precision better than 2% [123,132].

b. Photographic recording. The problems involves in photographic recording of the spectrum for purposes of quantitative analysis are considerable and it is probable that this method will rapidly fall out of use. A detailed discussion of photographic emulsions as ion detectors in quantitative mass spectrography has been given by Ahearn [121].

The number of emulsions suitable for ion detection is limited. The properties of the five most-used emulsions have been compared [133]. Of these, the Ilford Q emulsions are the most widely used. The latitude of an emulsion, that is the range between the threshold and saturated exposure levels, is not very great so that a series of exposures of the spectrum has to be made. Exposure time is measured in terms of the integrated total ion beam current. A series of exposures, begining with the shortest, in the ratio $1 : 3 : 10 : 30 : 100 : 300 :$ etc. are made and thirteen such exposures cover the concentration range of $1-10^6$.

The simplest method for estimating the ratio of concentration of an impurity element to that of the matrix element is to estimate the ratio of the minimum exposures in which an isotope of the impurity and an isotope of the matrix element are just detectable. The ratio, C_2, of the concentration of the impurity element to that of the matrix element is given by the expression

$$C_2 = K \frac{E_1}{E_2} \frac{A_1}{A_2} \frac{W_2}{W_1}$$

where E_1 and E_2 are the lowest exposures, in units of coulombs of total ion charge, at which the singly charged ions of the matrix and the impurity can be detected, A_1 and A_2 are the fractional abundances of the isotopes being used to evaluate the matrix and the impurity element, respectively, K is a proportionality constant which is given the value of unity when only an order of magnitude estimation is required, and W_1 and W_2 are the line widths of the matrix and impurity spectral lines, respectively. The ratio W_2/W_1 is taken as unity when only an order of magnitude estimation is required.

The value of K is dependent on the ionisation efficiency, the detection limit and the emulsion sensitivity. A practical relationship between K and atomic mass can be determined for a given spectrometer by analysis of a few standard samples. Alternatively, its value for the pair of isotopes under examination can be determined by analysis of one standard sample. For a mass spectrometer of the Mattauch—Herzog design, the line width, at perfect focus, is proportional to the square root of the m/e value and this relationship is usually employed to derive the value of W_2/W_1 [134].

In more refined methods of spectrographic analysis, the optical densities of the mass spectral lines are recorded using a densitometer. The developed plate is surveyed with a densitometer which records a plot of transmittance against distance along the plate. The distance axis can then be calibrated according to m/e values either visually by counting each peak at a unit mass number as discussed on p. 197, or with the aid of a computer after two or three of the main peaks have been identified by visual inspection. The strongest ions from a given isotope of mass m have a positive charge of one unit but doubly, triply and higher charged ions are found which give rise to peaks at $m/2$ and $m/3$ values which lie between the peaks at integral values of atomic mass units.

For accurate quantitative analysis, the emulsion must first be calibrated so that the relative densitometer transmission figures can be converted to relative ion beam intensities. The Churchill two-line method [135] is employed for this purpose.

The first step in the calibration process is to photograph a series of about twelve spectra of a selected element to yield a series of images ranging from just detectable to nearly saturated. The selected element should have two isotopes with relative abundances between 1.2 and 3.0. The plate is then developed and the mass spectra measured on a densitometer in the same manner as that in which the sample spectra is to be treated. A first curve is constructed by plotting for each spectrum the percentage transmission for the stronger line as y co-ordinate and the percentage transmission for the weaker line as x-co-ordinate on log—log graph paper. Data for the calibration curve are then obtained from this first curve. A point labelled $n = 0$ is chosen on the first curve at the *largest percentage transmission* value to be used in subsequent analyses. Its co-ordinates x_0 and y_0 are inserted in a table. The table is made up from other points with $n = 1, 2, 3$, etc. and their values of x_n and y_n bearing the relationship $y_n = x_{n-1}$. Values of r^n are computed for the final column of this table where r is the ratio of the more abundant to the less abundant isotope. The important quantities in this table are the corresponding values of y_n and r^n. These values of r^n are a series of relative ion exposure values with the value of 1 arbitrarily assigned to the lightest exposure and with each successive exposure a factor of r heavier than its predecessor. Values of y_n list the percentage transmission values corresponding to the relative exposure given by the value of r^n. A calibration curve is obtained by plotting on log—log co-ordinates the percentage transmission values (y_n) as ordinate against the relative exposure values (r^n) as abscissa.

264

In the unknown sample, it is desired to find the ratio between the concentrations of two elements. A suitable line on the mass spectrograph for one isotope of each element is selected and the percentage transmission determined using the densitometer. These figures are converted to relative exposures using the calibration curve. The ratio of these relative exposures is then corrected, if necessary, for the ratio of the exposures in which the two lines were measured to give a final figure for the ratio of the two ion beam intensities, I_1/I_2. If the ratio of the abundances of the two isotopes chosen is A_1/A_2, then the ratios of the concentrations of the two elements, C_1/C_2, is given by

$$\frac{C_1}{C_2} = K \frac{I_1}{I_2} \frac{A_2}{A_1} \frac{W_1}{W_2}$$

where W_1 and W_2 are the two line widths and K is the relative sensitivity of the mass spectrometer to the two ions measured.

The value of W_1/W_2 can be found from the densitometer data. In approximate work, K is taken as unity. For more accurate work, the value of K can be derived by analysis of a standard sample. The highest precision reported using a photographic plate detector is ±2—5% when strict control of all parameters was exercised [136].

Bibliography

Introductory texts

H.C. Hill, Introduction to mass spectrometry, Heyden, London, 2nd edn., 1972.
R.A.W. Johnstone, Mass Spectrometry for Organic Chemists, Cambridge University Press, London, 1972.
R.W. Kiser, Introduction to Mass Spectrometry and its Applications, Prentice-Hall, New Jersey, 1965.
J.R. Majer, The Mass Spectrometer, Wykeham Publications, London, 1977.
J. Roboz, Introduction to Mass Spectrometry, Interscience, New York, 1968.
S.R. Shrader, Introductory Mass Spectrometry, Allyn and Bacon, Boston, 1971.

Advanced texts. General

G.P. Barnard, Modern Mass Spectrometry, The Institute of Physics, London, 1953.
H.D. Beckey, Field Ionisation Mass Spectrometry, Pergamon Press, Oxford, 1971.

C. Brunnée and H. Voshage, Massenspektrometrie, Verlag Karl Thiemig, Munich, 1964.

A.L. Burlingame (Ed.), Topics in Mass Spectrometry, Wiley-Interscience, New York, 1970.

J.R. Chapman, Computers in Mass Spectrometry, Academic Press, New York, 1976.

R.G. Cooks, J.H. Beynon, R.M. Caprioli and G.R. Lester, Metastable Ions, Elsevier, Amsterdam, 1973.

H.E. Duckworth, Mass Spectrometry, Cambridge University Press, London, 1958.

F.H. Field and J.L. Franklin, Electron Impact Phenomena, Academic Press, New York, 1957.

P.F. Knewstubb, Mass Spectrometry and Ion Molecule Reactions, Cambridge University Press, London, 1969.

H. Knof, Massenspektrometrie von Kondensationkeimen in der Gasphase, Physik Verlag, Weinheim, 1974.

C.A. McDowell (Ed.), Mass Spectrometry, McGraw-Hill, New York, 1963.

A. MacColl (Ed.), MTP International Review of Science, Ser. 1, Vol. 5, Butterworths, London, 1972.

B.J. Millard, Quantitative Mass Spectrometry, Heyden, London, 1977.

G.W.A. Milne (Ed.), Mass Spectrometry: Techniques and Application, Wiley-Interscience, New York, 1971.

R.I. Reed, Ion Production by Electron Impact, Academic Press, London, 1962.

R.I. Reed (Ed.), Mass Spectrometry, Academic Press, London, 1965.

R.I. Reed (Ed.), Modern Aspects of Mass Spectrometry, Plenum Press, New York, 1968.

H.M. Rosenstock, K. Draxl, B.W. Steiner and J.T. Herron, Energetics of Gaseous Ions, National Bureau of Standards, Washington, 1977.

W.T. Wipke, S.R. Heller, R.J. Feldmann and E. Hyde (Eds.), Computer Representation and Manipulation of Chemical Information, Wiley, New York, 1974.

Advances in Mass Spectrometry, Institution of Petroleum, London, Vol. 1, 1961, Vol. 7, 1978.

Dynamic Mass Spectrometry, Heyden, London, Vol. 1, 1971, Vol. 5, 1978.

Specialist Periodical Report — Mass Spectrometry, The Chemical Society, London, Vol. 1, 1971, Vol. 4, 1977.

Advanced texts. Organic

J.H. Beynon, Mass Spectrometry and its Applications to Organic Chemistry, Elsevier, Amsterdam, 1960.

J.H. Beynon, R.A. Saunders and A.E. Williams, The Mass Spectra of Organic Molecules, Elsevier, Amsterdam, 1968.

K. Biemann, Mass Spectrometry and its Organic Chemical Applications, McGraw-Hill, New York, 1962.

H. Budzikiewicz, C. Djerassi and D.H. Williams, Interpretation of Mass Spectra of Organic Compounds, Holden-Day, San Francisco, 1964.

H. Budzikiewicz, C. Djerassi and D.H. Williams, Structural Elucidation of Natural Products by Mass Spectrometry. Vol. 1, Alkaloids, Vol. 2, Steroids, Terpenoids, Sugars and Miscellaneous Natural Products, Holden-Day, San Francisco, 1964.

H. Budzikiewicz, C. Djerassi and D.H. Williams, Mass Spectrometry of Organic Compounds, Holden-Day, San Francisco, 1967.

A.L. Burlingame (Ed.), Topics in Organic Mass Spectrometry, Wiley-Interscience, New York, 1970.

S.E. Drews (Ed.), Chroman and Related Compounds. Progress in Mass Spectrometry, Verlag Chemie, Weinheim, 1974.

R. Hague and F.J. Birds (Eds.), Mass Spectrometry and NMR Spectroscopy in Pesticide Chemistry, Plenum Press, New York, 1974.

M. Hesse and H.O. Bernhard, Alkoloide-ausser Indol-, Triterpen- und Steroidalkoloide, Verlag Chemie, Weinheim, 1975.

F.W. McLafferty, Mass Spectrometry of Organic Ions, Academic Press, New York, 1963.

F.W. McLafferty, Interpretation of Mass Spectra, Benjamin, Reading, Mass., 2nd edn., 1973.

S.P. Markey, W.G. Urban and S.P. Levine, Mass Spectra of Compounds of Biological Interest, U.S.A. E.C. National Technical Information Service, Springfield, Virginia.

R.I. Reed, Applications of Mass Spectrometry to Organic Chemistry, Academic Press, New York, 1966.

S. Safe and O. Hutzinger, Mass Spectrometry of Pesticides and Pollutants, CRC Press, Cleveland, Ohio, 1973.

J. Seibl, Massenspektrometrie, Akademie Verlag, Frankfurt-am-Main, 1970.

W. Simon and T. Clerc, Structural Analysis of Organic Compounds by Spectroscopic Methods, MacDonald, London, 1971.

G. Spiteller, Massenspectrometrische Strukturanalyse Organischer Verbindungen, Verlag Chemie, Weinheim, 1966.

G.R. Waller (Ed.), Biochemical Applications of Mass Spectrometry, Wiley-Interscience, New York, 1972.

D.H. Williams and N.G. Howe, Principles of Organic Mass Spectrometry, McGraw-Hill, London, 1972.

Advanced texts. Inorganic and organometallic

A.J. Ahearn (Ed.), Mass Spectrometric Analysis of Solids, Elsevier, Amsterdam, 1966.

J Charalambous (Ed.), Mass Spectrometry of Metal Compounds, Butterworth, London, 1975.

M.R. Litzow and T.R. Spalding, Mass Spectrometry of Inorganic and Organometallic Compounds, Elsevier, Amsterdam, 1973.

Advanced texts. GLC—mass spectrometry

A. Frigerio (Ed.), Proceedings of the International Symposium on Gas Chromatography—Mass Spectrometry, Tamburini Editore, Milano, 1972.

B.J. Gudzinowicz, M.J. Gudzinowicz and H.F. Martin, Fundamentals of Integrated Gas Chromatography—Mass Spectrometry, Marcel Dekker, Basel, 1976.

W.H. McFadden (Ed.), Techniques of Confined Gas Chromatography—Mass Spectrometry: Applications in Organic Analysis, Wiley-Interscience, New York, 1973.

O.A. Mamer, W.J. Mitchell and C.R. Scriver (Eds.), Application of Gas Chromatography—Mass Spectrometry to the Investigation of Human Disease, McGill University, Montreal, 1974.

Tables of mass spectra and related data

American Petroleum Institute and the Manufacturing Chemists Association, Catalogue of Mass Spectral Data, Chemical Thermodynamic Properties Center, Texas A.&M. University, College Station, TX.

American Society for Testing Materials Committee E-14, Index of Mass Spectral Data, ASTM, Philadelphia, 1963.

J.H. Beynon, R.A. Saunders and A.E. Williams, Tables of Metastable Transitions for use in Mass Spectrometry, Elsevier, Amsterdam, 1965.

J.H. Beynon and A.E. Williams, Mass and Abundance Tables for use in Mass Spectrometry, Elsevier, Amsterdam, 1963.

R. Binks, J.S. Littler and R.L. Cleaver, Tables for use in High Resolution Mass Spectrometry, Heyden, London, 1970.

A. Cornu and R. Massot, Compilation of Mass Spectral Data, Vols. 1 and 2, Heyden, London, 1975.

R.S. Gohlke, Uncertified Mass Spectral Data, Dow Chemical Co., Midland, Michigan, 1963.

J. Lederberg, Computation of Molecular Formulae for Mass Spectrometry, Holden-Day, San Francisco, 1964.

Mass Spectrometry Data Centre, Collection of Mass Spectra, AWRE, Aldermaston, Gt. Britain.

B.S. Middleditch and J.A. McCloskey, A Guide to Collections of Mass Spectral Data, American Society for Mass Spectrometry, Baylor College of Medicine, Houston, 1974.

M. Spiteller and G. Spiteller, A Collection of Mass Spectra of Solvents, Pollutants, Column Coatings and Simple Aliphatic Compounds, Springer Verlag, Vienna, 1973.

A. Tatematsu and T. Tsuchiya, Structure of Indexed Literature of Organic Mass Spectra, Academic Press of Japan, Tokyo, 1968.

D.D. Tunnicliff, P.A. Wadsworth and D.O. Shcissler, Mass and Abundance Tables, Shell Development Co., Emeryville, CA, 1965.

References

1 E. Goldstein, Sitzungsber. K. Preuss. Akad. Wiss., 39 (1886) 691; W. Wien, Ann. Phys. Chem. N.F., 65 (1898) 440.

2 J.J. Thomson, Rays of Positive Electricity and their Application to Chemical Analysis, Longmans, Green and Co., London, 1st edn., 1913.

3 F.W. Aston, Phil. Mag., 38 (1919) 514; Mass Spectra and Isotopes, Arnold, London, 1st edn., 1933.
4 A.J. Dempster, Phys. Rev., 11 (1918) 316.
5 W. Bleakney, Phys. Rev., 34 (1929) 157.
6 A.O. Nier, Rev. Sci. Instrum., 11 (1940) 212; 18 (1947) 398.
7 E.M. Clarke, Can. J. Phys., 32 (1954) 764.
8 R.E. Fox, W.M. Hickam, D.J. Grove and T. Kjeldaas, Rev. Sci. Instrum., 26 (1955) 1101.
9 M.R. Andrews, J. Phys. Chem., 27 (1923) 270.
10 F.H. Field, Acc. Chem. Res., 1 (1968) 42.
11 H.M. Fales, G.W.A. Milne and T. Axenrod, Anal. Chem., 42 (1970) 1432.
12 G.P. Arsenault, J.J. Dolhun and K. Biemann, Anal. Chem., 43 (1971) 1720.
13 R. Gomer and M.G. Inghram, J. Chem. Phys., 22 (1954) 1279.
14 H.D. Beckey, Field Ionisation Mass Spectrometry, Pergamon, Oxford, 1971.
15 J.N. Damico and R.P. Barron, Anal. Chem., 43 (1971) 17.
16 N.W. Reid, Int. J. Mass Spectrom. Ion Phys., 6 (1971) 1.
17 J.R. Woolston and R.E. Honig, Rev. Sci. Instrum., 39 (1962) 84.
18 J.G. Gorman, E.J. Jones and J.A. Hipple, Anal. Chem., 23 (1951) 438.
19 B. Chakravarty, V.S. Venkatasubramanian and H.E. Duckworth, in R.M. Elliott (Ed.), Advances in Mass Spectrometry, Pergamon, Oxford, 1963, Vol. 2, p. 128.
20 F.A. White, T.L. Collins and F.M. Rourke, Phys. Rev., 101 (1956) 1786; M.G. Inghram and W.A. Chupa, Rev. Sci. Instrum., 24 (1953) 518.
21 A.D. Nier, Rev. Sci. Instrum., 11 (1940) 212.
22 J.A. Hipple, J. Appl. Phys., 13 (1942) 551.
23 E.G. Johnson and A.O. Nier, Phys. Rev., 91 (1953) 10.
24 J. Mattauch and R.F.K. Herzog, Z. Phys., 89 (1934) 786.
25 (a) H. Sommer, H.A. Thomas and J.A. Hipple, Phys. Rev., 82 (1951) 697.
    (b) J.D. Baldeschwieler, Science, 159 (1968) 263.
26 W.Paul and H. Steinwedel, Z. Naturforsch. Teil A, 8 (1953) 448.
27 E.J. Bonnelli, M.S. Story and J.B. Knight, Dyn. Mass Spectrom., 2 (1971) 177; P.H. Dawson and N.R. Whetton, Adv. Electron. Electron Phys., 27 (1969) 59.
28 H.M. Rosenstock, M.B. Wallenstein, A.L. Wahrhaftig and H. Eyring, Proc. Natl. Acad. Sci. U.S.A., 38 (1952) 667; H.M. Rosenstock, Adv. Mass Spectrom., 4 (1968).
29 H. Budzikiewicz, C. Djerassi and D.H. Williams, Mass Spectrometry of Organic Compounds, Holden-Day, San Francisco, 1967.
30 T.W. Bentley and R.A.W. Johnstone, Adv. Phys. Org. Chem., 8 (1970) 151.
31 C. Krier, J.C. Lorquet and A. Berlingin, Org. Mass Spectrom., 8 (1974) 387.
32 L.A. Konig, J.H.E. Mattauch and A.H. Wapstra, Nucl. Phys., 31 (1962) 18.
33 D. Strominger, J.M. Hollander and G.T. Seaborg, Rev. Mod. Phys., 30 (1958) 585.
34 R.C. Weast (Ed.), Handbook of Chemistry and Physics, Chemical Rubber Co., Cleveland.
35 J.H. Beynon and A.E. Williams, Mass and Abundance Tables for use in Mass Spectroscopy, Elsevier, Amsterdam, 1963.

269

36 A. Cornu and R. Massot, Analyse Organique par Spectrométrie de Masse à Haute Résolution, Atlas de Raies, Press Universitaires de France, Paris, 1964.
37 J.H. Beynon, Mass Spectrometry and its Applications to Organic Chemistry, Elsevier, Amsterdam, 1960, p. 578.
38 Tables of Mass Ratios and Relative Peak Heights for Perfluorotributylamine Fragments, Associated Electrical Industries Ltd., Manchester, 1965.
39 T. Aczel, Anal. Chem., 40 (1968) 1917.
40 K.S. Quisenberg, T.T. Scolman and A.O. Nier, Phys. Rev., 102 (1956) 1071.
41 G.P. Moss, Org. Mass Spectrom., 5 (1971) 353.
42 Am. Pet. Inst. Manuf. Chem. Assoc., Catalogue of Mass Spectral Data, Chemical Thermodynamic Properties Center, Texas A.&M. University, College Station, TX, Spectrum No. 132.
43 Am. Pet. Inst. Manuf. Chem. Assoc., Catalogue of Mass Spectral Data, Chemical Thermodynamic Properties Center, Texas A&M University, College Station, TX, Spectrum No. 335.
44 Am. Pet. Inst. Manuf. Chem. Assoc., Catalogue of Mass Spectral Data, Chemical Thermodynamic Properties Center, Texas A&M University, College Station, TX, Spectrum No. 336.
45 A.C. Cope, U. Axen, E.P. Burrows and J. Weinlich, J. Am. Chem. Soc., 88 (1966) 4228.
46 Am. Pet. Inst. Manuf. Chem. Assoc., Catalogue of Mass Spectral Data, Chemical Thermodynamic Properties Center, Texas A&M University, College Station, TX, Spectrum No. 128.
47 Am. Pet. Inst. Manuf. Chem. Assoc., Catalogue of Mass Spectral Data, Chemical Thermodynamic Properties Center, Texas A&M University, College Station, TX, Spectrum No. 129.
48 Am. Pet. Inst. Manuf. Chem. Assoc., Catalogue of Mass Spectral Data, Chemical Thermodynamic Properties Center, Texas A&M University, College Station, TX, Spectrum No. 977.
49 Am. Pet. Inst. Manuf. Chem. Assoc., Catalogue of Mass Spectral Data, Chemical Thermodynamic Properties Center, Texas A&M University, College Station, TX, Spectrum No. 494.
50 Am. Pet. Inst. Manuf. Chem. Assoc., Catalogue of Mass Spectral Data, Chemical Thermodynamic Properties Center, Texas A&M University, College Station, TX, Spectrum No. 462.
51 Am. Pet. Inst. Manuf. Chem. Assoc., Catalogue of Mass Spectral Data, Chemical Thermodynamic Properties Center, Texas A&M University, College Station, TX, Spectrum No. 1060.
52 R.A. Friedel, J.C. Shultz and A.G. Sharkey, Anal. Chem., 28 (1956) 926.
53 Am. Pet. Inst. Manuf. Chem. Assoc., Catalogue of Mass Spectral Data, Chemical Thermodynamic Properties Center, Texas A&M University, College Station, TX, Spectrum No. 1059.
54 W. Benz and K. Biemann, J. Am. Chem. Soc., 86 (1964) 2375; S. Meyerson and L.C. Leitch, J. Am. Chem. Soc., 86 (1964) 2555.
55 F.W. McLafferty, Anal. Chem., 29 (1957) 1782.
56 Am. Pet. Inst. Manuf. Chem. Assoc., Catalogue of Mass Spectral Data, Chemical Thermodynamic Properties Center, Texas A&M University, College Station, TX, Spectrum No. 820.

57 Am. Pet. Inst. Manuf. Chem. Assoc., Catalogue of Mass Spectral Data, Chemical Thermodynamic Properties Center, Texas A&M University, College Station, TX, Spectrum No. 830.

58 Am. Pet. Inst. Manuf. Chem. Assoc., Catalogue of Mass Spectral Data, Chemical Thermodynamic Properties Center, Texas A&M University, College Station, TX, Spectrum No. 1132.

59 H. Audier, S. Bory, M. Fetizon, P. Longevialle and R. Toubiana, Bull. Soc. Chim. Fr., (1964) 3034.

60 J.A. McCloskey and M.J. McClelland, J. Am. Chem. Soc., 87 (1965) 5090; R.E. Wolff, G. Wolff and J.A. McCloskey, Tetrahedron, 22 (1966) 3093.

61 R.A. Friedel, J.L. Shultz and A.G. Sharkey, Anal. Chem., 28 (1956) 926.

62 J. Seibl and T. Gaümann, Z. Anal. Chem., 197 (1963) 33; Helv. Chim. Acta, 46 (1963) 2857.

63 F.W. McLafferty, Anal. Chem., 34 (1962) 2.

64 A.M. Duffield, S.D. Sample and C. Djerassi, Chem. Commun., (1966) 193.

65 F.W. McLafferty, Anal. Chem., 34 (1962) 16.

66 Am. Pet. Inst. Manuf. Chem. Assoc., Catalogue of Mass Spectral Data, Chemical Thermodynamic Properties Center, Texas A&M University, College Station, TX, Spectrum No. 540.

67 H. Budzikiewicz, J.I. Brauman and C. Djerassi, Tetrahedron, 21 (1965) 1855.

68 Am. Pet. Inst. Manuf. Chem. Assoc., Catalogue of Mass Spectral Data, Chemical Thermodynamic Properties Center, Texas A&M University, College Station, TX, Spectrum No. 1211.

69 Am. Pet. Inst. Manuf. Chem. Assoc., Catalogue of Mass Spectral Data, Chemical Thermodynamic Properties Center, Texas, A&M University, College Station, TX, Spectrum No. 1179.

70 W. Benz and K. Biemann, J. Am. Chem. Soc., 86 (1964) 2375.

71 S. Meyerson and L.C. Leitch, J. Am. Chem. Soc., 86 (1964) 2555.

72 C. Djerassi and C. Fenselau, J. Am. Chem. Soc., 87 (1965) 5747.

73 C. Djerassi and C. Fenzelau, J. Am. Chem. Soc., 87 (1965) 5752.

74 S.D. Sample and C. Djerassi, J. Am. Chem. Soc., 88 (1966) 1937.

75 E.M. Emery, Anal. Chem., 32 (1960) 1495.

76 K. Biemann, Angew. Chem., 74 (1962) 102.

77 F.W. McLafferty and R.S. Gohlke, Anal. Chem., 31 (1959) 2076.

78 S. Meyerson, I. Puskas and E.K. Fields, J. Am. Chem. Soc., 88 (1966) 4974.

79 F.W. McLafferty, Anal. Chem., 31 (1959) 82.

80 F.W. McLafferty, Chem. Commun., (1966) 78.

81 R.J. Liedke and C. Djerassi, J. Am. Chem. Soc., 91 (1969) 6814.

82 D.G. Earnshaw, F.G. Doolittle and A.W. Decora, Org. Mass Spectrom., 5 (1971) 801; K.K. Mayer and C. Djerassi, Org. Mass Spectrom., 5 (1971) 817.

83 J.A. Hipple and E.U. Condon, Phys. Rev., 68 (1945) 54.

84 A.H. Jackson, G.W. Kenner, K.M. Smith, R.T. Aplin, H. Budzikiewicz and C. Djerassi, Tetrahedron, 21 (1965) 2913; R.A.W. Johnstone, B.J. Millard, F.W. Dean and A.W. Hill, J. Chem. Soc. C, (1966) 1712; J. Siebl, Helv. Chim. Acta, 50 (1967) 263.

85   J.H. Benyon and A.E. Fontaine, Z. Naturforsch. Teil A, 22 (1967) 334.
86   (a) R. Jennings, J. Chem. Phys., 43 (1965) 4176; J.H. Futrell, K.R. Ryan and L.W. Sieck, J. Chem. Phys., 43 (1965) 1832.
     (b) J.H. Beynon, J.W. Amy and W.E. Baitinger, Chem. Commun., (1969) 723; J.H. Beynon, W.E. Baitinger and J.W. Amy, Int. J. Mass Spectrom. Ion Phys., 3 (1969) 55.
87   (a) N.R. Daly, A. McCormick and R.E. Powell, Rev. Sci. Instrum., 39 (1968) 1163; N.R. Daly, A. McCormick and R.E. Powell, Org. Mass Spectrom., 1 (1968) 167.
     (b) A.F. Weston, K.P. Jennings, S. Evans and R.M. Elliot, Int. J. Mass Spectrom. Ion Phys., 20 (1976) 317.
     (c) D.S. Millington and J.A. Smith, Org. Mass Spectrom., 12 (1977) 264.
88   T.W. Bentley, R.A.W. Johnstone and J. Grimshaw, J. Chem. Soc. C, (1967) 2234.
89   F.H. Field, in MTP International Review of Science, Physical Chemistry Ser. 1, Vol. 5, Butterworth, London, 1973, p. 133; G.P. Arsenault, in G.R. Waller (Ed.), Biochemical Applications of Mass Spectrometry, Wiley-Interscience, New York, 1972, p. 817; L. Foltz, Lloydia, 35 (1972) 344.
90   G.A. Olah, Chem. Br., 8 (1975) 281.
91   F.H. Field, M.S.B. Munson and D.A. Becker, Adv. Chem. Ser. 58 (1966) 167.
92   F.H. Field and M.S.B. Munson, J. Am. Chem. Soc., 89 (1967) 4272.
93   F.H. Field, J. Am. Chem. Soc., 90 (1968) 5649.
94   M.S.B. Munson and F.H. Field, J. Am. Chem. Soc., 89 (1967) 1047.
95   S.G. Lias, A. Viscomi and F.H. Field, J. Am. Chem. Soc., 96 (1974) 359.
96   F.H. Field, American Chemical Society 155th Meeting, 1968.
97   (a) Am. Pet. Inst. Manuf. Chem. Assoc., Catalogue of Mass Spectral Data, Chemical Thermodynamic Properties Center, Texas A&M University, College Station, TX, Spectrum No. 880.
     (b) F.H. Field, J. Am. Chem. Soc., 92 (1970) 2672.
98   H.M. Fales, H.A. Lloyd and G.W.A. Milne, J. Am. Chem. Soc., 92 (1970) 1590.
99   M.S.B. Munson and F.H. Field, J. Am. Chem. Soc., 88 (1966) 4337.
100 H.D. Beckey, in G.R. Waller (Ed.), Biochemical Applications of Mass Spectrometry, Wiley-Interscience, New York, 1972, p. 795.
101 H.D. Beckey, E. Hilt, A. Maas, M.O. Migahed and E. Ochterbeck, Int. J. Mass Spectrom. Ion Phys., 3 (1969) 161.
102 H.D. Beckey, Int. J. Mass Spectrom. Ion Phys., 2 (1969) 500.
103 P. Brown and G.R. Pettit, Org. Mass Spectrom., 3 (1970) 67.
104 H. Krone and H.D. Beckey, Org. Mass Spectrom., 2 (1969) 427.
105 S.E.J. Johnsen, Anal. Chem., 19 (1947) 305.
106 H. Hoover and H.W. Washburn, Am. Inst. Min. Metall. Eng. Tech. Publ. No. 1205, 1940; H.W. Washburn, H.F. Wiley and S.M. Rock, Anal. Chem., 15 (1943) 541; H.M. Washburn, H.F. Wiley, S.M. Rock and C.E. Berry, Anal. Chem., 17 (1945) 74.
107 F.W. McLafferty, Anal. Chem., 28 (1956) 306.
108 E.E. Hughes and W.D. Dorks, Anal. Chem., 40 (1968) 866.

109 J.A. Dorsey, R.H. Hunt and M.J. O'Neal, Anal. Chem., 35 (1963) 511.
110 C. Brunnée, L. Jenckel and K. Kronenberger, Z. Anal. Chem., 189 (1962) 50.
111 R. Ryhage, Anal. Chem., 36 (1964) 759.
112 E.J. Bonelli, M.S. Story and J.B. Knight, Dyn. Mass Spectrom., 2 (1971) 177.
113 E.W. Becker, in H. London (Ed.), Separation of Isotopes, Newnes, London, 1961, p. 360.
114 J.T. Watson and K. Biemann, Anal. Chem., 36 (1964) 1135; 37 (1965) 844.
115 P.M. Krueger and J.A. McCloskey, Anal. Chem., 41 (1969) 1930.
116 M. Blumer, Anal. Chem., 40 (1968) 1590; M.A. Grayson and J.J. Bellina, Anal. Chem., 45 (1973) 487.
117 C. Brunnée, H.J. Bultemann and G. Kappus, 17th Annual Conference on Mass Spectrometry and Allied Topics, Dallas, 1969, Paper No. 46.
118 S.R. Lipsky, C.G. Horvath and W.J. McMurray, Anal. Chem., 38 (1966) 1585; M.A. Grayson and C.J. Wolf, Anal. Chem., 39 (1967) 1438.
119 D.P. Lucero and F.C. Haley, J. Gas Chromatogr., 6 (1968) 477; P.G. Simmonds, G.R. Schoemake and J.E. Lovelock, Anal. Chem., 42 (1970) 881.
120 J.E. Lovelock, K.W. Charlton and P.G. Simmonds, Anal. Chem., 41 (1969) 1048.
121 A.J. Ahearn, Mass Spectrometric Analysis of Solids, Elsevier, Amsterdam, 1966.
122 R.J. Conzemius and H.J. Svec, Talanta, 16 (1969) 365; C.A. Evans, R.J. Guidoboni and F.D. Leipziger, Appl. Spectrosc., 24 (1970) 85.
123 J.S. Gorman, J.A. Hipple and J.E. Jones, Anal. Chem., 23 (1951) 438.
124 J.S Halliday, P. Swift and W.A. Wolstenholme, Adv. Mass Spectrom., 3 (1966) 143.
125 W.A. Wolstenholme, Appl. Spectrosc., 17 (1963) 51.
126 R. Brown and W.A. Wolstenholme, Nature (London), 201 (1964) 598; W.A. Wolstenholme, Nature (London), 203 (1964) 1284.
127 W.L. Harrington, R.K. Skogerboe and E.H. Morrison, Anal. Chem., 37 (1965) 1480.
128 R.L. Heath, in J.W. Guthrie (Ed.), Table of Atomic Masses, Sandia Corporation, Albuquerque, 1961.
129 E.B. Owens and A.M. Sherman, Mass Spectrographic Lines of the Elements, Lincon Lab. Tech. Rep. 265, Massachusetts Institute of Technology, 1962.
130 A. Cornu, R. Massot and J. Terrier, Atlas de Raies, Centre d'Etudes Nucleaires, Grenoble, 1963.
131 J.R. Woolston and R.W. Honig, Rev. Sci. Instrum., 35 (1964) 69.
132 R.A. Bingham and R.M. Elliott, Anal. Chem., 43 (1971) 43.
133 W. Rudloff, Z. Naturforsch. Teil A, 17 (1962) 414.
134 E.B. Owens and N.A. Giardino, Anal. Chem., 35 (1963) 1172.
135 J.R. Churchill, Anal. Chem., 16 (1944) 653.
136 G.D. Nicholls, A.L. Graham, E. Williams and M. Wood, Anal. Chem., 39 (1967) 584; P.F.S. Jackson, P.G.T. Vossen and J. Whitehead, Anal. Chem., 39 (1967) 1737.

Chapter 3

Ion selective electrodes

W.E. VAN DER LINDEN

1. Introduction

Ion selective electrodes (ISEs) are electrochemical sensors that respond selectively to the activity of ionic species. As, in practice, no electrode is sensitive to one particular ion to the exclusion of all others, the term *selective* is preferred to the term *specific*. In this chapter, discussion will be confined to membrane electrodes that operate on a potentiometric principle: that is, a change in potential is sensed at constant current. In most cases, this current is virtually zero, which means that the electromotive force (EMF) is measured. An example of such an ISE with which every chemist is familiar is the hydrogen ion selective glass electrode.

Our present knowledge of membrane potentials dates back to Nernst [1] and Planck [2] who introduced concepts of the diffusion potential and the potential difference at the liquid—liquid interface. Another fundamental contribution was also made at the turn of the 19th century when Ostwald [3] studied the potential difference across semipermeable membranes. Only a few years later, Cremer [4] discovered the hydrogen ion response potentialities of some glasses. The subject was studied in more detail by Haber and Klemensiewicz [5].

In the decades following the introduction of the glass electrode, the research of membranes was mainly focused on biological membranes. In this period, a noteworthy contribution was made by Donnan [6]. He developed a theory for the electrical potential difference between two electrolytes separated by a membrane that completely prevents diffusion of at least one kind of ion present in one of the two electrolytes. An important extension of this concept was given by Teorell [7] and by Meyer and Sievers [8] in the 1930s. By combination of the two Donnan potentials at both electrolyte—membrane interfaces and the diffusion potential across the membrane, they could derive a quantitative expression for the membrane potential of membranes that are easily permeable for ions with a certain charge but poorly permeable for ions with opposite charge.

At the same time, progress was made in the study of different glasses with respect to the relation between chemical structure and electrochemical behaviour. Special attention was paid to the selectivity. The work carried out by Lengyel and Blum [9] has to be mentioned in this connection. Assuming an ion-exchange process at the solution—glass interface, Nicolsky and Tolmacheva [10] derived a simple equation for the response of the glass electrode in a solution containing both hydrogen ions and any one alkali metal ion.

The first experiments with other solid-state electrodes were also published in the thirties. Kolthoff and Sanders [11] investigated silver halide membrane electrodes for the selective determination of halide ions. They found a reduced sensitivity to the influence of redox systems present in the solution as compared with the ordinary silver—silver halide electrodes of the second kind. Another attempt was made by Tendeloo [12] who embedded, for instance, barium sulfate micro crystals in a paraffin matrix in order to prepare a sulfate selective membrane.

During the 1940s, little was published on membranes and membrane potentials, but an enormous development took place in the field of instrumentation. Thanks to this development, direct-reading mV-meters with high input impedances have become available which facilitate the EMF measurements to a considerable extent. In the late fifties, interest was focused on glass electrodes again with the systematic study by Eisenman and his colleagues (see ref. 13). This has led to the development of glass compositions particularly selective to certain monovalent ions such as Na^+, K^+, Ag^+, NH_4^+, Tl^+, Li^+, and Cs^+.

From that time on, a rapid succession of new results can be observed. Pungor et al. introduced heterogeneous solid-state electrodes based on active materials dispersed in silicone rubber (see ref. 14). Shortly thereafter, Ross [15] described a liquid membrane electrode selective to calcium which strongly stimulated research in the field of· liquid membranes. Special mention must be made of the work of Simon and his co-workers [16] on neutral carriers. Another impulse to the renewed interest in the search for ISEs was the discovery of the fluoride selective electrode by Frant and Ross [17]. This homogeneous solid state electrode consists of a single crystal lanthanum fluoride membrane. Since solid-state electrodes have, in general, low response times, long operative lifetimes, and good resistance to corrosive media, and are relatively free from interference by redox systems, other single crystals have been examined. However, due to the lack of suitable single crystals or because of the high cost of preparing them, the majority of homogeneous solid-state electrodes commercially available nowadays consists of polycrystalline membranes. Mixed metal sulfide—silver sulfide precipitates are mainly used as far as metal selective electrodes are concerned.

The tremendous growth in interest of ISEs is best illustrated by the number of publications in recent years. In his review article on the theory and applications of ISEs, Koryta [18] listed more than 700 relevant references covering the literature until the beginning of 1972. Five years later, the second part of this review appeared [19] including another 1250 references. Also, the number of symposia devoted exclusively to ISEs can be considered to be an indication of this interest [20—24]. Progress in the field to the present time is summarized in the biennial reviews in *Analytical Chemistry* [25].

In view of the considerable impact of ISEs on analytical chemistry, a chapter on this subject cannot be omitted in a series such as *Comprehensive Analytical Chemistry*, even though several other monographs on ISEs have been published in the last decade [18,20, 26—30].

2. List of symbols and abbreviations

a_i	activity of species i
$\bar{c}$	"concentration" of fixed sites in ion-exchange membrane
d	thickness of membrane
f	titration parameter
j_i	electrical current density of species i
k_i	distribution coefficient of species i
k_{MN}^{pot}	potentiometric selectivity coefficient
(m)	membrane phase
pX	negative logarithm of concentration (activity) of X
(s)	solution phase

t_i	transference number of species i
t_{90}	response time referred to 90% of the total potential shift
$\bar{u}$	mean ionic mobility
x	linear coordinate
z_i	electrical charge of ion i
C_i	concentration of species i
D_i	diffusion constant of species i
D^*	mean diffusion coefficient
E	potential of electrode
E^0	standard potential
F	Faraday constant
G	Gibbs free energy
I	ionic strength
J_i	mass flux of species i
K_a	dissociation constant of acid
K_{ex}, K'_{ex}	exchange equilibrium constants
K_{MN}	ion-exchange constant
K_{MS} (..)	stability constant of complex MS in phase (..)
K_{so}^{MX}	solubility product of MX
R	gas constant
T	absolute temperature
U_i	particle mobility of species i
β_n	overall stability constant of complex MS_n
γ_i	activity coefficient for species i
$\gamma_\pm$	mean activity coefficient
δ	thickness of Nernst diffusion layer
λ	Donnan distribution coefficient
μ_i	chemical potential of species i
μ_i^0	standard chemical potential
$\bar{\mu}_i$	electrochemical potential of species i
ϕ	inner potential
ϕ_D	Donnan potential
$\Delta\phi_L$	liquid junction potential
$\Delta\phi_m$	membrane potential
CFA	continuous flow analysis
EDTA	ethylenediaminetetraacetic acid
EGTA	ethyleneglycol(2-aminoethylether)tetraacetic acid
EMF	electromotive force

DCTA	1,2-diaminocyclohexanetetraacetic acid
FET	field effect transistor
FIA	flow injection analysis
ISE	ion selective electrode
ISFET	ion sensitive field effect transistor
MOSFET	metal oxide semiconductor field effect transistor
NTA	nitrilotriacetic acid
SEM	scanning electron microscope
TETREN	tetraethylenepentamine
TISAB	total ionic strength adjustment buffer
TRIEN	triethylenetetramine

3. Membrane potentials

(A) INTRODUCTION

In colloquial speech, the term *membrane* is generally associated with a sheet-like and often elastic separating wall. Particularly in chemistry, however, a more restricted definition is used, i.e. a region of space that separates two solutions and is (semi) permeable. As membrane potentials are little affected by the geometrical shape of the membrane [31], it is justifiable to accept a somewhat broader definition in this chapter, viz. a membrane is a body of any shape

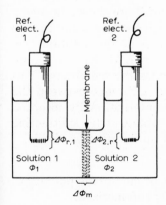

Fig. 1. Schematic diagram of an electrochemical cell.

and structure that separates two liquid phases and is permeable for at least one component of these phases.

If a membrane is equally permeable for all species in the solution, including the solvent molecules, it is usually called a *diaphragm*. In that case, it can only be used to prevent direct convective mixing of the two solutions. If the various components have different rates of transport through the membrane, it is called *semipermeable*. The term *permselectivity* is reserved for membranes which allow the permeation of ions with charges of uniform sign, but which are almost completely impermeable for ions with charges of opposite sign.

To measure the potential difference across a membrane, an electrochemical cell has to be formed. This can be accomplished by the immersion of an electrode in each of the two solutions (Fig. 1). Graphically, such a cell can be represented as

| (reference) electrode 1 | solution 1 ϕ_1 $x=0$ | membrane | solution 2 $x=d$ ϕ_2 | (reference) electrode 2 |

where ϕ_1 and ϕ_2 are the so-called *inner potentials* and x is the linear coordinate. It is assumed throughout the further treatment of the subject that the possible change in composition of the membrane is directed along this x axis only.

The EMF of this cell can be divided into three parts, (a) the potential difference between the reference electrode 1 and solution 1, $\Delta\phi_{r,1}$; (b) the potential difference across the membrane, $\Delta\phi_m$; and (c) the potential difference between solution 2 and the reference electrode 2, $\Delta\phi_{2,r}$. When the electrodes 1 and 2 are chosen in such a way that the potentials with regard to the respective solutions are constant and independent of the composition of the solution (i.e. one is dealing with reference electrodes), the EMF will vary only with the change in potential difference across the membrane. This membrane potential can be expressed as the difference between the inner potentials ϕ_1 and ϕ_2. According to the internationally accepted IUPAC convention [32], the difference is obtained by subtracting the potential on the left-hand side from that on the right-hand side.

$$\Delta\phi_m = \phi_2 - \phi_1 \qquad (1)$$

Before giving a more quantitative treatment of membrane potentials, some thermodynamic concepts will be briefly summarized.

For ionic species, equilibrium conditions are described in terms of the *electrochemical potential*, $\bar{\mu}$.

$$\bar{\mu}_i = \mu_i + z_i F \phi \tag{2}$$

where $\mu_i = (\partial G / \partial n_i)_{p,T}$ and denotes the *chemical potential*, i.e. the Gibbs free energy change per mole of component i, z_i is the charge of species i, ϕ is the inner potential, and F the Faraday constant. Because it is extremely difficult to assess all interactions in a certain system, it is virtually impossible to determine or calculate the absolute value of μ. Therefore, the chemical potential is always related to some standard state (^{0}). For an arbitrary species i, it can be expressed by

$$\mu_i = \mu_i^0 + RT \ln a_i \tag{3}$$

where μ_i^0 is the *standard* chemical potential, a_i is the *activity* of component i, R is the gas constant, and T is the absolute temperature. For dilute solutions, which can be taken as "ideal", the activity of a component can be replaced by its concentration.

Equilibrium between different phases (α, β, γ, etc.) for a component i exists when the condition

$$\bar{\mu}_i(\alpha) = \bar{\mu}_i(\beta) = \bar{\mu}_i(\gamma) = \ldots \tag{4}$$

is satisfied. This will lead to a potential difference at the phase boundary of two otherwise homogeneous phases α and β. Substitution of eqns. (2) and (3) in eqn. (4) yields

$$\bar{\mu}_i(\alpha) = \bar{\mu}_i(\beta)$$

or

$$\mu_i^0(\alpha) + RT \ln a_i(\alpha) + z_i F \phi(\alpha) = \mu_i^0(\beta) + RT \ln a_i(\beta) + z_i F \phi(\beta) \tag{5}$$

Hence

$$\Delta \phi_{\alpha,\beta} \equiv \phi(\beta) - \phi(\alpha) = \frac{\mu_i^0(\alpha) - \mu_i^0(\beta)}{z_i F} + \frac{RT}{z_i F} \ln \frac{a_i(\alpha)}{a_i(\beta)} \tag{6}$$

With regard to membrane equilibria, two cases can be distinguished.

(1) Equilibrium exists at the interface of two adjacent phases (in this case the two solution—membrane interfaces) and the composition of the membrane is the same throughout. The total membrane potential is then composed of the two boundary phase potentials only.

A special situation is observed when the solutions cannot be considered as homogeneous because of, for instance, the continuous dis-solution of the membrane material. Under certain conditions, however, a steady state can be attained in which the concentration of the active component at the surface remains constant, leading to a constant membrane potential.

(2) Equilibrium is established at the two interfaces but the composition inside the membrane is not constant in the x direction. Consequently, an electrochemical potential *gradient* exists which is the driving force for mass transport. If ions are involved, this mass transport is accompanied by charge transport. In the case of different mobilities of the various ions, this will cause a tendency for charge separation and the development of a so-called *diffusion potential* [33]. Across a membrane of this type, the total potential difference is composed of two boundary phase potentials and this diffusion potential.

The pure diffusion potential is obtained when two solutions of differing concentrations are in direct contact or separated by a diaphragm. In that case, no real phase boundary occurs. The solution just inside the pores of the diaphragm is not different from the solution outside the membrane, so in eqn. (6), $\mu_i^0(\alpha) = \mu_i^0(\beta)$ and $a_i(\alpha) = a_i(\beta)$. Hence, the phase boundary potential, $\Delta\phi_{\alpha,\beta}$, is absent. The potential thus obtained is often called the *liquid junction potential.* The use of a diaphragm merely restricts the interphase region between both solutions and prevents the disturbance of this interphase by mechanical or other convective influences.

For the quantitative evaluation of membrane potentials in the presence of diffusion potentials, the relation between mass transport and the electrochemical potential gradient has to be examined in more detail. The rate of mass transport is usually expressed as the mass flux, J, i.e. the number of moles passing a unit area per unit time. In the absence of any convection and gradients of pressure and activity coefficients, the flux can be represented by

$$J_i = -U_i C_i \frac{d\bar{\mu}_i}{dx} \tag{7}$$

where U_i is the particle mobility of i and C_i is the local concentration. Substitution of eqns. (2) and (3) in eqn. (7) leads to

$$J_i = -U_i RTC_i \frac{d \ln a_i}{dx} - z_i FU_i C_i \frac{d\phi}{dx} \tag{8}$$

282

For dilute ("ideal") solutions, this can be replaced by

$$J_i = -U_i RT \frac{dC_i}{dx} - z_i FU_i C_i \frac{d\phi}{dx}$$

$$= -D_i \frac{dC_i}{dx} - z_i FU_i C_i \frac{d\phi}{dx} \tag{9}$$

where D_i denotes the diffusion coefficient.

Equation (9) is known as the *Nernst—Planck relation* and is valid for all mobile species provided that they show ideal behaviour, and that the solutions are so diluted that the influence of osmotic pressure can be disregarded. The first term on the right-hand side represents the pure diffusion contribution and is identical with the first law of Fick. The second term accounts for the (electro)migration or pure conduction.

For a proper understanding of the real response mechanisms, a profound knowledge of the structure of the solution—membrane interface on a molecular scale will be indispensable. This would demand a complete treatment of the *electrical double layer*, etc. [33]. It is beyond the scope of this chapter, however, to deal with all these aspects. Therefore, the discussion will be primarily confined to interfaces at which a rapid, reversible, and faradaic exchange of ions can take place. In that case, equilibrium conditions will, at least locally, prevail and a purely thermodynamic approach is appropriate. The reader who is interested in so-called blocked interfaces at which the potential differences are not generated by faradaic but by non-faradaic processes such as, for example, adsorption, is referred to the publications by Buck [34,35].

On the basis of the aforementioned thermodynamic concepts, formulae for the potential—concentration relationships for the various types of membrane can be derived.

In the following discussion, a distinction will be made between membrane potentials in the absence and in the presence of internal diffusion potentials according to the classification used by Koryta [18,19].

(B) MEMBRANE POTENTIALS IN THE ABSENCE OF INTERNAL DIFFUSION POTENTIAL

(1) Solid-state membranes

a. A formula for the membrane potential. Solid-state membranes consist of sparingly soluble inorganic salts or solid ion-exchangers. In this section, the discussion will be confined to the inorganic salt membranes. Such membranes are available either as single crystals, non-porous pellets pressed from microcrystalline precipitates, or as microcrystals embedded in some suitable polymeric matrix. For reasons of simplicity, both anions and cations will be supposed to be monovalent in the subsequent discussions unless stated otherwise.

If such a membrane made of MX is in contact with solutions containing M^+ or X^-, an equilibrium situation can be established at the surface. Graphically the membrane can be represented as

$$\text{(solution 1)} \ M^+; X^- \quad \bigg| \quad M^+ X^- \quad \bigg| \quad M^+; X^- \quad \text{(solution 2)}$$
$$a_M(1); a_X(1) \quad \bigg| \quad a_M(m); a_X(m) \quad \bigg| \quad a_M(2); a_X(2)$$
$$\qquad\qquad\qquad x=0 \qquad\qquad\quad x=d$$

Then

$$\bar{\mu}_M(1)_{x=0} = \bar{\mu}_M(m)_{x=0} \qquad \bar{\mu}_M(m)_{x=d} = \bar{\mu}_M(2)_{x=d}$$
$$\bar{\mu}_X(1)_{x=0} = \bar{\mu}_X(m)_{x=0} \qquad \bar{\mu}_X(m)_{x=d} = \bar{\mu}_X(2)_{x=d} \tag{10}$$

Substitution of eqns. (2) and (3) leads to

$$\mu_M^0(1) + RT \ln a_M(1)_{x=0} + F\phi(1)_{x=0} = \mu_M(m) + F\phi(m)_{x=0} \tag{11a}$$

$$\mu_X^0(1) + RT \ln a_X(1)_{x=0} - F\phi(1)_{x=0} = \mu_X(m) - F\phi(m)_{x=0} \tag{11b}$$

Of course, similar equations hold at the interface membrane—solution 2. The potential differences at the phase boundary can be given by

$$\phi(m)_{x=0} - \phi(1)_{x=0} = \Delta\phi_{x=0} = \frac{1}{F}\{\mu_M^0(1) - \mu_M(m)\} + \frac{RT}{F} \ln a_M(1)_{x=0} \tag{12a}$$

or, similarly

$$\Delta\phi_{x=0} = \frac{1}{F}\{\mu_X(m) - \mu_X^0(1)\} - \frac{RT}{F} \ln a_X(1)_{x=0} \tag{12b}$$

284

Combination of eqns. (12a) and (12b) gives

$$\mu_M(m) + \mu_X(m) - \mu_M^0(1) - \mu_X^0(1) = RT \ln \{a_M(1)a_X(1)\} = RT \ln K_{so}^{MX}$$

(13)

where K_{so}^{MX} denotes the solubility product of MX.

Since the membrane consists of the pure substance MX and therefore

$$\mu_M(m) + \mu_X(m) = \mu_{MX}(m) = \mu_{MX}^0$$

eqn. (13) can be written as

$$\mu_{MX}^0 - \mu_M^0(1) - \mu_X^0(1) = RT \ln K_{so}^{MX}$$

(14)

Equations similar to (12a) and (12b) apply to the other interface. As the membrane is homogeneous, the following equations are valid if no net electrical current passes the membrane.

$$a_i(m)_{x=0} = a_i(m)_{x=d} \quad \text{and} \quad \phi(m)_{x=0} = \phi(m)_{x=d}$$

Using the assumption that solutions 1 and 2 differ only in the concentration of M^+ and X^-, $\mu^0(1)$ may be taken equal to $\mu^0(2)$ and the total potential difference across the membrane can be written as

$$\Delta\phi_m = \Delta\phi_{x=0} + \Delta\phi_{x=d} = \frac{RT}{F} \ln \frac{a_M(1)_{x=0}}{a_M(2)_{x=d}} = \frac{RT}{F} \ln \frac{a_X(2)_{x=d}}{a_X(1)_{x=0}}$$

(15)

When, in the solutions, the concentrations at the surface are equal to the respective bulk concentrations, the subscripts $x = 0$ and $x = d$ can be omitted in eqn. (15).

Keeping the solution constant at one side of the membrane (e.g. solution 1), the membrane potential will be a function of the activity of the other solution. Thus

$$\Delta\phi_m = \text{constant} - \frac{RT}{F} \ln a_M(2)$$

(16a)

$$\Delta\phi_m = \text{constant} + \frac{RT}{F} \ln a_X(2)$$

(16b)

b. *Limit of detection; interference; selectivity.* At low concentrations of M^+ or X^- in sample solutions, the linear relationship between the potential difference across the membrane and the logarithm of the respective activities is no longer observed due to the dissolution of the membrane material. Starting from the equations

$$a_X = \gamma_X \{[X^-] + [X^-]_m\} = \gamma_X[X^-]_t \qquad (17)$$

where γ_X is the activity coefficient of X and

$$K_{so}^{MX} = \gamma_\pm^2[M^+][X^-]_t \qquad (18)$$

where $[X^-]_m$ denotes the contribution because of the dissolution process and $\gamma_\pm$ is the mean activity coefficient $(\gamma_\pm = \{\gamma_M\gamma_X\}^{1/2})$, it can easily be shown that

$$\Delta\phi_m = \text{constant} + \frac{RT}{F} \ln\left\{\frac{1}{2}\gamma_X\left[[X^-] + \left([X^-]^2 + \frac{4 K_{so}^{MX}}{\gamma_\pm^2}\right)^{1/2}\right]\right\} \qquad (19)$$

provided that the solution at one side of the membrane is kept constant.

If $[X^-] \gg 2(K_{so}^{MX}/\gamma_\pm^2)^{1/2}$

$$\Delta\phi_m = \text{constant} + \frac{RT}{F}\ln \gamma_X[X^-] \qquad (20)$$

which is equal to eqn. (16b) and if $[X^-] \ll 2(K_{so}^{MX}/\gamma_\pm^2)^{1/2}$

$$\Delta\phi_m = \text{constant} + \frac{RT}{F}\ln\{\gamma_X(K_{so}^{MX}/\gamma_\pm^2)^{1/2}\} \qquad (21)$$

The *limit of detection* is expressed by the concentration or activity at which the two extrapolated linear segments of the calibration line, given by eqns. (20) and (21), respectively, intercept. Hence

$$[X^-]_{lim.ofdet.} = (K_{so}^{MX}/\gamma_\pm^2)^{1/2} \qquad (22)$$

or

$$(a_X)_{lim.ofdet.} = (K_{so}^{MX}\gamma_X/\gamma_M)^{1/2} \qquad (23)$$

Equations similar to eqns. (17)—(23) hold for the cation M^+.

In Fig. 2, an illustration is given of the theoretical response according to eqn. (19) for X^- and the corresponding one for M^+.

Any disturbance of the equilibrium at the interface will affect the membrane potential. As far as disturbances by chemical substances are concerned, two cases will be considered: (a) a compound Y will react with MX forming a soluble complex MY_n and (b) an ion Y^- will react with MX forming an insoluble compound MY.

An example of this latter type of interference is a silver chloride

286

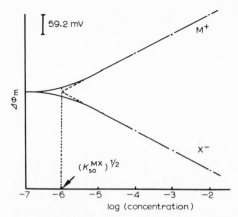

Fig. 2. Dependence of potential across a membrane made of MX on the logarithm of $[M^+]$ and $[X^-]$. $K_{so}^{MX} = 10^{-12}$; $\gamma_\pm = 1$; $2.303RT/F = 59.2$ mV at $25°C$.

membrane in contact with iodide ions. As long as

$$a_I < \frac{K_{so}^{AgI}}{K_{so}^{AgCl}} a_{Cl} \qquad (24)$$

there is no formation of AgI and eqn. (16b) will apply and

$$\Delta\phi_m = \text{constant} + \frac{RT}{F} \ln a_{Cl}(2) \qquad (25)$$

If the iodide activity exceeds the right-hand side of eqn. (24) a new phase will be formed at the surface.

$$AgCl + I^- \rightleftharpoons AgI + Cl^-$$

The membrane thus formed can be represented schematically as

soln. (1) $a_{Cl}(1); a_{Ag}(1) \underset{x=0}{|} \ AgCl \mid AgI \underset{x=d}{|} \ a_{Ag}(2); a_{Cl}(2); a_I(2)$ soln. (2)

Under equilibrium conditions

$$\overline{\mu}_{Ag}(1) = \overline{\mu}_{Ag}(AgCl) = \overline{\mu}_{Ag}(AgI) = \overline{\mu}_{Ag}(2) \qquad (26)$$

or

$$\mu_{Ag}^0(1) + RT \ln a_{Ag}(1) + F \ \phi_{x=0} = \mu_{Ag}^0(2) + RT \ln a_{Ag}(2) + F \ \phi_{x=d} \quad (27)$$

The assumption $\mu_{Ag}^0(1) = \mu_{Ag}^0(2)$ together with the substitution of the two solubility products leads to

$$\Delta\phi_m = \frac{RT}{F}\left\{\ln\frac{K_{so}^{AgCl}}{K_{so}^{AgI}} + \ln[a_I(2)/a_{Cl}(1)]\right\} \tag{28}$$

If solution 1 is again kept constant, this equation can be reduced to

$$\Delta\phi_m = \text{constant} + \frac{RT}{F}\ln a_I(2) \tag{29}$$

which means that an iodide-responsive membrane is obtained.

The transition between these two limiting cases, expressed by eqns. (25) and (29), respectively, is a gradual one. On the basis of the supposition of a fixed number of available sites at the surface of the membrane for chloride and iodide ions, expressed as a concentration of silver ions, $C_{Ag}(m)$, Buck [36] obtained a relation between the potential at the phase boundary and the concentration of the primary ion (Cl⁻) and the interfering ion (I⁻). Buck starts from the reaction

$$Cl^-(m) + I^-(2) \rightleftharpoons I^-(m) + Cl^-(2)$$

with the corresponding equilibrium constant

$$K_{Cl,I} = \frac{a_I(m)a_{Cl}(2)}{a_{Cl}(m)a_I(2)} \tag{30}$$

Assuming $C_{Ag}(m) = C_{Cl}(m) + C_I(m)$ and using the general relation $a_i = \gamma_i C_i$ for the membrane phase as well, the iodide "concentration" adsorbed at the surface of the membrane can be calculated to be

$$C_I(m) = \frac{K_{Cl,I}\{\gamma_{Cl}(m)/\gamma_I(m)\}\{a_I(2)/a_{Cl}(2)\}C_{Ag}(m)}{1 + K_{Cl,I}\{\gamma_{Cl}(m)/\gamma_I(m)\}\{a_I(2)/a_{Cl}(2)\}} \tag{31}$$

If, apart from chloride, the equilibrium condition for iodide is also applicable

$$\bar{\mu}_I(m) = \bar{\mu}_I(2) \tag{32}$$

the use of eqn. (3) for the membrane phase leads to

$$\Delta\phi_{x=d} = \phi(2) - \phi(m) = \frac{1}{F}\left\{\mu_I^0(2) - \mu_I^0(m) + RT\ln\frac{a_I(2)}{a_I(m)}\right\} \tag{33}$$

Substitution of eqn. (31) yields

$$\Delta\phi_{x=d} = \frac{1}{F}\{\mu_I^0(2) - \mu_I^0(m)\} - \frac{RT}{F}\ln \gamma_{Cl}(m)K_{Cl,I}C_{Ag}(m) +$$

$$\frac{RT}{F}\ln[a_{Cl}(2) + \{\gamma_{Cl}(m)/\gamma_I(m)\}K_{Cl,I}a_I(2)] \tag{34}$$

This can be reduced to

$$\Delta\phi_{x=d} = \text{constant} + \frac{RT}{F}\ln[a_{Cl} + k_{Cl,I}^{pot}a_I] \tag{35}$$

where $k_{Cl,I}^{pot}$ is the *potentiometric selectivity coefficient*. In this case, this coefficient is equivalent to $\{\gamma_{Cl}(m)/\gamma_I(m)\}K_{Cl,I}$. It is easily verified that $K_{Cl,I}$ is equal to the quotient of the solubility products

$$K_{Cl,I} = \frac{K_{so}^{AgCl}}{K_{so}^{AgI}} \tag{36}$$

Obviously, the smaller the value of the selectivity coefficient, the more selective the membrane will be.

Recently, Hulanicki and Lewenstam [37] have discussed the same kind of interference based on exchange reaction of the interfering ion with the primary ion in the membrane material. They assumed the reaction to be chemically reversible and fast in comparison with the diffusional transport in the system. As a result of their model, the selectivity coefficient depends, not only on the value of the equilibrium constant, but also on the values of the respective diffusion coefficients.

If a membrane made of MX is brought in contact with a complexing agent Y, which can form a stable and soluble complex MY_n, the following exchange reaction can occur at the surface.

$$MX_{solid} + n\,Y^- \rightleftharpoons MY_n^{(n-1)-} + X^-$$

the equilibrium constant being

$$K'_{ex} = \frac{a_{MY_n}a_x}{a_Y^n} \tag{37}$$

At constant ionic strength, this equilibrium constant can be replaced by

$$K_{ex} = K'_{ex}\frac{\gamma_Y^n}{\gamma_{MY_n}\gamma_X} = \frac{[MY_n][X]}{[Y]^n} \tag{38}$$

in which charges are omitted.

When K_{ex} is large and the reaction rate is sufficiently high, a continuous dissolution of X^- will take place. Under constant experimental conditions, a steady state can be attained at which the concentrations at the surface of the membrane remain constant, governed by the transport rates of the different participating species.

In the steady state, the following flux equations must hold at the surface.

$$J_Y + n\, J_{MY_n} = 0 \text{ (mass balance)} \tag{39}$$

and

$$J_X + (n-1)\, J_{MY_n} + J_Y = 0 \text{ (charge balance)} \tag{40}$$

Combination of both equations leads to

$$J_X = -\frac{1}{n} J_Y \tag{41}$$

In the presence of a relatively large amount of other ions, the transport of the relevant species is mainly governed by diffusion through the (assumed) stagnant layer adjacent to the membrane surface [38]. In most cases, it is permissible to consider the concentration profile within this layer to be linear, leading to a simple expression for the concentration gradient, viz.

$$\left(\frac{dC_i}{dx}\right)_{x=0} = \frac{C_i - C_i(0)}{\delta_i} \tag{42}$$

where δ_i is the thickness of this stagnant layer, the so-called *Nernst diffusion layer*, C_i is the concentration in the *bulk* of the solution $(x \geqslant \delta_i)$, and $C_i(0)$ is the concentration at the surface.

Application of eqn. (42) to eqn. (41) leads to

$$[X]_0 = [X] + \frac{1}{n}\frac{D_Y}{D_X}\frac{\delta_X}{\delta_Y}[Y] \tag{43}$$

The value of δ_i will depend on the way in which the diffusion layer is formed, e.g. by natural convection or by forced hydrodynamic conditions. In general, δ_i can be expressed as

$$\delta_i = \text{constant} \times D_i^\alpha \tag{44}$$

290

Hence

$$[X]_0 = [X] + \frac{1}{n} \frac{D_Y^{(1-\alpha)}}{D_X^{(1-\alpha)}} [Y] \tag{45}$$

After substitution in eqn. (16b), the equation

$$\Delta\phi_m = \text{constant} + \frac{RT}{F} \ln \gamma_X \left\{ [X] + \frac{1}{n} \frac{D_Y^{(1-\alpha)}}{D_X^{(1-\alpha)}} [Y] \right\} \tag{46}$$

for the membrane potential is obtained. This concept was originally introduced by Jaenicke [39] in his discussion of the dissolution rate of salts in the presence of complexing agents. This principle underlies the use of silver iodide membranes for the determination of cyanide.

When K_{ex} is small or the reaction rate is low, no noticeable attack of the membrane will occur, so $[X]_0$ may be taken equal to $[X]_{bulk}$ and

$$\Delta\phi_m = \text{constant} + \frac{RT}{F} \ln \gamma_X [X]_{bulk} \tag{47}$$

Mixed silver sulfide—copper sulfide membranes used for the determination of copper(II) ions show, in some cases, an anomalous behaviour in the presence of polyaminopolycarboxylic acids such as EDTA and NTA [40]. Especially in solutions containing these complexing agents in excess, the electrodes seem to respond to the free ligand concentration instead of the free copper(II) concentration in the bulk of the solution. Surprisingly, such a discrepancy is not observed in the presence of polyamines like TRIEN and TETREN, although, at least at higher pH values, these compounds form copper complexes of comparable stability. The membranes that show this behaviour are prepared under such conditions that monovalent copper sulfides can be formed [actually ternary copper(I)—silver sulfides are formed] and, consequently, the reaction

$$Cu_2S \rightleftharpoons CuS + Cu^{2+} + 2e$$

has to be considered. This reaction, which proceeds at the surface of the membrane, is affected in different ways by polyaminopolycarboxylic acids and polyamines. A quantitative explanation is not available, but it is important to stress that some consequences must be taken into account for the use of these membranes in compleximetric titrations and particularly in calibration procedures by means of metal buffer solutions, as will be discussed in Sect. 6.(B).

c. Kinetics; response time. When the composition of a solution in contact with a membrane is changed, a new equilibrium has to be established. The length of time necessary to reach the "final" equilibrium potential within a given limit is denoted as the *response time*. In the report on recommendations for nomenclature of ion selective electrodes [41], a IUPAC commission has recommended the use of the *practical response time*. In this version, the given limit is taken to be 1 mV. If the relation between potential and activity (concentration) can be described according to a Nernstian-type equation, this means that the activity at that time has reached the eventual value within 4% as far as monovalent ions are concerned. For n-valent ions, the corresponding percentage is $100\{1 - \exp[-nF/RT]\}$. This definition is particularly suitable in practical analysis, where, in general, the relative error in the result of the determination is the relevant factor. However, in most papers dealing with the origin of transient potentials, the definition of response time is related to the time needed to get a potential shift which is a fixed percentage of the total potential shift. As, in many cases, an approximately exponential relationship between potential shift and time is observed, the response time is sometimes referred to 50% of the total shift $(t_{1/2})$, analogously to other exponential relations such as radioactive decay. Other authors have used 95% (t_{95}) or 99% (t_{99}) of the total shift. In a later provisional recommendation, the IUPAC commission has abandoned the original definition based on a fixed mV-deviation and has recommended the use of the 90% (t_{90}) limit [41a].

The response time of ion selective membranes is of great importance from the point of view of the analyst and should be considered as one of the chief characteristics of these membranes. The purely thermodynamic approach to the establishment of potential differences at phase boundaries, and so across membranes, cannot yield a deeper understanding with respect to the origin of transient phenomena. This is obvious because only the energetic conditions of the initial and the final states play a role in thermodynamic treatments and neither the path along which this final state is reached nor the time needed is involved. Transient potentials should therefore be discussed on the basis of a kinetic approach. To do so, appropriate mechanistic models are needed. Unfortunately, relatively little attention has been paid to this aspect compared with the total number of papers dealing with ISEs. According to Shatkay [42], this stems from two causes; first, the treatment is fairly complicated and secondly, in general, it is difficult to test any theory experimentally. At present, no completely satisfactory model is available to explain the potential—time behaviour for solid-state membranes. The existing models can be classified into those based on the existence of an energy barrier at the surface of the electrode [43,44] and those based on the diffusion through a stagnant layer [45,46]. In the energy barrier model, the passage of ions from the solution into the membrane phase, and vice versa, is considered to be the essential rate-determining process. Starting with general expressions for the ion fluxes across an energy barrier, an expression can be found for the net charge added to the membrane phase. With the assumption that this charge accumulates in a thin layer near the surface, forming an electrical condenser with a constant capacity, the transfer of charge can be related to the potential difference across the interface.

$$\Delta\phi_t = \Delta\phi_\infty \left(1 - \exp[-Kt]\right) \tag{48}$$

where $\Delta\phi_t$ is the phase boundary potential at an arbitrary time, t, $\Delta\phi_\infty$ is the final steady-state potential, and K is a "constant" which depends on the concentration of the solution with which the membrane is brought into contact, and on the temperature.

The model based on diffusion through a stagnant layer can be considered to be a further elaboration of the concept introduced in Sect. 3.(B)(1)b. It is assumed that the membrane senses the activity of

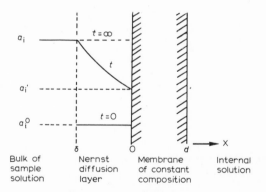

Fig. 3. Diffusion model used in the case of membranes of constant composition. The course of the activity profile in the Nernst diffusion layer after a stepwise change in the sample activity at $t = 0$ is illustrated [46]. (By kind permission of *Analytical Chemistry*.)

the relevant ion which is in immediate contact with it, and that it responds instantaneously to any change in this activity. Following the approach by Morf et al. [46], the diffusion problem illustrated in Fig. 3 must be solved. The somewhat simplified solution is

$$(a_i' - a_i^0) = (a_i - a_i^0)(1 - \exp[-t/\tau']) \qquad (49)$$

in which a_i', a_i^0 and a_i are the activities of i at the surface at times $t = t$, $t = 0$ and $t = \infty$, respectively. The time constant, τ', is approximately equal to $\delta^2/2D'$, where δ is the thickness of the Nernst diffusion layer and D' is the diffusion constant. In Fig. 4, the potential—time curves are shown calculated according to eqns. (16a) and (16b) making use of relation (49). The figure clearly illustrates that, on changing the activity from higher to lower values, the response times are much larger than in the case of changes in the opposite direction. This behaviour is always observed in real experiments. A careful analysis of the experimental response time curves reveals that, in many cases, more than one single process is involved, each separate process having its own time constant [42]. Moreover, a hyperbolic relationship seems to fit the experimental results better in some cases as was shown for the fluoride selective lanthanum fluoride crystal membrane [47]. Other factors that have to be taken into account are the fact that, shortly after a stepwise change of the activity in the bulk of the solution, the electrode surface might be in contact with

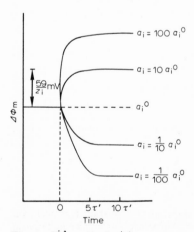

Fig. 4. Theoretical $\Delta\phi_m$ response vs. time profiles calculated according to eqn. (49) [46]. (By kind permission of *Analytical Chemistry*.)

294

different activities and the fact that, at the surface, inhomogeneities might be present in the membrane phase. The first was discussed by Lindner et al. [48] while Morf [49] dealt with the latter. The sluggish response behaviour in the presence of interfering ions was ascribed to such inhomogeneities. Recently, Buffle and Parthasarathy [50] have proposed a model in which dissolution or crystallization and charge transfer at the membrane surface are the predominant rate-determining factors. Based on a general formula for the rate of growth or dissolution of salt crystals, together with the Butler—Volmer equation, which describes the current—voltage relationship at an electrode, they could derive a relation which, after some simplification, led to a hyperbolic current—time relation.

Mechanistic information concerning the operation of solid-state ion selective membranes was also obtained by purely electrochemical methods. In this connection, the work of Brand and Rechnitz [51] and Camman [52,53] must be mentioned. The first two authors have used electrode impedance measurements to derive equivalent electrical circuit models for ion selective membranes. An example of such a circuit is shown in Fig. 5. In this circuit, R_{M_1} and R_{M_2} account for the rate of the exchange reactions at both surfaces. The capacitances in parallel with both resistances can be attributed to the electrical double layers present at the phase boundaries. The other resistances may account for the limited transport rates of the relevant species inside and outside the membrane. Camman has tried to measure exchange current densities, among other methods by means of current—voltage curves and by current—time and voltage—time curves [52,54]. According to his findings, a high correlation is observed between the selectivity of the electrode towards a certain ion and the corresponding standard exchange current density. This discussion is facilitated by making use of so-called Evans diagrams [55],

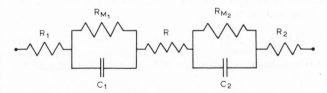

Fig. 5. Equivalent electrical circuit model for ISE with ion transport across solid-state membrane (e.g. silver chloride). R, R_1, R_2, R_{M_1} and R_{M_2} are resistances; C_1 and C_2 are capacitances.

an approach developed for the discussion of corrosion processes at metal electrodes. The occurrence of parallel electrode reactions results in mixed potentials (corrosion potentials). It is pointed out that high standard exchange current densities are essential for a proper functioning of the electrode as an ideal Nernstian sensor, because it allows the measurements of the potential, even if some current is drawn, without a potential drift in consequence of a disturbance of the equilibrium (direct reading mV-meters use currents of 10^{-10} to 10^{-14} A).

(2) Semipermeable membranes; Donnan potential

Semipermeable membranes are characterized by the property that the various components diffuse through such a membrane with different velocities. If the membrane prevents the diffusion of at least one kind of ion completely, a so-called *Donnan potential* is formed [6], which prevents the other ions crossing the membrane. To illustrate the phenomenon, equal volumes of two solutions (1 and 2) will be considered. Solution 1 contains a salt MX in a concentration C_1 and solution 2 contains the salt MR in a concentration C_2. The solutions are separated by a membrane that allows the transport of M^+ and X^- ions, but is impermeable to R^- ions. It is obvious that, irrespective of the values of C_1 and C_2, X^- ions will tend to diffuse from solution 1 to solution 2. To maintain electroneutrality at both sides of the membrane, an equivalent amount of positive charges, i.e. M^+ ions, has to be transferred. This leads to the following situation after establishment of the equilibrium.

solution 1; M^+	X^+	membrane	solution 2; M^+	X^-	R^-
$c_1 - \Delta c$	$c_1 - \Delta c$		$c_2 + \Delta c$	Δc	c_2

At thermodynamic equilibrium, the electrochemical potentials for M^+ should be equal at both sides of the membrane. Of course, the same applies to X^-, but not to R^-.

$$\bar{\mu}_{M^+}(1) = \bar{\mu}_{M^+}(2); \bar{\mu}_{X^-}(1) = \bar{\mu}_{X^-}(2) \qquad (50)$$

For sufficiently dilute solutions, pressure effects can be disregarded leading to

$$\mu^0_{M^+}(1) + RT \ln a_{M^+}(1) + F \phi(1) = \mu^0_{M^+}(2) + RT \ln a_{M^+}(2) + F \phi(2)$$

$$(51)$$

296

If the same solvent is used on both sides of the membrane, the standard electrochemical potentials μ^0 are equal and it is possible to write eqn. (51) as

$$\phi_D = \phi_2 - \phi_1 = \frac{RT}{F} \ln \frac{a_{M^+}(1)}{a_{M^+}(2)} \tag{52a}$$

and, similarly

$$\phi_D = \frac{RT}{F} \ln \frac{a_{X^-}(2)}{a_{X^-}(1)} \tag{52b}$$

where ϕ_D is the Donnan potential.
Combination of both equations gives

$$\frac{a_{M^+}(1)}{a_{M^+}(2)} = \frac{a_{X^-}(2)}{a_{X^-}(1)} \tag{53}$$

More generally

$$\frac{a_{M^+}(1)}{a_{M^+}(2)} = \left\{\frac{a_{M^{2+}}(1)}{a_{M^{2+}}(2)}\right\}^{1/2} = \left\{\frac{a_{M^{3+}}(1)}{a_{M^{3+}}(2)}\right\}^{1/3} = .. = \frac{a_{X^-}(2)}{a_{X^-}(1)} = \lambda \tag{54}$$

where λ denotes the *Donnan distribution coefficient*.
Substitution of eqn. (54) into eqns. (52a) and (52b) yields [56]

$$\phi_D = \frac{RT}{F} \ln \lambda \tag{55}$$

If activities can be replaced by concentrations, application of eqn. (54) to the example described above leads to

$$\lambda = \frac{C_1 - \Delta C}{C_2 + \Delta C} = \frac{\Delta C}{C_1 - \Delta C} \rightarrow \Delta C = \frac{C_1^2}{C_2 + 2C_1} \tag{56}$$

It has to be emphasized that eqns. (50)—(55) also apply to systems in which equilibrium conditions are restricted to regions near the interfaces, provided that the proper local activities or concentrations are used.

(C) LIQUID JUNCTION POTENTIAL

If two solutions of electrolytes are brought into contact with each other, there will be a region in which the composition varies gradually from that of the one solution to that of the other. Generally, such an interphase region is very liable to mechanical or other convective disturbances and therefore a diaphragm is often used to stabilize the interphase region. As long as the compositions of both solutions are mutually different, there will exist a tendency for the transport of ions across the interphase. Differences in the mobilities of the various kinds of ions will cause a separation of charge, leading to the production of a *liquid junction potential.*

In many electrochemical cells, such liquid junction potentials have to be taken into account. Since irregularities in the total EMF values of cells can often be attributed to the instability of these junction potentials, a thorough knowledge of them is of great importance.

The liquid junction potential is caused by the mass-fluxes of ions, which are governed by the Nernst—Planck equation [eqn. (9)]. The mass-flux of each kind of ion is accompanied by transfer of charge and will contribute to the *electrical current density*, j, i.e. the electrical current per unit area.

$$j_i = z_i F_i J_i \tag{57}$$

In most potentiometric measurements, the total current density is very small so that

$$j = \sum j_i = \sum z_i F J_i \approx 0 \tag{58}$$

Substitution of eqn. (9) leads to

$$\frac{\mathrm{d}\phi}{\mathrm{d}x} = -\frac{RT}{F} \frac{\sum z_i U_i \mathrm{d}C_i/\mathrm{d}x}{\sum z_i^2 U_i C_i} = -\frac{RT}{F} \frac{\sum z_i U_i C_i \mathrm{d} \ln C_i/\mathrm{d}x}{\sum z_i^2 U_i C_i} \tag{59}$$

Introduction of the *transference number*, t, which is defined as

$$t_i = \frac{z_i^2 U_i C_i}{\sum z_i^2 U_i C_i} \tag{60}$$

yields

$$\frac{\mathrm{d}\phi}{\mathrm{d}x} = -\frac{RT}{F} \sum \frac{t_i}{z_i} \frac{\mathrm{d} \ln C_i}{\mathrm{d}x} \tag{61}$$

Integration across the complete interphase of thickness d gives the liquid junction potential, $\Delta\phi_L$.

$$\Delta\phi_L = -\frac{RT}{F} \int_0^d \sum \frac{t_i}{z_i} \mathrm{d} \ln C_i \tag{62}$$

298

Obviously, a knowledge of the concentration profiles inside the interphase is required to evaluate the integral presented in eqn. (62). A solution was presented by Henderson [57] who arbitrarily assumed linear concentration profiles for all ions within the diffusion layer, i.e.

$$\frac{dC_i(x)}{dx} = \frac{C_i(d) - C_i(0)}{d} = \frac{\Delta C_i}{d}; \; C_i(x) = C_i(0) + \frac{\Delta C_i}{d}x \tag{63}$$

Substitution in eqn. (62) and integration yields the well-known Henderson formula

$$\Delta\phi_L = -\frac{RT}{F} \frac{\sum z_i U_i \Delta C_i}{\sum z_i^2 U_i \Delta C_i} \ln \frac{\sum z_i^2 U_i C_i(d)}{\sum z_i^2 U_i C_i(0)} \tag{64}$$

An exact, but more complicated, solution of eqn. (62) was presented earlier by Planck [2] and a modification of his solution was recently given in a new form by Morf [58].

$$\Delta\phi_L = -\frac{RT}{F} \frac{\bar{u}_m - \bar{u}_x}{|z_m|\bar{u}_m + |z_x|\bar{u}_x} \ln \frac{\sum C_i(d)}{\sum C_i(0)} \tag{65}$$

in which the index m applies to cations and the index x to anions; $\bar{u}_i$

TABLE 1

Liquid junction potential values at 25°C calculated from the Planck theory and the Henderson equation [58]

Sample solution	$\Sigma C_i(d)/\Sigma C_i(0)$	$\Delta\phi_L$ (mV) [eqns. (65) and (66)]	$\Delta\phi_L$ (mV) [eqn. (64)]
KCl	10^{-3}	0.00	0.00
	10^{-2}	0.01	0.01
	10^{-1}	0.05	0.05
	1	0.18	0.18
NaCl	10^{-2}	0.20	0.20
	10^{-1}	1.11	1.11
	1	4.60	4.60
HCl	10^{-4}	−0.04	−0.04
	10^{-3}	−0.32	−0.28
	10^{-2}	−2.07	−1.73
	10^{-1}	−9.40	−8.31
	1	−26.73	−26.77
	10	−52.84	−57.58

References pp. 385–392

are mean mobilities which are defined by the relationship

$$\bar{u}_i = \frac{\sum U_i C_i(d) \exp\{z_i F \Delta \phi_L / RT\} - \sum U_i C_i(0)}{\sum C_i(d) \exp\{z_i F \Delta \phi_L / RT\} - \sum C_i(d)} \tag{66}$$

$\Delta\phi_L$ can be calculated by means of an iterative process. At first any arbitrary value of $\Delta\phi_L$ is inserted in eqn. (66). The mean mobilities thus obtained are the first-order approximations to be used in eqn. (65). The value of $\Delta\phi_L$ then calculated is used to obtain the next value of $\bar{u}_i$, etc. Due to a restriction introduced in the derivation, the Morf equations (65) and (66) apply only to solutions containing anions of the same charge and cations of the same charge.

To give an idea of the order of magnitude of liquid junction potential values, some values are given in Table 1.

(D) MEMBRANE POTENTIALS IN THE PRESENCE OF INTERNAL DIFFUSION POTENTIAL

In the previous paragraphs, two extreme cases of membrane potential have been considered. At first, a dicussion was given of membrane potentials in the absence of any contribution by an inner diffusion potential; subsequently, a discussion of the pure diffusion potential in the absence of any boundary potential was presented. In the following paragraphs, intermediate cases will be considered. The general model that will be adopted comprises at least three different contributions: two boundary (Donnan) potentials at both solution—membrane interfaces and a liquid junction potential inside the membrane. This three-regions concept was first introduced by Teorell [7], and Meyers and Sievers [8] (see also ref. 59). Different expressions for the boundary potential will be introduced, depending on the type of equilibrium existing at the interface. Also, the inner diffusion potential will be expressed in different ways depending on the composition of the membrane and the way in which the relevant ions are transported through it. For the sake of simplicity and in order to obtain equations that are explicit in the relevant parameters, Henderson's instead of Planck's relation will be used. A basic condition that always has to be fulfilled inside the membrane, is that of electroneutrality.

(1) Solid-state ion-exchange membranes

a. A formula for the membrane potential. Let us first consider a system of a solid-state ion-exchange membrane with fixed negative charges separating two solutions of the same salt MX with different concentrations. Schematically, this system can be represented in the following way.

where $\bar{c}$ denotes the "concentration" of fixed negative sites inside the membrane. At low concentration levels of MX, activities can be replaced by concentrations. Then, the following relations must hold true according to eqn. (53).

$$C_M(1)C_X(1) = C_M(m)_{x=0} C_X(m)_{x=0} \tag{67}$$

$$C_M(2)C_X(2) = C_M(m)_{x=d} C_X(m)_{x=d}$$

Electroneutrality conditions applied to the membrane phase lead to

$$C_M(m)_{x=0} = C_X(m)_{x=0} + \bar{c} \atop C_M(m)_{x=d} = C_X(M)_{x=d} + \bar{c}, \tag{68}$$

Substitution into Henderson's equation, eqn. (64), yields the inner diffusion potential contribution

$$\phi_{(x=d)} - \phi_{(x=0)} = -\frac{RT}{F}\frac{U_M - U_X}{U_M + U_X}\ln\frac{(U_M + U_X)C_M(m)_{x=d} - U_X\bar{c}}{(U_M + U_X)C_M(m)_{x=0} - U_X\bar{c}} \tag{69}$$

Together with the two Donnan potentials [N.B. $\mu^0(1) = \mu^0(2)$]

$$\Delta\phi_D(x=0) = \frac{RT}{F}\ln\frac{C_M(1)}{C_M(m)_{x=0}} + \frac{\mu_M^0(1) - \mu_M^0(m)}{F} \tag{70a}$$

and

$$\Delta\phi_D(x=d) = \frac{RT}{F}\ln\frac{C_M(m)_{x=d}}{C_M(2)} + \frac{\mu_M^0(m) - \mu_M^0(2)}{F} \tag{70b}$$

the total membrane potential can be represented as

$$\Delta\phi_m = \frac{RT}{F}\left[\ln\frac{C_M(1)C_M(m)_{x=d}}{C_M(2)C_M(m)_{x=0}} - \frac{U_M - U_X}{U_M + U_X}\right.$$

$$\left.\ln\left\{\frac{(U_M + U_X)C_M(m)_{x=d} - U_X\overline{c}}{(U_M + U_X)C_M(m)_{x=0} - U_X\overline{c}}\right\}\right] \tag{71}$$

If the membrane is really permselective, the membrane with fixed negative charges will completely prevent the permeation of co-ions, i.e. X^-. Hence, $U_X \simeq 0$ and eqn. (71) reduces to

$$\Delta\phi_m = \frac{RT}{F}\ln\frac{C_M(1)}{C_M(2)}$$

which is identical to the pure Donnan potential as given by eqns. (52a) and (52b).

b. Interference; selectivity. If, apart from M^+, another cation N^+ is present in the solution, a competition between both cations will be observed. This can be expressed by means of the exchange reaction

$$N^+(s) + M^+(m) \rightleftharpoons N^+(m) + M^+(s)$$

where the symbol (s) denotes ions in the solution and (m) denotes those in the membrane. If, furthermore, the membrane with fixed negative sites can be considered to be virtually permselective, i.e. $U_X \simeq 0$, the Henderson relation can be written as

$$\phi(x = d) - \phi(x = 0) = -\frac{RT}{F}\ln\frac{U_M C_M(m)_{x=d} + U_N C_N(m)_{x=d}}{U_M C_M(m)_{x=0} + U_N C_N(m)_{x=0}}$$

$$= -\frac{RT}{F}\ln\frac{C_M(m)_{x=d} + (U_N/U_M)C_N(m)_{x=d}}{C_M(m)_{x=0} + (U_N/U_M)C_N(m)_{x=0}} \tag{72}$$

The total membrane potential is obtained by summation of eqn. (72) and both Donnan potentials at the solution—membrane interfaces.

$$\Delta\phi_m = \frac{RT}{F}\ln\frac{C_N(m)_{x=d}C_N(1)}{C_N(m)_{x=0}C_N(2)} - \frac{RT}{F}\ln\frac{C_M(m)_{x=d} + (U_N/U_M)C_N(m)_{x=d}}{C_M(m)_{x=0} + (U_N/U_M)C_N(m)_{x=0}}$$

$$\tag{73}$$

Introduction of the ion-exchange constant

$$K_{MN} = \frac{C_N(m)_{x=0}C_M(1)}{C_N(1)C_M(m)_{x=0}} = \frac{C_N(m)_{x=d}C_M(2)}{C_N(2)C_M(m)_{x=d}} \qquad (74)$$

into eqn. (73) leads to

$$\Delta\phi = \frac{RT}{F}\ln\frac{C_M(1) + (U_N/U_M)K_{MN}C_N(1)}{C_M(2) + (U_N/U_M)K_{MN}C_N(2)} \qquad (75)$$

If solution 1 is kept constant, this equation can be written as

$$\Delta\phi = \text{constant} - \frac{RT}{F}\ln[C_M(2) + (U_N/U_M)K_{MN}C_N(2)]$$

$$= \text{constant} - \frac{RT}{F}\ln[C_M + k_{M,N}^{pot}C_N] \qquad (76)$$

where

$$k_{M,N}^{pot} = (U_N/U_M)K_{MN}. \qquad (77)$$

This means that the potentiometric selectivity coefficient depends not only on an equilibrium constant, but also on the ratio of the particle mobilities of both metal ions [13].

c. Kinetics; response time. A constant membrane potential implies the establishment of a steady state. Apart from the diffusion in the membrane, the diffusion in the two Nernst layers adjacent to the membrane has to be taken into account. As long as the permeability through the membrane is small, the diffusion in the Nernst layers is of minor importance for the overall reaction rate. This applies to non-electrolytes and co-ions. For the counter ions, M^+ in the case of a membrane with fixed negative sites, the rate-determining step depends on the diffusion coefficients, the thickness of the membrane and the ratio of the concentrations inside and outside the membrane. The smaller the diffusion coefficient inside the membrane and the thicker the membrane, the more important the diffusion in the membrane will be in comparison with the diffusion in the Nernst layers. Helfferich [60] has derived a formula for the response time for the case in which diffusion in the membrane is the rate-determining step. He has shown that the flux has approached its steady state value to within 3% in the time

$$t_{97} = 0.42\frac{d^2}{D} \qquad (78)$$

in which d is the membrane thickness and D the diffusion coefficient in the interior of the membrane. For a typical ion-exchanger membrane ($D \sim 2.10^{-6}$ cm^2 sec^{-1}; $d \sim 0.1$ cm) the response time, t_{97}, is, according to Helfferich, of the order of 30 min.

(2) Liquid membranes with dissolved electroneutral carrier

a. A formula for the membrane potential. Liquid membranes are distinguished from solid-state membranes by the fact that the ion selective sites, S, are mobile. Such membranes consist of a solvent, immiscible with the two adjacent aqueous solutions, in which a *carrier* is dissolved. A carrier is a compound that is able to form complexes selectively with an ion, e.g. the cations M^{+z_m}, N^{+z_n}. The membrane potential of liquid membranes in contact with these cations, as well as with anions X^{-z_x} can be obtained by combination of eqn. (65) and the two boundary potentials. After some algebra, the following general relationship was found by Morf [58,61].

$$\Delta\phi_m = (1 - t_x)\frac{RT}{z_m F}\ln\frac{\sum k_M a_M(1)}{\sum k_M a_M(2)} + (t_x)\frac{RT}{z_x F}\ln\frac{\sum k_X a_X(1)}{\sum k_X a_X(2)} \tag{79}$$

where k_M and k_X are overall distribution coefficients and t_x is the integral anionic transference number defined as

$$t_x = \frac{|z_x|\overline{u}_x}{|z_m|\overline{u}_m + |z_x|\overline{u}_x} \tag{80}$$

The mean ionic mobilities introduced in eqn. (80) may be expressed as

$$\overline{u}_i = \frac{\sum U_i k_i a_i(d)\exp\{z_i F\Delta\phi_m/RT\} - \sum U_i k_i a_i(0)}{\sum k_i a_i(d)\exp\{z_i F\Delta\phi_m/RT\} - \sum k_i a_i(0)} \tag{81}$$

If the carrier itself is electrically neutral and the membrane solvent has a low dielectrical constant, the membrane appears to be almost completely permselective to cations, i.e. $t_x \simeq 0$, and the last term of the right-hand side of eqn. (79) will vanish. Morf and Simon [61] assume that the complexation of M^{+z_m} proceeds in three consecutive steps (1) transfer of free carriers from the membrane into the boundary layers, (2) formation of the complexes $MS_n^{+z_m}$, and (3) transfer of complexes into the membrane phase. k_M can then be expressed as

$$k_M = \sum_n (\beta_n)_{aq} k_{MS_n}\left(\frac{C_S}{k_S}\right)^n \tag{82}$$

304

where $(\beta_n)_{aq}$ is the stability constant of the complex MS_n^{+zm} in the aqueous phase, k_{MS_n} is the distribution coefficient of MS_n^{+zm}, C_S is the concentration of the carrier in the membrane phase, and k_S is the distribution coefficient of S.

b. Interference; selectivity. If a virtually permselective membrane is in contact with two monovalent cations, M^+ and N^+, and it is assumed that only 1 : 1 complexes can be formed with the electroneutral carrier S, then eqn. (79) can be written as

$$\Delta\phi_m = \frac{RT}{F} \ln \frac{k_M a_M(1) + k_N a_N(1)}{k_M a_M(2) + k_N a_N(2)} \tag{83}$$

If the solution at one side of the membrane is kept constant, eqn. (83) can be simplified to

$$\Delta\phi = \text{constant} + \frac{RT}{F} \ln\{a_M + (k_N + (k_N/k_M)a_N\}$$

$$= \text{constant} + \frac{RT}{F} \ln\{a_M + k_{M,N}^{pot} a_N\} \tag{84}$$

$k_{M,N}^{pot}$ corresponds to the equilibrium constant of the ion exchange reaction

$$N^+(s) + MS^+(m) \rightleftharpoons NS^+(m) + M^+(s)$$

where (s) denotes the solution and (m) the membrane phases. Suitable neutral carriers are, in general, macrocyclic compounds. Since the geometrical dimensions and therefore the distribution coefficients of these complexes are approximately independent of the nature of the central ion, the potentiometric selectivity coefficient can be simply expressed as the quotient of the stability constants in aqueous medium, provided that the complexes are solvated in the same way in the interior of the membrane phase.

$$k_{M,N}^{pot} = \frac{(\beta_{NS})_{aq}}{(\beta_{MS})_{aq}} \tag{85}$$

It is interesting to notice that, in this case, the selectivity is not influenced by the kind of solvent used for the membrane. From eqn. (82), it is evident that such a simple relationship for the selectivity coefficient cannot be expected when the interfering ion, N, forms complexes with S of a stoichiometry which differs from that of the complexes of M with S. In that case, the concentration of free ligand in the membrane becomes important. If the interfering ion has a valency different from that of the primary ion, it is difficult to derive an explicit relation for the selectivity coefficient. It was pointed out by Morf and Simon [61] that, for instance, the mono-valent—divalent cation selectivity varies with the carrier concentration and the polarity of the membrane solvent as described by the dielectric constant. So, preference of a membrane for monovalent ions is improved when reducing the polarity, whereas an increase of polarity favours the selectivity towards divalent cations. However, an increase of the polarity of the membrane solvent enhances the solubility and permeability of lipophilic anions such as thiocyanate and perchlorate. This may lead to the situation that the last term of the right-hand side of eqn. (79) can no longer be neglected and an anion interference has to be expected. As polar solvents are a pre-requisite for liquid membranes selective to divalent cations, and a decrease of the permeability is not suitable, an alternative method for reducing the anion interference was suggested by Morf et al. [62]. By dissolving a highly lipophilic anion such as tetraphenyl borate, they incorporated permanent anionic sites into the mem-brane which excluded the penetration of anions from the sample. Thus, a diminution of the anion interference was obtained. Figure 6 illustrates the favourable influence of tetraphenyl borate.

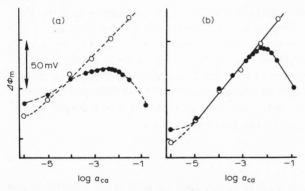

Fig. 6. $\Delta\phi_m$ response of neutral carrier membrane to Ca^{2+} in the presence of chloride (O) and thiocyanate (●) [62]. (a) Membrane solvent without tetra-phenyl borate (TPB); (b) the same membrane with TPB (equivalent to 100 mole % of the ligand concentration). (By kind permission of *Analytical Letters*.)

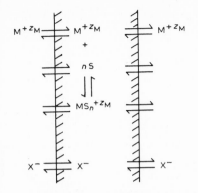

Fig. 7. Schematic representation of a neutral carrier membrane.

c. Kinetics; response time. The kinetics of a neutral carrier membrane will be discussed on the basis of the schematic representation depicted in Fig. 7. As long as the concentration of cationic complexes is much smaller than the concentration of free carrier, the latter concentration can be considered to remain constant throughout the membrane. In that case, the following relationship holds.

$$\frac{C_{MS_n}(m)_{x=0}}{a_M(1)} = \frac{C_{MS_n}(m)_{x=d}}{a_M(2)} = K \quad (K \ll 1) \tag{86}$$

Since diffusion in the membrane is, in general, slow in comparison with diffusion in the aqueous Nernst layers, the establishment of the steady state inside the membrane is the rate-determining step. The same assumption was introduced in the discussion of the response of solid-state ion-exchange membranes.

Figure 8 illustrates the diffusion problem that has to be solved [46]. The activity $a_M(2)$ is kept constant and it is assumed that the activity is the same throughout the whole solution at the right-hand side of the membrane. On the left-hand side, the activity in the bulk is changed stepwise from $a_M(1)_{t=0} = a_M^0(1)$ to $a_M(1)$. When equilibrium is again attained at this side of the membrane (i.e. for $t \to \infty$), the activity at the surface will have obtained the value $a_M(1)$ as well, but in the transient period $a_M(1)_{x=0}(t = t) = a_M'(1)$ will be smaller leading to a flux of $M^{+z}m$ across the Nernst diffusion layer.

$$J_M(t) = D_M \frac{a_M(1) - a_M'(1)}{\delta} \tag{87}$$

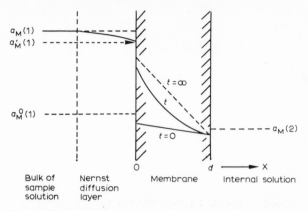

Fig. 8. Diffusion model for neutral carrier membrane. The establishment of a steady-state concentration profile within the membrane is assumed to be rate-determining [46]. (By kind permission of *Analytical Chemistry*.)

Because of the continuity of fluxes

$$J_M(t) = J_M(m)(x = 0, t) \tag{88}$$

where $J_M(m)$ is the flux inside the membrane.

Carslaw and Jaeger [63] have solved a similar problem for the conduction of heat. Transformation of their result to the diffusion problem leads to

$$J_M(m)(x = 0, t) = D*K \frac{a'_M(1) - a_M(1)}{\sqrt{\pi D*t}} \left(1 + 2\sum_n^\infty \exp[-n^2 d^2/D*t]\right.$$
$$\left. - \frac{\sqrt{\pi D*t}}{d}\right) + D*K \frac{a'_M(1) - a_M(2)}{d} \tag{89}$$

where D^* signifies the mean diffusion coefficient of the ions ($MS_n^{+z_M}$ and X^-) in the membrane. For the first period (t is small), $D*t \ll d^2$ and the term between brackets in eqn. (89) can be simplified, leading to

$$J_M(m)(x = 0, t) = D*K \frac{a'_M(1) - a_M^0(1)}{\sqrt{\pi D*t}} \tag{90}$$

Combination of eqns. (87), (88) and (90) leads to

$$a'_M(1) - a_M(1) = [a_M(1) - a_M^0(1)](\sqrt{t/\tau})(1 + \sqrt{t/\tau})^{-1} \tag{91}$$

308

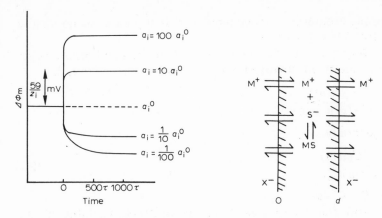

Fig. 9. Theoretical $\Delta\phi_m$ response vs. time profiles for neutral carrier membranes according to eqn. (91) [46]. (By kind permission of *Analytical Chemistry*.)

Fig. 10. Schematic representation of a membrane with dissolved charged ligands.

in which

$$\tau = \frac{D^* K^2 \delta^2}{D_m^2 \pi} \tag{92}$$

The time response constant, τ, can decrease not only by reduction of D^* and K, which are properties of a particular membrane, but also by thoroughly stirring the sample solution and so decreasing δ, the Nernst diffusion layer thickness. In Fig. 9, theoretical EMF versus time curves are presented calculated on the basis of eqn. (91).

(3) Liquid membranes with dissolved charged ligands

a. A formula for the membrane potential. When a membrane composed of a water-immiscible solvent, interposed between two aqueous solutions, contains an appreciable concentration of a charged ligand, S^-, it can function as a liquid ion-exchange membrane. Unlike solid ion-exchangers, the sites are mobile and will contribute to the charge transfer across the membrane. Such a membrane can be represented as shown in Fig. 10. Sandblom et al. [64] (see also ref. 13) have discussed the potential response based on the following assumptions.

(1) Only 1 : 1 complexes will be considered.

(2) The membrane is homogeneous and is supposed to be ideally permselective, i.e. $C_X(m) = 0$.

(3) At every point in the membrane, chemical equilibrium exists and the concentrations involved obey the law of mass action, i.e.

$$K_{MS}(m) = \frac{C_{MS}(m)}{C_M(m)C_S(m)}$$

or, more generally

$$K_{iS}(m) = \frac{C_{iS}(m)}{C_i(m)C_S(m)} \tag{93}$$

where $i = M^+, N^+$ etc.

(4) S^- and the complexes iS are completely insoluble in the aqueous phase, hence the total concentration of the ligand can be written

$$C_S^{tot}(m) = C_S(m) + \sum C_{iS}(m) \tag{94}$$

(5) The condition of electroneutrality must hold inside the membrane. In the case of monovalent ions, this leads to

$$C_S(m) = \sum C_i(m) \tag{95}$$

The Nernst—Planck relation for all the species present in the solution yields the following set of equations.

$$J_i = -U_i RT \frac{dC_i(m)}{dx} - FU_i C_i(m) \frac{d\phi}{dx} \tag{96a}$$

$$J_S = -U_S RT \frac{dC_S(m)}{dx} + FU_S C_S(m) \frac{d\phi}{dx} \tag{96b}$$

$$J_{iS} = -U_{iS} RT \frac{dC_{iS}(m)}{dx} \tag{96c}$$

When no S^- or iS is leaving the membrane, the total flux of S must be zero in the steady state.

$$J_S^{tot} = J_S + \sum J_{iS} = 0 \tag{97}$$

Under such conditions that no electrical current flows through the cell, the summation of the fluxes of positively charged species must just counterbalance the flux of the anionic sites, S^-, i.e.

$$J_S - \sum J_i = 0 \tag{98}$$

310

After substitution of eqns. (96a) and (96b) in eqn. (98), the following result is obtained.

$$\frac{d\phi}{dx} = -\frac{RT}{F} \frac{1}{\sum U_i C_i(m) + U_S C_S(m)} \frac{d}{dx} \{ \sum U_i C_i(m) - U_S C_S(m) \} \quad (99)$$

This equation is essentially equal to the formula for the liquid junction potential [eqn. (59)]. Before eqn. (99) can be integrated, the term $dC_S(m)/dx$ must be eliminated. Therefore, another equation containing this term has to be available. Combination of eqns. (96b), (96c) and (97) yields, after introduction of the equilibrium constant [eqn. (93)], the following relation.

$$\frac{-RTU_S dC_S(m)}{dx} + \frac{FU_S C_S(m)d\phi}{dx} - \frac{RT \sum U_{iS} K_{iS}(m) C_S(m) dC_i(m)}{dx}$$
$$- \frac{RT \sum U_{iS} K_{iS}(m) C_i(m) dC_S(m)}{dx} = 0 \quad (100)$$

Together with eqn. (99) and after some rearrangement, a relation is obtained for the steady state at zero current.

$$\frac{d\phi}{dx} = -\frac{RT}{F} \frac{\sum U_i dC_i(m)/dx}{\sum U_i C_i(m)} -$$
$$t \left[\frac{\sum U_{iS} K_{iS}(m) dC_i(m)/dx}{\sum U_{iS} K_{iS}(m) C_i(m)} - \frac{\sum U_i dC_i(m)/dx}{\sum U_i C_i(m)} \right] \quad (101)$$

where

$$t = \frac{U_S C_S(m)}{[\{ U_S C_S(m)/\sum U_{iS} C_{iS}(m) \} + 1] \sum U_i C_i(m) + U_S C_S(m)} \quad (102)$$

On integration of eqn. (101) between the limits $x = 0$ and $x = d$, an expression for the diffusion potential is obtained. Addition of the two Donnan potentials will lead to the total membrane potential

$$\Delta\phi_m = -\frac{RT}{F} \ln \frac{\sum U_i C_i(m)_{x=d}}{\sum U_i C_i(m)_{x=0}} - \int_0^d t d \ln \left[\frac{\sum U_{iS} K_{iS}(m) C_i(m)}{\sum U_i C_i(m)} \right]$$
$$+ \Delta\phi_D(x = 0) + \Delta\phi_D(x = d) \quad (103)$$

The two Donnan potential equations (70a) and (70b) can be written in a modified form.

$$\Delta\phi_D(x = 0) = \frac{RT}{F} \ln \frac{a_i(1)k_i}{C_i(m)_{x=0}} \tag{104a}$$

$$\Delta\phi_D(x = d) = \frac{RT}{F} \ln \frac{C_i(m)_{x=d}}{a_i(2)k_i} \tag{104b}$$

where the distribution coefficient, k_i, is defined as

$$k_i = \exp \frac{\mu_i^0 - \mu_i^0(m)}{RT} \tag{105}$$

According to eqn. (104a)

$$\frac{a_M(1)k_M}{C_M(m)_{x=0}} = \frac{a_N(1)k_N}{C_N(m)_{x=0}} = \ldots = \frac{a_i(1)k_i}{C_i(m)_{x=0}} = \frac{\sum a_i(1)k_i}{\sum C_i(m)_{x=0}} \tag{106}$$

Because of the electroneutrality condition represented by eqn. (95), eqn. (106) can be rewritten as

$$C_i(m)_{x=0} = C_S(m)_{x=0} \frac{a_i(1)k_i}{\sum a_i(1)k_i} \tag{107a}$$

and, similarly

$$C_i(m)_{x=d} = C_S(m)_{x=d} \frac{a_i(2)k_i}{\sum a_i(2)k_i} \tag{107b}$$

These results can be substituted in eqns. (104a) and (104b), respectively, and subsequently combined with eqn. (103).

$$\Delta\phi_m = -\frac{RT}{F} \ln \left[\frac{\sum U_i k_i a_i(2)}{\sum U_i k_i a_i(1)} \right] - \int_0^d t \, d \ln \left[\frac{\sum U_{iS} K_{iS}(m)C_i(m)}{\sum U_i C_i(m)} \right] \tag{108}$$

Two limiting cases can now be distinguished, (1) the complexes are almost completely dissociated, i.e. $C_{iS} \rightarrow 0$; (2) the association is almost complete, i.e. $C_{iS} \gg C_S$.

Strong dissociation. In this case, the term $[U_S C_S(m)]/[\Sigma U_{iS} C_{iS}(m)]$ and so the whole denominator of eqn. (102) becomes very large and t will tend to zero. Equation (108) will then simply result in

$$\Delta\phi_m = -\frac{RT}{F} \ln \left[\frac{\sum U_i k_i a_i(2)}{\sum U_i k_i a_i(1)} \right] \tag{109}$$

312

Strong association. Now, the term $[U_S C_S(m)]/[\Sigma U_{iS} C_{iS}(m)]$ is much smaller than unity and the parameter t can be expressed as

$$t = \frac{U_S C_S(m)}{\sum U_i C_i(m) + U_S C_S(m)} \tag{110}$$

in which t can be identified as the transference number of free sites. If the discussion is restricted to the situation where two different kinds of cations, M^+ and N^+, are present, the electroneutrality condition demands that

$$C_S(m) = C_M(m) + C_N(m) \tag{111}$$

In this case, the total membrane potential, eqn. (108), yields

$$\Delta\phi_m = -\frac{RT}{F}\ln\left[\frac{\sum U_i k_i a_i(2)}{\sum U_i k_i a_i(1)}\right] - \int_0^d \left\{\frac{U_S[C_M(m)/C_N(m)] + 1}{(U_M + U_S)[C_M(m)/C_N(m)] + U_N + U_S}\right\}$$

$$\times \, \text{dln}\left[\frac{U_{MS}K_{MS}(m)[C_M(m)/C_N(m)] + U_{NS}K_{NS}(m)}{U_M[C_M(m)/C_N(m)] + U_N}\right] \tag{112}$$

After integration, Sandblom et al. [64] obtained a general expression

$$\Delta\phi_m = -\frac{RT}{F}\left[(1-\tau)\ln\frac{(U_M + U_S)k_M a_M(1) + (U_N + U_S)k_N a_N(1)}{(U_M + U_S)k_M a_M(2) + (U_N + U_S)k_N a_N(2)}\right.$$

$$\left. + \tau\ln\frac{U_{MS}K_{MS}(m)k_M a_M(1) + U_{NS}K_{NS}(m)k_N a_N(1)}{U_{MS}K_{MS}(m)k_M a_M(2) + U_{NS}K_{NS}(m)k_N a_N(2)}\right] \tag{113}$$

where

$$\tau = \frac{U_S(U_{NS}K_{NS}(m) - U_{MS}K_{MS}(m))}{(U_M + U_S)U_{NS}K_{NS}(m) - (U_N + U_S)U_{MS}K_{MS}(m)} \tag{114}$$

b. Interference; selectivity. The influence of interfering cations is included in the derivation of the equations in Sect. 3.(D)(3)a. An expression for the selectivity towards the primary ion, M^+, with respect to any other cation, N^+, can be found by considering just these two cations.

In the case of almost complete dissociation, this leads, according to eqn. (109), to

$$
\Delta\phi_m = -\frac{RT}{F} \ln \frac{U_M k_M a_M(2) + U_N k_N a_N(2)}{U_M k_M a_M(1) + U_N k_N a_N(1)}
$$

$$
= -\frac{RT}{F} \ln \frac{a_M(2) + \{(U_N k_N)/(U_M k_M)\} a_N(2)}{a_M(1) + \{(U_N k_N)/(U_M k_M)\} a_N(1)} \tag{115}
$$

When solution (1) is kept constant, this results in

$$
\Delta\phi_m = -\frac{RT}{F} \ln\{a_M + k_{M,N}^{pot}\} \tag{116}
$$

in which

$$
k_{M,N}^{pot} = \frac{(U_N k_N)}{(U_M k_M)} \tag{117}
$$

$$
\Delta\phi_m = -\frac{RT}{F} \ln \frac{(U_M + U_S)k_M a_M(1) + (U_N + U_S)k_N a_N(1)}{(U_M + U_S)k_M a_M(2) + (U_N + U_S)k_N a_N(2)}
$$

In the case of strong association, a distinction can be made depending on the value of the mobility of the free sites. When U_S is very small, τ tends to zero, and eqn. (113) reduces to

$$
\Delta\phi_m = -\frac{RT}{F} \ln \frac{a_M(1)(U_N + U_S)k_N/(U_M + U_S)k_M a_N(1)}{a_M(2)(U_N + U_S)k_N/(U_M + U_S)k_M a_N(2)} \tag{118}
$$

When solution (1) is kept constant again, eqn. (118) reduces to eqn. (116) in which now

$$
k_{M,N}^{pot} = \frac{(U_N + U_S)k_N}{(U_M + U_S)k_M} \tag{119}
$$

Furthermore, when $U_S \ll U_M$, U_N, the selectivity coefficient can be reduced to

$$
k_{M,N}^{pot} = \frac{U_N k_N}{U_M k_M}
$$

which is essentially the same result as obtained in the case of almost complete dissociation [see eqn. (117)].

If the mobility of the free sites is very large, the value of τ approaches unity and eqn. (113) will have the form

$$
\Delta\phi_m = -\frac{RT}{F} \ln \frac{a_M(1) + [U_{NS} K_{NS}(m) k_N/U_{MS} K_{MS}(m) k_M] a_N(1)}{a_M(2) + [U_{NS} K_{NS}(m) k_N/U_{MS} K_{MS}(m) k_M] a_N(2)} \tag{120}
$$

or, when the composition of the solution (1) is not variable

$$\Delta\phi_m = -\frac{RT}{F} \ln\{a_M + k_{M,N}^{pot} a_N\}$$

where

$$k_{M,N}^{pot} = \frac{U_{NS}K_{NS}(m)k_N}{U_{MS}K_{MS}(m)k_M} \tag{121}$$

The quotient $K_{NS}(m)k_N/K_{MS}(m)k_M$ can be identified as the ion-exchange equilibrium constant, K_{MN}, of the reaction

$$N^+(s) + MS(m) \rightleftharpoons NS(m) + M^+(s)$$

The value of K_{MN} and so the selectivity coefficient will, in this case of large mobilities of the free sites, depend on the properties of the charged ligand as well as on the characteristics of the solvent, whereas, in the case of small mobilities of the free sites, the selectivity coefficient will predominantly depend on the solvent alone according to eqn. (119).

If M^+ forms much stronger complexes than N^+, i.e. $K_{MS}(m) \gg K_{NS}(m)$, and if the mobilities of both complexes are also approximately equal, then eqn. (114) reduces to

$$\tau = \frac{U_S}{U_N + U_S} \tag{122}$$

Since, in this case, the electrical charge transport through the membrane is mainly governed by the ions M^+ and S^-, τ can be identified as the transference number of the most dissociated cation.

c. Kinetics; response time. In the literature, no special attention has been paid to the theoretical aspects of kinetics and response times of membranes with dissolved charged ligands. However, it can be expected that two cases can be distinguished. If the mobility of charged species in the membrane is relatively large, the establishment of steady Nernst diffusion layers will be the rate-determining step. On the other hand, if the mobility is small, this diffusion is of minor importance and the overall reaction time will depend on the time necessary to establish a steady concentration profile in the membrane. The first case was discussed in Sect. 3(B)(1)c; an exponential potential—time curve has to be expected. In the second case, the situation is similar to the one discussed in Sect. 3.(D)(2)c and eqn. (91) will be applicable.

(E) SURVEY OF PARAMETERS DETERMINING THE SELECTIVITY COEFFICIENTS

In the previous paragraphs, various types of membranes have been discussed. It appeared to be possible to express the total membrane potential in a general form, viz.

$$\Delta\phi_m = \frac{RT}{F}\ln\{a_M + k^{pot}_{M,N}a_N\}$$

in which M^+ is considered to be the primary ion and N^+ is the interfering ion. The selectivity coefficient, however, is dependent on the type of membrane. In Table 2, the expressions for the selectivity coefficients are summarized.

Although the general form given above was theoretically derived

TABLE 2

Theoretical relationships of selectivity coefficients to physiochemical parameters for the various types of membrane

Type of membrane	$k^{pot}_{M,N}$	Equation
solid membrane		
Sparingly soluble inorganic salts	K^{MX}_{so}/K^{NX}_{so}	36
Ion-exchange membrane	$(U_N/U_M)K_{MN}$	7·7
Liquid membrane		
Charged ion-exchangers dissociated complexes	$(U_N k_N)/(U_M k_M)$	117
strongly associated complexes mobility of free sites is small	$(U_N + U_S)k_N/(U_M + U_S)k_M$	119
or when also $U_S \ll U_M$; U_N mobility of free sites is large	$(U_N k_N)/(U_M k_M)$	
	$(U_{NS}K_{NS}(m)k_N/(U_{MS}K_{MS}(m)k_M$	121
or	$(U_{NS}/U_{MS})K_{MN}$	
electroneutral carrier	k_N/k_M	84
when $U_{NS} \simeq U_{MS}$	$(\beta_{NS})_{aq}/(\beta_{MS})_{aq}$	85

K^{MX}_{so}, K^{NX}_{so} = solubility products; U_i = mobility of the particle i in the membrane; K_{MN} = ion-exchange equilibrium constant for the reaction $N^+(s) + M..(m) \rightleftharpoons N..(m) + M^+(s)$; k_M, k_N = distribution coefficients; $K_{MS}(m)$, $K_{NS}(m)$ = stability constants of the complexes in the membrane phase; $(\beta_{MS})_{aq}$, $(\beta_{NS})_{aq}$ = stability constants of the complexes in the aqueous phase.

for monovalent primary and interfering ions, it has been proved experimentally that it is permissible to generalize this equation as

$$\Delta\phi_m = \frac{RT}{z_M F} \ln[a_M + \sum_{i \neq M} k_{M,i}^{pot}(a_i)^{z_M/z_i}] \tag{123}$$

4. Construction and performance of ion selective electrodes

(A) INTRODUCTION; CLASSIFICATION; CHARACTERISTIC PARAMETERS

The membranes discussed so far can be used for the construction of ion selective electrodes. Such electrodes consist of a suitable reference electrode, an inner solution, and the active membrane. If an ISE is in contact with a sample solution containing the ion(s) for which the electrode is selective, and a reference electrode is placed in the same sample solution, an electrochemical cell is formed. The EMF value measured will be equal to

$$E = E_{ref(inner\ solution)} + \Delta\phi_m + E_{ref(sample\ solution)} \tag{124}$$

or, in combination with eqn. (123)

$$E = E^0 + \frac{RT}{z_M F}\ln[a_M + \sum_{i \neq M} k_{M,i}^{pot} a_i^{z_M/z_i}] \tag{125}$$

As long as the composition of the inner solution is kept unaltered and the potential difference at the reference electrode—sample solution boundary remains virtually constant, E^0 is constant and the EMF of the cell will depend only on the composition of the sample solution.

In addition to the classification of ISEs according to the properties of the membranes with respect to aspects such as solid or liquid state, the presence or the absence of diffusion in the membrane, and the kind of ligand in the membrane when liquid membranes are concerned, it is necessary to consider the preparation of the membranes and the construction of the electrode as well.

As far as membranes made of sparingly soluble (inorganic) salts are concerned, a subdivision can be made in single crystal and polycrystalline membranes. The latter membranes can be obtained either homogeneous, by pelletizing microcrystalline precipitates, or heterogeneous in which the precipitate is dispersed in some inert polymeric matrix like silicone rubber, PVC, or polyethylene. Many of the homogeneous solid-state membranes are not composed of one single compound but are formed from mixtures of compounds, e.g. $AgCl-Ag_2S$, $CdS-Ag_2S$, etc. In the construction of electrodes, the inner solution is sometimes replaced by a direct metallic contact with the membrane. For instance, in many membranes containing silver sulfide, electrical contact is accomplished by pressing a silver rod directly on the pellet or by means of an electrically conductive adhesive containing finely divided silver powder.

Such electrodes, which are often referred to as *all solid-state electrodes*, have the advantage of being very robust. This type of electrode has proved to be very useful in practice, but there is still some discussion about the thermodynamic aspects because of the transition of mainly ionic conduction in the membrane to purely electronic conduction in the metal wires [65—69]. In that case, the potential difference at the membrane—metal wire interface, and so the value E^0 of eqn. (125), will depend very strongly on the actual composition of the membrane phase. In this connection, not only the presence of stoichiometric compounds has to be considered but also the presence of non-stoichiometric ones (e.g. it was found that besides CuS and Cu_2S, $Cu_{1.86}S$ and $Cu_{1.98}S$ can also be present in copper sulfide precipitates [68,69]. In the case of mixed precipitates, the formation of ternary compounds, e.g. $Ag_{1.5}Cu_{0.5}S$ [40] has to be taken into account.

It was suggested by Růžička and Lamm [70] that an all solid-state electrode can also be obtained by pressing the electroactive membrane material directly on a hydrophobized graphite surface. However, if the explanation of the coupling between ionic processes in the membrane and the electronic conduction in metal contacts already gives rise to problems, still greater difficulties are to be expected in the case of graphite as the electronic conductor.

Other electrodes that can be classified as solid-state electrodes are those made of glass membranes and those made of solid ion-exchanger membranes. A very interesting sensor that should also be classified in this category, but which is not included in the IUPAC classification, is the so-called ISFET. In this device, introduced by Bergveld [71], the metal gate of a conventional metal oxide semiconductor field effect transistor (MOSFET) is replaced by an ion selective membrane.

318

With regard to liquid membrane electrodes, there is little more difference between the electrodes than the various possibilities in which the liquid membrane is kept in place. Originally, these electrodes were often made by impregnating a glass frit or some other porous material, but now dispersion of the liquid in a neutral polymeric matrix is often recommended. The first category can be considered as homogeneous, the second category as heterogeneous. Also in the case of liquid membrane electrodes, the inner solution has been replaced by a direct metallic contact (e.g. platinum) [72], but, again, it is not well understood how the ionic processes in the membrane influence the inner potential in the metal [73].

Apart from the electrodes mentioned thus far, which can be classified under the heading *primary electrodes* (see classification in ref. 41), another main group, specified as *sensitized ion selective electrodes*, has to be considered. This group can be divided into gas-sensing electrodes and enzyme (substrate) electrodes. According to the definitions recommended by the IUPAC commission [41], gas-sensing electrodes are defined as ". . . sensors composed of an indicating and a reference electrode which use a gas-permeable membrane or an air-gap to separate the sample solution from a thin film of an intermediate solution, which is either held between the gas membrane and the ion-sensing membrane, or placed on the surface of the electrode using a wetting agent (e.g. air-gap electrode). This intermediate solution interacts with the gaseous species in such a way as to produce a change in a measured value (e.g. pH) of this intermediate solution. This change is then sensed by the ISE and is proportional to the partial pressure of the gaseous species in the sample".

The enzyme electrodes are defined as ". . . sensors in which an ion selective electrode is covered with a coating containing an enzyme which causes the reaction of an inorganic or organic substance (substrate) to produce a species to which the electrode responds. Alternatively, the sensor could be covered with a layer of substrate which reacts with the enzyme to be assayed".

In Table 3, a survey is given of the various types of ISE. In the following sections, some practical aspects of all these electrodes will be reviewed.

With regard to the performance of the electrodes, attention will be paid to the linear range and the slope of the calibration curve, interferences and selectivity, the practical limit of detection, the practical

TABLE 3

Classification of ion selective electrodes

Class of electrode	Type of active material	Membrane
Primary electrodes		
Solid-state electrodes	Crystalline, single crystal	
	polycrystalline	Homogeneous, heterogeneous
	mixed crystals	Homogeneous, heterogeneous
	Glass membrane	
	Solid ion-exchanger membrane	
Liquid membrane electrodes	Charged ligands	Homogeneous, heterogeneous
	Neutral ligands	Homogeneous, heterogeneous
ISFETs		
Sensitized ISEs		
Gas-sensing electrodes	Gas-permeable membrane	
	Air-gap	
Enzyme electrodes		

response time, drift, stability and reproducibility, and maintenance and lifetime. Most of these characteristics or terms have already been introduced in the preceding sections, but some still need a clear definition. In order to further the standardization with respect to the terminology, the nomenclature and definitions recommended by IUPAC [41a] will be adopted.

A *calibration curve* is a plot of the EMF of a given ion selective cell assembly vs. the logarithm of the ionic activity (or concentration) of a given species. It is recommended that the potential value be plotted on the ordinate with the more positive potential values at the top of the graph and that the negative value of the logarithm of the activity (pa_i) or the concentration (pC_i) is plotted on the abscissa with increasing activity to the right (Fig. 11).

By analogy with definitions adopted in other fields, the limit of detection should be defined as that activity (or concentration) for which, under the specific conditions, the potential E deviates from the average potential in region I by some arbitrary multiple of the standard error (deviation) of a single measurement of the potential in region I. However, in the present state of the art, and for the sake

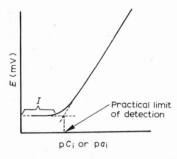

Fig. 11. General picture of a calibration curve.

of practical convenience, a simpler definition is recommended. The *practical limit of detection* may be taken as the activity (or concentration) of i at the point of intersection of the extrapolated linear branches. The term *practical response time* has already been discussed in Sect. 3.(B)(1)c. With regard to the slope of the calibration curve, the term *Nernstian response* is often used. An ISE is said to exhibit Nernstian response over a given range of activities (or concentrations) if the calibration curve in this range is linear with a slope of $2.303RT/z_MF$ mV per decade change in activity (or concentration).

In many cases, the slope is less than the theoretically expected value: the electrode is then said to have a *sub-Nernstian response*. Also cases of larger slopes, so-called *super-Nernstian response*, have been observed. The term *drift* is used in its normal way as the slow non-random change of the signal (i.e. potential) with time.

(B) SOLID-STATE ELECTRODES

(1) Single crystal membrane electrodes

Pungor and Tóth [74] are formally right when they state that a subdivision of precipitate-based solid-state electrodes in single crystal, polycrystalline and mixed crystalline membranes as well as the distinction in homogeneous and heterogeneous membranes is not relevant from a theoretical point of view. However, the unambiguity of their membrane composition and the absence of any grain boundary effects as, for example, observed by van de Leest and Geven [75], place the single crystal membrane electrodes in a special position. Furthermore, no problems are encountered with single crystals with respect to the possibilities of compressing the precipitates to compact pellets which are mechanically stable and non-porous. In fact, one of the most successful ISEs, apart from the pH glass electrode, is such a single crystal membrane electrode, i.e. the lanthanum-(III) fluoride membrane electrode introduced by Frant and Ross [17] in 1966.

a. Lanthanum(III) fluoride membrane electrode. LaF_3 is sparingly soluble and forms crystals with a hexagonal structure. In this lattice, each lanthanum(III) ion is surrounded by five fluoride ions, whereas the six next closest neighbours are fluoride ions as well. The total lattice shows layers of LaF_2^+ with adjacent layers of F^- ions at both sides. The transport of charge across the membrane is governed by the F^- ions only by means of a so-called lattice defect conduction mechanism in which the mobile ion, i.e. the F^- ion, moves from a fixed position into an adjacent vacant fluoride position in the lattice creating a new vacancy. The incorporation of some europium(II) fluoride was found to result in an increase of the conductivity [76], probably by the production of extra fluoride vacancies in the lattice. The fluoride electrode exhibits Nernstian response in the concentration range of $1-10^{-5}$ or 10^{-6} M in neutral or slightly acidic media. The practical limit of detection is about 10^{-6} to 10^{-7} depending on the ionic strength of the solution. This experimentally found limit is of the order of magnitude to be expected from the solubility product of lanthanum(III) fluoride. Since only a few anions have such a dimension and electrical charge that they can replace the fluoride ions in the lattice, the membrane potential should be little affected by other anions and the selectivity should be very large. This is indeed observed. However, interferences can be expected if chemical reactions occur at the surface of the membrane which lead to a modification of the surface layers and/or to an increase of the local fluoride concentration. Although the hydroxyl ion has an ionic radius comparable with the fluoride ion, it is assumed that the interference by hydroxyl ions (see Fig. 12) is caused by the reaction

$$3\ OH^-(s) + LaF_3(m) \rightleftharpoons 3\ F^-(s) + La(OH)_3(m)$$

At high pH values, the interference by OH^- ions is even more complicated than suggested by this reaction because of the formation of stable lanthanum(III) hydroxide complexes such as, for example,

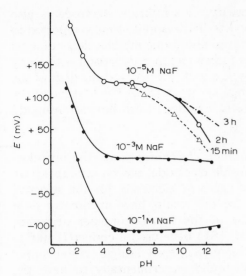

Fig. 12. Effect of pH on the potential of the LaF$_3$ membrane electrode in NaF solutions of various concentrations [77]. (By kind permission of *Zeitschrift für Analytische Chemie*.)

LaOH^{2+} [78]. Whereas the potential values show a negative deviation at larger pH values, at lower pH a positive deviation is observed (Fig. 12). These deviations are not caused by interference with the membrane material but are entirely due to a decrease of the fluoride ion activity in the bulk of the solution as a result of the protonation of the F$^-$ ions. The formation of HF and HF$_2^-$ starts below pH 5.

Complex formation with metal ions like iron(III) or aluminum(III) will also affect the fluoride ion activity. If one is interested in the determination of the total fluoride concentration, it is necessary to decomplex the various metal fluoride complexes in the solution. This is usually accomplished by the addition of strongly complexing agents such as citrate or DCTA. Furthermore, the pH must be adjusted to between pH 5 and 7 and the ionic strength should be kept at a constant level. For this purpose Frant and Ross [79] have recommended the use of a Total Ionic Strength Adjustment Buffer (TISAB) containing a buffer (acetic acid/sodium acetate) of pH 5, a large amount of sodium chloride to maintain the constant ionic strength, and some citrate or DCTA as masking agents for Fe(III) and Al(III) ions.

Apart from its large selectivity, the fluoride electrode is also almost ideal in many other respects. The response time is reported to be in the range of a few seconds to about one minute, depending on the concentration level. This is rather fast in comparison with many other ISEs. The reproducibility is about 0.2 mV and the drift of the potential is limited to about 2 mV per week [80]. The lifetime of the electrode is very long; with careful handling, it can function properly for several years.

b. Other single crystal membrane electrodes. Since the introduction of the very successful fluoride electrode, several attempts have been made to produce similar types of electrode for the selective determination of ions. However, the results have apparently been rather poor until now, in view of the limited number of papers published so far. Hulanicki et al. [81] have used a copper(I) sulfide single crystal membrane and Veselý [82] used a single crystal copper selenide membrane which was found experimentally to have the composition $Cu_{1.8}Se$. Both electrodes show no particular advantage over the copper selective electrodes based on membranes pressed from mixed copper sulfide—silver sulfide precipitates [see Sect. 4.(B)(3)]. Single crystals of silver chloride and silver bromide have been prepared by the Monokrystaly Research Institute of Single Crystals in Czechoslovakia. The principal disadvantages are associated with the rather high electrical resistance and with the fact that both materials exhibit photoelectric potentials which give rise to fluctuations in the EMF values with changes in the illumination conditions [83]. Although annealing at $320°C$ seems to improve the electrode properties [84], this type of membrane is superseded almost completely by heterogeneous polycrystalline membranes or by membranes compressed from mixed silver halide—silver sulfide precipitates.

(2) Polycrystalline membrane electrodes

Pure polycrystalline precipitates are, in general, much easier to prepare than single crystals. Therefore, it is not surprising that so much effort had already been spent on the preparation of ion selective electrodes from polycrystalline materials during the decades preceding the time at which the first single crystal membrane electrode became available. The oldest work in this field goes back to that of Tendeloo [12] on barium sulfate and that of Kolthoff and

324

Sanders [11] on the use of silver halide discs in the thirties. This research was continued by Tendeloo and Krips [85] in the late fifties. They embedded calcium oxalate and calcium stearate into paraffin in order to prepare calcium-selective membranes. Important contributions were subsequently made by Pungor et al. [86] by the introduction of silicone rubber as the inert matrix for barium sulfate and silver iodide polycrystalline precipitates.

It was not by accident, that, at the outset, the heterogeneous membranes received more attention than the homogeneous, because most of the precipitates which would be useful as the active material for the preparation of ISEs are difficult to compress to give compact, non-porous pellets. The compressibility of a large number of precipitates, the mechanical stability of the pellets obtained, and the functioning of the pellets as the membrane in an ISE were extensively studied by Karh [87]. In their monograph on ISEs, Moody and Thomas [88] present ". . . some stringent requirements which any material must satisfy before it can act as a successful membrane in a solid-state electrode device". It must be completely non-porous, exhibit no photoelectric response, have good mechanical resistance, have good conductive properties, and have a small solubility product.

To get membranes which respond selectively to one of the constituent ions, e.g. M^+ or X^- in the case of the insoluble salt MX, the solubility products of other salts formed either by the reaction of M^+ with an interfering anion Y^- or by the reaction of X^- with a cation N^+, should be much larger than the solubility product of MX.

Since only a few precipitates meet all these requirements, they are often mixed with another very insoluble compound which should improve one or more of the less favourable properties. In many cases, silver sulfide is used for this purpose because it combines good mechanical properties with a large ionic conductivity, whereas it does not affect the electrode potential determined by the pure precipitate because of its very low solubility ($K_{so}^{Ag_2S} \sim 10^{-51}$). At present, electrodes based on membranes prepared from pure silver halide and prepared from pure silver sulfide are commercially available.

a. Silver halide membrane electrodes. Silver halide crystals are ionic conductors in which the silver ion is the charge carrier. The mobility of Ag^+ ions is generally explained by the *Frenkel mechanism* [89]. In this mechanism, it is assumed that an ion leaves its regular site in the lattice to occupy what is known as an *interstitial*

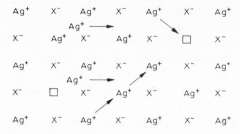

Fig. 13. Schematic representation of the transport of Ag^+ in a silver halide lattice according to the Frenkel mechanism. The number of interstitial silver ions is the same as the number of unoccupied sites.

position in the space between the fixed lattice points. The vacancy left after such a transition can be filled up again by another ion of the same kind. Since cations are usually smaller than anions and can therefore move more readily in the free spaces in the lattice, it is quite common that the cations are responsible for the electrical conductivity. This process is depicted schematically in Fig. 13. Most of the work on silver halide membranes was originally done by Pungor and Tóth [90,91] with silicone rubber-based heterogeneous membranes. For the chloride and bromide electrodes, a linear response was reported between 10^{-1} and 10^{-4} M. For the less soluble silver iodide membrane, the linear range was 10^{-1} to 10^{-5} M. In general, the slope was found to be somewhat sub-Nernstian (approximately 56 mV/pX; $X = Cl^-$, Br^-, or I^-). The practical limit of detection corresponds, for all the three different electrodes, with the respective values calculated according to eqns. (22) and (23). Although it is very difficult to compare response times obtained by various authors, because the values depend very strongly on the experimental conditions, it can be stated that the electrodes respond rather quickly, as do most solid-state electrodes. Reported t_{95} values vary between some seconds and about one minute.

All three halide electrodes exhibit an almost Nernstian response to silver ions (linear range pAg 1—4). No interference by other cations was observed as long as the halide ion activity in the solution is not affected by complexation. As for the selectivity to other anions forming insoluble silver salts, it has been found that the selectivity coefficients compare quite well with the ratios of the corresponding solubility products as could be expected theoretically (see Table 2).

To avoid a chemical transformation of the surface, the following condition must be fulfilled.

$$a_X < \frac{K_{so}^{AgX}}{K_{so}^{AgX^*}}\, a_{X^*} \tag{126}$$

in which X is the interfering ion and X^* is the primary halide ion [see Sect. 3.(B)(1)a].

Because silver ions are not very liable to hydrolysis and halide ions are normally not involved in any protolytic reaction, the electrodes can be used over quite a large pH range, viz. $1 < pH < 11$.

An interesting interference is observed in the presence of cyanide. In that case, the reaction

$$AgX(m) + 2\,CN^-(s) \rightleftharpoons Ag(CN)_2^-(s) + X^-(s)$$

can proceed in which the soluble silver cyanide complex is formed. The results obtained in Sect. 3.(B)(1)c are applicable which means that, according to eqn. (46), the electrode potential can be written as

$$E = E^0 + \frac{RT}{F} \ln \gamma_X \left[[X^-] + \frac{1}{2} \frac{[CN^-]D_{Ag(CN)_2}^{(1-\alpha)}}{D_X^{(1-\alpha)}} \right] \tag{127}$$

When the concentration of X^- in the solution is negligible, eqn. (127) can be simplified to

$$E = \text{constant} + \frac{RT}{F} \ln[CN^-] \tag{128}$$

This means that the halide electrodes can function as a cyanide sensitive electrode [90]. Since HCN is a very weak acid, the potential will depend very strongly on the pH and a strict buffering of the solution is a prerequisite for proper functioning and for obtaining reproducible results.

   b. Silver sulfide membrane electrodes. Crystalline silver sulfide is an almost ideal membrane material. It combines a very low solubility with a large conductivity at room temperatures. Below $176°C$, the stable silver sulfide crystal structure is the monoclinic β-modification, acanthite. In this modification, the conductivity has to be attributed to mobile silver ions. Above $176°C$, the cubic α-modification, argentite, is stable. Argentite is an electronic conductor. The compressibility at room temperature is very good and the pellets thus obtained are mechanically stable. Furthermore, the photoelectric response is almost completely absent. The silver sulfide membrane electrode responds in a Nernstian way to silver as well to sulfide ions: 29.6 mV/pS^{2-} and 59.2 mV/pAg$^+$. The practical limit of detection, based on a calibration curve obtained with solutions made by successive dilution, corresponds to about 10^{-7} to 10^{-8} M which is much higher than the value of about 10^{-17} M to be expected on the basis of the solubility. According to Morf et al. [93], this effect should be due to the transference to the solution of interstitial silver ions produced by the Frenkel mechanism. Crombie et al. [94] found, however, that the limit of detection is really determined by the solubility product and Buck [34] pointed out that the phenomena observed by Morf et al. could also be attributed to the leaching of soluble silver salts occluded during the precipitation procedure. Moreover, Camman [95] has stressed the general point that, for electrodes provided with very insoluble membranes, the detection limit can be due to the instability of dilute solutions (concentrations $<10^{-7}$ M) because of problems connected with adsorption, desorption, and contamination. This latter problem can be overcome by the use of, for example, metal buffers [see Sect. 6.(B)]. In that case, the electrode was found to exhibit a Nernstian response down to about 10^{-20} M for silver as well as for sulfide ions. The Ag$_2$S membrane is very selective to Ag$^+$ ions and mercury(II) ions are the only cationic interferent observed. The electrode is virtually insensitive to any anionic interference; only at very large concentrations will cyanide give a contribution to the response [96].

Recently, Gulens and Ikeda [97] have studied the super-Nernstian response observed with older solid-state electrodes at low concentration levels. They concluded that this phenomenon has to be attributed to the accumulation of metallic silver at the membrane—solution interface, which will lead to the measurement of mixed potentials. The appearance of such a film of silver was examined by means of a scanning electron microscope (SEM). Polishing of the electrode surface will remove the tarnish and will restore the Nernstian response. The internal silver contact was supposed to be the source of the metallic silver.

(3) Mixed crystal membrane electrodes

During the last decades, numerous insoluble compounds have been examined for their usefulness as membrane material. The most extensive study has been made by Kahr [87] who tested over one hundred pure compounds or binary mixtures of compounds with special attention to their compressibility to mechanically stable pellets and to the functioning of the electrode. In general, agreement was found with the results obtained by a large number of other research workers in this field. Of the pure precipitates, only the silver halides, silver sulfide and, to a lesser extent, copper(I) sulfide yielded suitable membranes. All the other compounds, such as Cu_2Se, $CuTe$, CuS, $CuSe$, $CuTe$, PbS, $PbSe$, $PbTe$, PbI_2, $CdTe$, $CdSe$, Cds, HgS, $HgSe$, $HgTe$, Hg_2I_2, Sb_2S_3, and Bi_2S_3, were either not compressible to stable pellets or the pellets showed a limited response, or no response at all, to the activity of the corresponding metal ion in solution. Better results were obtained with the mixed precipitates. Those of copper sulfide, cadmium sulfide, and lead sulfide with silver sulfide, especially, appeared to be suitable substances for the preparation of membranes selective to copper(II), cadmium(II) or lead(II), respectively. To some extent, mercury sulfide—silver sulfide can also be used, but it has appeared to be difficult to prepare membranes with a reproducible behaviour.

In the subsequent part of this section some remarks will first be made on the mixed silver halide—silver sulfide membrane electrodes, followed by a brief discussion of the mixed metal sulfide—silver sulfide membrane electrodes.

a. Silver halide—silver sulfide membrane electrodes. As mentioned before, the presence of silver sulfide improves the electrical and mechanical properties of the membrane. Moreover, in the case of silver chloride (and silver bromide)—silver sulfide precipitates, the presence of silver sulfide results in a reduced photoelectric effect and an increased conductivity; silver iodide is a good conductor on its own. Intimate mixtures of the two components are obtained by co-precipitation from aqueous solution, adding an excess of silver nitrate to a solution containing sodium sulfide and sodium chloride. During the pressing procedure, the bromide and iodide precipitates are subject to a partial transformation to Ag_3SBr and Ag_3SI. The conductivity of these two compounds is even larger than that of pure Ag_2S [98].

Van de Leest [99] has utilized these good ionic conductors as supporting materials for membranes which were rendered ion selective by covering them with a thin layer of pure silver halide. Since the coverage procedure was accomplished under thermodynamically well-defined conditions, electrodes were obtained which showed an excellent reproducibility.

Sorrentino and Rechnitz [100] have observed that a silver iodide—silver sulfide precipitate-based electrode ceased to respond in a Nernstian way to iodide in a concentrated solution of potassium iodide. This could be partly explained by the dissolution of the membrane with formation of polyiodo argentate(I) complexes ($AgI_n^{-(n-1)}$) [101].

b. Copper sulfide—silver sulfide membrane electrodes. Since Ross and Frant [83,102] introduced the use of mixed metal sulfide—silver sulfide membranes for the construction of ISEs, many methods have been published for the preparation of copper(II) selective electrodes. Two main aspects of the various electrodes can be distinguished: the procedure for the preparation of the electroactive material and the way in which this material is applied, i.e. by compressing to compact homogeneous pellets or by dispersing in an inert polymeric matrix.

For the simultaneous precipitation, three reagents have been examined: sodium sulfide [102—107], hydrogen sulfide [103,104, 107], and thioacetamide [104,107]. Although the precipitation with hydrogen sulfide, and particularly the homogeneous precipitation in slightly acidic medium with thioacetamide, will result in materials which suffer less from contamination by occlusion, it was found by Heijne et al. [107] that mechanically stable and homogeneous pellets of good performance could only be obtained from precipitates prepared by sodium sulfide precipitation in alkaline medium. X-Ray powder diffractograms have shown that, in the case of the hydrogen sulfide or the thioacetamide procedure, mixtures of Ag_2S (acanthite) and CuS (covellite) are formed, whereas with the sodium sulfide procedure, ternary copper(I) sulfide—silver sulfides are the main components, mainly $Ag_{1.5}Cu_{0.5}S$ (jalpaite) [108]. Furthermore, it was found that, in the first two procedures, the precipitation proceeds .stepwise; first, silver sulfide is formed, followed by the precipitation of copper(II) sulfide in the second stage. The partial coverage of the Ag_2S particles formed in the first stage by CuS might be responsible for the fact that, just as for pure CuS, no stable non-porous pellets could be pressed from these mixed precipitates. It must be emphasized that, in jalpaite, copper is in its monovalent state so that, at the surface, the reaction

$$4\ Ag_{1.5}Cu_{0.5}S \rightleftharpoons 3\ Ag_2S + CuS + Cu^{2+} + 2\ e$$

330

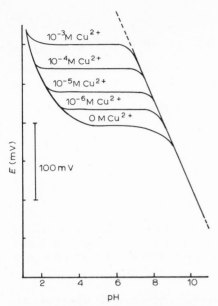

Fig. 14. Response of the copper(II) ISE vs. pH in 0.1 M potassium nitrate and various concentrations of copper(II).

can proceed. This reaction might account for the enhanced solubility and hence the high detection limit, particularly in the presence of oxidizing agents.

Copper(II) selective electrodes made of membranes which contain monovalent copper exhibit a linear response over about five orders of magnitude ($1 < pCu < 6$) at pH values between 3 and 6. At higher pH values, the formation of $Cu(OH)_2$ will reduce the upper limit of determination, as can be seen from Fig. 14. The practical (lower) limit of detection is much higher than can be expected from the solubility product of either CuS or Cu_2S, but it can be markedly decreased by the addition of complexing agents. In the presence of an excess of a suitable ligand such as triethylenetetramine (TRIEN), this limit was found to be as low as 10^{-20} M. However, in the presence of other complexing agents such as the polyaminopoly-carboxylic acids EDTA and NTA, the electrode response was found to depend on the free ligand concentration [109]. Ishibashi and his co-workers [110] have suggested masking the ligands that show such an abnormal behaviour by adding an excess of polyamines. With

TABLE 4

Selectivity coefficients, $k_{Cu,N}^{pot}$, of Cu ISE at pH 4.7 (acetate buffer) $C_{Cu} = 10^{-3}$ M; $C_N = 10^{-3}$ M [111].

Metal N	$\log k_{Cu,N}^{pot}$ after contact with [a]	
	free Cu(II)	free EDTA
Zn	−3.6	−4.1
Cd	−3.5	−4.0
Ni	−3.7	−4.0
Mn	−3.8	−4.2
Pb	−2.7	−2.9
Fe(II) under N_2	−3.3	−3.4

[a] Since it makes some difference how the electrode is pretreated, selectivity coefficients have been determined with the electrode preconditioned in a 10^{-3} M copper(II) solution and with the electrode preconditioned in a 10^{-3} M solution of EDTA.

membranes that contain divalent copper sulfide, such discrepancies have not been observed [104]. As for the response to copper(II), the slope of the linear part of the calibration curve is almost ideally Nernstian; the response to S^{2-} ions is often somewhat sub-Nernstian.

Observed response times depend very strongly on the experimental conditions and on the history of the electrode. Values between some seconds and several minutes have been reported at 10^{-5} M concentration level. Especially at pH above 7, a drifting of the potential is observed. In contrast to many other solid-state electrodes, polishing of the surface does not have a significant influence on the response times. Only old, tarnished surfaces could be improved in their performance by polishing.

The selectivity towards many other ions is very good. Alkali and alkaline earth metals do not interfere. The selectivity coefficients with respect to several transition metals are summarized in Table 4. Mercury(II) interferes because of the formation of the very insoluble HgS, whereas the interference by iron(III) is probably due to the oxidation of monovalent copper with the simultaneous production of copper(II) ions. The interference by chloride has been studied by several authors. Ross [83] suggests a reaction that might take place in solutions containing both copper(II) and chloride ions, viz.

$$Ag_2S(m) + Cu^{2+} + 2\,Cl^- \rightleftharpoons AgCl(m) + CuS(m)$$

332

which will be accompanied by an abrupt change in the electrode function from a response to copper(II) to a response towards chloride. Crombie et al. [112] have found evidence for this reaction but they have also found evidence for the fact that the formation of copper(II) chloride complexes is involved and Lanza [113] suggested the contribution of $AgCl_n^{-(n-1)}$ complexes.

Midgley [114] has examined the influence of several mineral acids. He has found a strong dependence on the age of the membrane surface. Oglesby et al. [115] have found that, in the presence of large amounts of sodium chloride (e.g. in sea water), the slope of the EMF vs. pCu curve was super-Nernstian; values up to $45\,mV/pCu$ have been measured in 1 M NaCl medium. No such effects were found in the case where potassium nitrate was the major background electrolyte indicating that this phenomenon can not be attributed to ionic strength effects. El-Taras et al. [116] have reported on the response caused by complexing agents in copper-free solutions. They demonstrated that their silicone rubber-based copper(II) selective electrode possessed some selectivity towards the various complexing agents. Similar results have been obtained by Sekerka and Lechner [117]. They have observed response times of about 5 min at 10^{-3} M level to 15 min at 10^{-6} M level for EDTA, NTA, etc. Exposure to high concentrations of EDTA, NTA, and EGTA ($>10^{-4}$ M) for prolonged periods of time led to a slow deterioration of the electrode function. Repolishing restored the proper functioning.

c. *Cadmium sulfide—silver sulfide membrane electrodes.* Membranes pressed from cadmium sulfide alone do not show an electrochemical response, but by co-precipitation of cadmium sulfide with silver sulfide, suitable electroactive materials were obtained. The various precipitation procedures mentioned for the mixed copper—silver sulfides have also been applied to cadmium [102,104—106, 118]. Just as for copper, the best results have been obtained with the procedure in which the solution of metal ions is added to an excess of sodium sulfide in alkaline medium. In order to obtain better membranes, it was found to be favourable to press the pellets at elevated temperatures ($150°C$; pressure 980 MPa) for 1—3 h. Gordievskii and co-workers [119] have compared these membranes with those obtained by crystallization of a melt of the two sulfides at $1200°C$. Membranes cut from a slowly cooled melt showed the same performance as the membranes prepared by hot pressing. Hirata and

Higashyima [120] have also used a mixture of the two sulfides but they compressed the material at higher temperatures and lower pressures: 200—500°C and 300—700 MPa). Although all authors found the compression at elevated temperature to be a prerequisite, X-ray diffractograms of the pellets and the starting material have not revealed any significant difference. In both cases, CdS (greenockite) and Ag_2S (acanthite) were the only compounds present. Apparently, the elevated temperatures are beneficial for the intimate "packing" of the microcrystalline particles.

The calibration curve is linear in the range $1 <$ pCd < 6, but the response is less reproducible than in the case of the copper(II) selective electrode. For different electrodes the slope of the calibration curve varies between 27 and 31 mV/decade. For one single electrode, variations of the slope of 0.2 mV/decade have been observed within one day; larger fluctuations will be found over periods of several days. Frequent calibration is therefore absolutely necessary.

At cadmium concentrations smaller than 10^{-3} M, the electrode can be applied between pH 3 and 8. At higher pH values, the formation of cadmium hydroxo complexes will occur, in particular the formation of $Cd(OH)_2$.

The practical (lower) limit of detection found for calibration curves obtained by serial dilution is about 10^{-7} M. In the presence of polyamines like TRIEN or TETREN, the limit can become as low as 10^{-13} M, which corresponds fairly well with the value expected from the solubility product ($pK_{so}^{CdS} \sim 26$—28). Although less pronounced in comparison with many copper(II) electrodes, an abnormal behaviour was observed in the presence of ligands such as EDTA, etc. [118]. In contact with oxidizing agents, a deterioration of the performance is found, while the membrane surface takes on a dull yellowish and tarnished appearance. The same can be observed after prolonged exposure to humid air. Some elemental sulfur is probably formed, which will yield an insulating layer. This oxidation process differs from that found for the copper sulfide where monovalent copper sulfide is oxidized to divalent. The formation of such an insulating layer might explain why the results found with the cadmium selective electrode are less reproducible and why the electrode has to be polished much more frequently than in the case of the copper selective electrode.

It has been found that, for a concentration change from 10^{-7} to 10^{-4} M, 95% of the total potential change was reached within 10 sec.

334

TABLE 5

Selectivity coefficients, $k_{Cd,N}^{pot}$, of Cd ISE at pH 6.5 (maleate buffer) $C_{Cd} = 10^{-3}$ M; $C_N = 10^{-3}$ M [118].

Metal N	$\log k_{Cd,N}^{pot}$	
	Electrode from ref. 118	Orion electrode No. 94-48
Pb	−1.5	−1.8
Zn	−3.1	−3.2
Mn	−3.4	−3.4
Ni	−3.4	−3.5
Fe(III)	−3.3	−4.0
Fe(II) under N_2	−2.7	−3.4

For the reversed concentration change (10^{-4} to 10^{-7} M), the response time was ten times as large.

The cadmium electrode does not respond to alkali and alkaline earth metals. The selectivity to some other metals is summarized in Table 5. It is interesting to notice that iron(III) hardly interferes, in contrast to the results with the CuISE. Apparently, iron(III) is too weak an oxidizing agent to oxidize sulfide to sulfur. Mercury(II) and copper(II) will interfere because of the much lower values of the solubility products for the corresponding sulfides. In the literature, little can be found about anion interferences, but a study by Brand et al. [121] suggests a large selectivity towards almost all anions. Because of the fact that the reproducibility is not completely satisfactory, the electrode is, in most applications, not used for direct measurements but as an indicating electrode in titrations.

d. Lead sulfide—silver sulfide membrane electrodes. Suitable mixed sulfides for the manufacture of lead selective electrodes were obtained by simultaneous precipitation of both metals in the presence of an excess of sodium sulfide in alkaline medium [122]. By X-ray analysis, it was shown that only acanthite and galena (PbS) were present in the membrane material. Compression to stable non-porous pellets should be accomplished at elevated temperatures (150°C), just as for the mixed cadmium—silver sulfide membrane, but, for lead the pressing time has to be considerably longer, at least 6 h. Several authors have mentioned oxidation problems [102,123,

124]. Therefore, one of the manufacturers of this type of lead(II) selective electrode recommends the addition of 10^{-3} M formaldehyde to the inner solution to retard oxidation [125]. Probably, the oxidation results in an oxide layer because a simultaneous increase of the membrane resistance has been observed [123] together with a drift of the potential, but also the possibility of the formation of elemental sulfur cannot be excluded. Frequently, polishing of the surface ensures the proper functioning of the PbISE.

The calibration curve is linear between pPb 4 and 6. Below $[Pb^{2+}] = 10^{-6}$ M, the slope begins to decrease and at pPb = 7, the practical (lower) limit of detection is reached. The linear part of the calibration curves has a slope varying between 27 and 30 mV/decade. For one single electrode, the variations within one day did not exceed 0.5 mV/decade. Above $[Pb^{2+}] = 10^{-4}$ M, super-Nernstian response occurs [122,126]. The origin of this deviation is not clear, but the fact that the response time simultaneously increases from 0.5 to 2 min in the linear part to over 5 min in the super-Nernstian part suggests a change in the response mechanism. The selectivity is worse than that of the CdISE and the CuISE. Not only do mercury(II), copper(II), and cadmium(II) strongly interfere, but also iron(III) was found to be an interferent, supporting the suggestion that at the surface it is not sulfide that is oxidized to sulfur but more likely lead is oxidized to lead oxide. In Table 6, selectivity coefficients are summarized. As for the anions, a 100-fold amount of chloride or nitrate interfere seriously.

The PbISE can be used between pH 3 and 7 for lead concentra-

TABLE 6

Selectivity coefficients, $k_{Pb,N}^{pot}$, of the Pb ISE at pH 4.7 (acetate buffer) $C_{Pb} = 10^{-3}$ M; $C_N = 10^{-3}$ M [122].

Metal N	$\log k_{Pb,N}^{pot}$
Zn	−3.6
Ni	−3.7
Mn	−4.0
Fe(II) under N_2	−3.3
Fe(III)	+0.6
Cd	−0.9

tions up to 10^{-3} M. The upper pH limit is due to the fact that lead hydroxo complexes or lead hydroxide precipitate will be formed. The lower pH limit is due to the enhanced dissolution of the sulfides in acidic medium. The lead(II) selective electrode has also been used in potentiometric titrations, e.g. the titration of sulfate with lead perchlorate solutions. To lower the solubility, titrations are often performed in dioxan—water mixtures. If the electrode is used in compleximetric titrations, it should be kept in mind that, in the case of polyamines, the EMF values correspond to the values expected on the basis of equilibrium conditions in the bulk of the solution, but that again, in the presence of EDTA, deviations have been observed.

   e. Mercury sulfide—silver sulfide membrane electrode. Little has been published on mixed mercury—silver sulfide membranes. Růžička and Lamm [70] have mentioned the preparation of a mercury(II) selective electrode by activating the surface of a hydrophobized carbon rod with pure mercury(II) sulfide or its admixture with silver sulfide. Anfält and Jagner [127] have prepared a HgISE by coating a silver rod with silver sulfide by anodic oxidation in sulfide medium followed by precipitation of mercury(II) sulfide on the surface. A similar procedure has been proposed by Van de Leest [128]. None of these electrodes showed such a behaviour that they could be used successfully for the determination of low concentrations of mercury.

   Some difficulties are encountered in the preparation of the sulfides. Kahr [87] has found that it is difficult to press good pellets from the black cubic HgS (metacinnabar) whereas the red hexagonal modification (cinnabar) cannot be pelletized at all. Pellets with fairly good mechanical properties can be obtained starting from solutions of mercury(I) nitrate and silver nitrate. This was confirmed by van der Linden and Oostervink [118]. From X-ray diffraction analysis, it can be concluded that, in this material, the major component is a ternary sulfide of unknown composition.

   The mercury(II) selective electrode exhibits a Nernstian response over a very small range of concentration ($2.5 < \text{pHg} < 4.5$). This limited range and the difficulty in producing reproducible electrodes are certainly the reasons why no mixed mercury—silver sulfide electrodes are commercially available, although the selectivity is rather good because of the low value of the solubility of mercury sulfide (e.g. $k_{\text{Hg,Cu}}^{\text{pot}} \sim 10^{-2}$ [118]).

   To conclude this section on mixed crystal membrane electrodes, it can be stated that this type of solid-state electrode did not fulfil all the expectations formulated at the end of the sixties. Apart from the silver halides—silver sulfide membrane electrodes, only the mixed copper—silver sulfide and, to a somewhat lesser extent, the mixed cadmium—silver sulfide and mixed lead—silver sulfide membrane electrodes have found application in analytical practice. The last two are used particularly in titrations, and the copper electrode is used in titrations and also for direct measuring of samples and continuous monitoring.

(4) Glass electrodes

During the last two decades, so many new membrane materials for the fabrication of ISEs have been developed that it is easily forgotten that most of the work on ion selective electrodes was originally directed to glass electrodes. Since the discovery of the pH-responsive properties of some glasses, about eighty years ago [4], several hundred papers have been published on glass membranes. The major part of these contributions is well covered by monographs [129—134]. In this section, only a brief outline will be presented, in the course of which the potentialities of glass electrodes responding to cations other than hydrogen ions will be emphasized.

Glasses can be considered to be supercooled liquids. Although, for instance, many oxides, sulfides, selenides, halides, borates, and even nitrates can form glasses, attention has always been focused on oxide glasses in particular. Essential for the formation of glasses is the presence of cations with a valency of three or higher, e.g. SiO_2, P_2O_5, B_2O_3, GeO_2, etc., as well as a small ratio of the cation to anion radii. These oxides can build up a three-dimensional network, as suggested by Zachariasen [135], in which the central atom is surrounded by three or more oxygen atoms. This is illustrated for silicate in Fig. 15 (I). The number of oxygen atoms is determined by the co-ordination valency of the cation. The assemblies of a cation surrounded by fixed

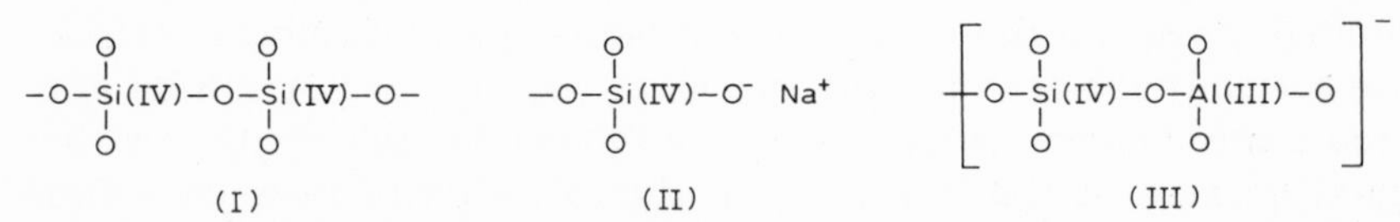

$$-O-\overset{\overset{\textstyle O}{|}}{\underset{\underset{\textstyle O}{|}}{Si(IV)}}-O-\overset{\overset{\textstyle O}{|}}{\underset{\underset{\textstyle O}{|}}{Si(IV)}}-O- \qquad -O-\overset{\overset{\textstyle O}{|}}{\underset{\underset{\textstyle O}{|}}{Si(IV)}}-O^-\ \ Na^+ \qquad \left[+O-\overset{\overset{\textstyle O}{|}}{\underset{\underset{\textstyle O}{|}}{Si(IV)}}-O-\overset{\overset{\textstyle O}{|}}{\underset{\underset{\textstyle O}{|}}{Al(III)}}-O\right]^-$$

(I)   (II)   (III)

Fig. 15. Schematic silicate structures. (I) Pure silicate; (II) silicate incorporating Na_2O; (III) silicate incorporating Al_2O_3.

number of oxygen atoms, e.g. tetrahedra in the case of silicate, are not arranged in a regular way as in crystals, but in a random way. Just as in liquids, only a short-range order exists.

By the introduction of a cation which can replace the central atom in the skeleton but which has a lower oxidation state, a negative site is created [Fig. 15 (III)]. The introduction of alkali oxides or alkaline earth oxides in the silicate will cause a disrupture of the structure, creating negative sites which show a high selectivity for hydrogen ions over other cations (Fig. 15 (II)]. A drawback of glass membranes has always been the large electrical resistance. Because, formerly, no mV-meters with a very high input impedance were available, the glass membranes had to be extremely thin in order to make the total resistance as low as possible. Nowadays, mV-meters with such high input impedances have come on the market that much thicker and mechanically stronger membranes are possible.

The conduction proceeds by means of a defect mechanism in which the cation (e.g. sodium ion in the case of a silicate containing sodium oxide) jumps from one interstitial position to another. The conductivity increases with the amount of monovalent or divalent oxide in the silicate, monovalent oxides being much more favourable than divalent oxides in this respect. The composition of silicate glasses is always presented in mole percentages of the constituent oxides. The first glass produced commercially for the fabrication of pH electrodes had the composition 22% Na_2O, 6% CaO, and 72% SiO_2. Typical sodium selective electrodes contain substantial amounts of Al_2O_3. Examples are

NAS 11-18 (11% Na_2O, 18% Al_2O_3, 71% SiO_2) [136]

NAS 10.6-10 (10.6% Na_2O, 10% Al_2O_3, 79.4% SiO_2) [137]

LAS 26.2-12.4 (26.2% Li_2O, 12.4% Al_2O_3, and 61.4% SiO_2) Beckmann No. 39278

These glasses exhibit a good selectivity for sodium over potassium; the selectivity coefficient, $k_{Na,K}^{pot}$, is of the order of 10^{-2} to 10^{-3}. Although the selectivity towards potassium can be increased by a decrease of the Al_2O_3 content, the value of $k_{K,Na}^{pot}$ never exceeds a value of about 0.1. Hence, these electrodes will allow the selective determination of potassium only if the ratio of the potassium concentration to the sodium concentration is larger than about 1000. Examples of these potassium selective glasses are

NAS 27-4 (27% Na_2O, 4% Al_2O_3, 69% SiO_2) [137]

KABS 20-5-9 (20% K_2O, 5% Al_2O_3, 9% B_2O_3, 66% SiO_2) [138]

The best lithium selective electrode has the glass composition
LAS 25-15 (25% Li_2O, 15% Al_2O_3, 60% SiO_2) [139]

All alkaline selective electrodes respond to hydrogen ions and silver ions and, in fact, show a large preference to these two ions. This can be seen from the selectivity coefficients. For instance, for NAS 11-18 $k_{Na,H}^{pot} \sim 10^3$ and $k_{Na,Ag}^{pot} \sim 5.10^2$. This means that, in general, the alkali selective electrodes can be used for the determination of alkali metals at above pH 7 only.

The Nernstian response and the large preference for hydrogen ions might suggest that the membrane potentials can be explained by the occurrence of Donnan potentials and transport of charge through the membrane by the hydrogen ion only. However, such a simple model cannot account for the selectivity to alkali metals. It was conclusively demonstrated by Schwabe and Dahms [140] by means of coulometric experiments that hydrogen ions do not penetrate through the membrane at all. Apparently, the charge transfer within the membrane is achieved by the small monovalent cations. The response to hydrogen ions and other cations proceeds through an ion-exchange mechanism at the solution—glass interface. For a better understanding of the processes going on at the membrane surface, it is desirable to examine the interface more closely.

It is well known that glass electrodes require a preconditioning by soaking in an aqueous solution for several hours. The solution has to contain the cation to which the electrode is supposed to respond in a concentration of about 0.1 M. During this preconditioning time, a hydrated gel layer is formed, the depth of which varies from 0.05 to 10 μm depending on the glass composition. Sodium silicate glasses, for instance, have a much thicker swollen layer than lithium glasses. The ability of a glass electrode to function satisfactorily is closely associated with the water content of the glass in equilibrium with water as was clearly demonstrated by Hubbard et al. [141]. They found a positive correlation between the hygroscopicity of glasses and the electrode function. A Nernstian response was observed for proper glasses used for electrodes, whereas a sub-Nernstian response was found for window glass which was found to have a water content in the gel layer that is about 10 times less than good responsive glasses. Chemically resistant glasses hardly absorb any water and, consequently, do not respond at all. Once formed, these gel layers should not be allowed to dry out. This means that the exposure to dehydrating and corrosive solutions should be limited to brief

340

periods. Because the hydration and dehydration processes are reversible, as long as the membranes have not been subjected to exhaustive drying, the electrode function can be restored by conditioning again.

The actual exchange of the monovalent cations in the glass with the ions to be measured occurs at the solution—hydrated layer interface. Because of this exchange reaction, the activities at this interface will differ from the activities at the hydrated glass layer—dry (bulk) glass layer interface. The ions to be measured will diffuse inward and the cations present in the glass will move to the solution. Hence, a diffusion potential is set up in the gel layer. Two other contributions to the total membrane potential have to be considered; first, the phase boundary potential at the surfaces of the dry glass layer, second, the diffusion potential across the bulk dry glass layer. Fortunately, both contributions are unaffected by changes in the solutions and so they may be considered constant. The selectivity of the membrane is then, in principle, governed by the ion-exchange process and a diffusion through the "solid-state" gel layer. This should imply that eqn. (77) for the selectivity coefficient can be applied. Eisenman [132] has verified this relation using a glass electrode responding to both potassium and sodium ions. A reasonably good agreement between calculated and experimental values was observed. The gel layer has been extensively studied. Concentration profiles for the cations involved in the exchange process have been measured by Boksay et al. [142] by means of fractional glass dissolution with HF. Such profiles give valuable information on how far the leaching process has proceeded. A very sophisticated technique for the determination of concentration profiles was introduced by Baucke and his colleagues [144]. They have applied ion-sputtering induced photon emission and could assess the lithium profile in the extremely thin surface layers as observed in modern lithium silicate-based glass electrodes for pH measurements. Wikby [143] has measured resistivity profiles within the glass. He showed that the gel layer front moves inward in the direction of the bulk of dry glass while, simultaneously, the solution—gel layer front moves in the same direction due to the dissolution of the silica skeleton. The velocity at which the inner gel front moves is determined by the rate of hydration which in turn depends on the glass composition. When the movement of the two fronts becomes equal, the thickness of the gel layer becomes time-invariant. Owing to this process, the membrane will very slowly become thinner which can ultimately limit the lifetime of the electrode. The chemical durability of glasses decreases almost exponentially with temperature, so the lifetime of electrodes is significantly shortened if they are used for longer times at elevated temperatures.

After this brief exposition of the complexity of the gel layer in which the essential rate-determining step takes place, it can hardly be expected that the two surfaces of a glass membrane behave in exactly the same way. A difference in behaviour will give rise to the so-called *asymmetry potential*. This is the potential observed if the solutions at both sides of the membrane are exactly identical. Kratz [130] has summarized some of the possible causes of this asymmetry.

(a) Loss of alkali from the outer surface in the flame during fabrication of the glass. This may lead to an alteration of the water sorptive capacity.

(b) Dehydration of the swollen outer surface layer by drying or immersion in dehydrating solutions.

(c) Mechanical or chemical destruction of the gel layer.

(d) Adsorption of foreign ions, formation of films of grease, proteins, etc. This will result in a decreased exchange capacity of the surface. Furthermore, differences in mechanical strain within the glass can contribute to the asymmetry. This latter contribution can be diminished by annealing.

The gradual change of the asymmetry, finding expression in a drift of the potential, demands a frequent standardization of the electrode.

Apart from the pH glass electrodes, only the sodium selective glass electrodes have been thoroughly investigated. An up-to-date comparison of the performance of some commercially available pNa electrodes has been published by Wilson et al. [145]. For most of the electrodes, the linear Nernstian range extends down to 10^{-4} to 10^{-5} M. Because it is very difficult to prepare solutions which contain very small amounts of sodium ions, or no sodium at all, it is impossible to assess the exact value of the practical limit of detection. The reproducibility of measurements with one single electrode is about 0.5 mV. Large variations can be observed between the E^0 values of different electrodes, especially between those from different manufacturers. This is mainly due to the differences in the composition of inner filling solutions.

As for the selectivity, it has already been mentioned that silver ions and hydrogen ions strongly interfere. This latter interference limits the pH range in which the sodium electrode can be used to the region pH < pNa + 3 (or 4)

The pNa electrode is about 10^3 times more sensitive towards sodium than to potassium. Although the electrode is also sensitive to other monovalent cations such as Rb^+, Cs^+, Li^+, Tl^+, and NH_4^+ and, to a lesser extent, to divalent cations such as Ca^{2+} and Mg^{2+}, these ions only interfere when present in large excess. The response time depends largely on the velocity at which the gel layer adapts itself to changes in the external solution. Therefore, the pretreatment of the membrane is of great importance. Values for t_{95} have been found ranging from several seconds to many minutes.

Other alkali selective glass electrodes have been developed. However, the lack of selectivity in the presence of sodium ions as well as the fact that suitable alternative electrodes have become available by the development of liquid membrane electrodes, explain why these electrodes have not found wide application.

From this brief introduction to glass electrodes, it is clear that water plays an important role in the response mechanism. Nevertheless, it is interesting to notice that the electrode can also be used successfully in other solvents as long as the swollen layer remains intact. The electrode function, which will show a tendency to gradual deterioration because of dehydration, can be restored by soaking the membrane in an aqueous solution of the ion to be determined.

(5) Solid ion-exchange membrane electrodes

Solid ion-exchange membranes have proved to be very useful in separation processes on both the laboratory and industrial scale. They have also been used in many fundamental studies, particularly as models for biological membranes. An extensive account of the properties and applications of these membranes is given in monographs [60,146,147]. Although these solid ion-exchange membranes have, in general, a low electrical resistance and should be suitable for the construction of selective electrodes, hardly any such electrodes have been reported. Only a very limited number of papers deal with the analytical application of these electrodes. The most extensive study is that by IJsseling and van Dalen [148] who used a home-made solid ion-exchange membrane electrode for the potentiometric indication of the end-point in precipitation titrations.

The limited success of these membranes for the construction of selective electrodes must be due to the fact that they do not meet simple analytical standards with respect to selectivity, response time, etc., while, at the same time, attractive alternatives are on the market in the form of solid-state electrodes or liquid membrane electrodes. This latter type of electrode will be discussed in the next section.

(C) LIQUID MEMBRANE ELECTRODES

In this paragraph, attention will be focused on membranes with mobile ion selective sites. To distinguish the electrodes equipped with these membranes from the solid-state electrodes which have fixed sites, the collective term *liquid membrane electrodes* will be used. Originally, these membranes were really liquid in nature, the ion-exchanger being dissolved in an organic solvent which was immiscible with water. This solution was fixed by means of an inert porous plug or wick (glass filter, porous plastic membrane, ceramic plug, etc.). Nowadays, the ion-exchanger is often incorporated in an inert matrix such as PVC, polythene, or silicone rubber. This latter design [fig. 16(a)] should have the advantage of avoiding any problems that can attend the handling of fluid membranes. However, an improved construction of the electrodes [Fig. 16(b)] equipped with

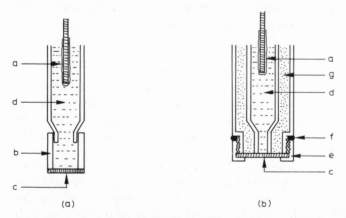

Fig. 16. Construction of ion selective liquid membrane electrodes. (a) Membrane consists of an inert matrix in which the ion-exchanger is incorporated; (b) membrane consists of porous plug which fixes the ion-exchanger. a, Silver—silver chloride electrode; b, PVC tube; c, membrane; d, inner filling solution; e, membrane cap; f, "O"-ring; g, ion-exchanger solution.

344

fluid membranes has eliminated the drawbacks, so that, at present, a fair comparison between the two designs can be made. According to Bailey [149] the most fundamental difference between the two types arises from the quantity of active material present. In some cases, the mobility of the species, particularly charged species, is lower within plastic matrices. Because all ion-exchangers are slightly soluble in water, one should ensure continuous replenishment of the active material. In this respect, the electrodes with a real liquid membrane in contact with a reservoir containing the ion-exchanger have the advantage over membranes in which the ion-exchanger is incorporated in an inert matrix because, once the ion-exchanger has dissolved from the inert matrix, such an electrode has to be discarded. On the other hand, a large quantity of active material is unfavourable when high concentrations of interfering species are present in the sample solution because a substantial amount of these interferents will then enter the membrane and it will take quite a long time to restore the proper response when the electrode is immersed again in a solution that is free from the interferent. This results in a prolonged drift of the potential. So, in fact, the choice between the real liquid type of membrane and a membrane with an inert matrix is a choice between longer lifetimes and shorter response times in the case of varying concentrations of interfering substances.

One of the first successful liquid membrane electrodes was the calcium selective electrode introduced by Ross [15]. This electrode was based on charged ligands derived from diesters of phosphoric acid. A study on a potassium selective liquid membrane electrode was published by Simon and co-workers at the same time [150,151]. Successively, liquid membranes with charged ligands and neutral carriers will be discussed more systematically in the next two sections.

(1) Liquid membrane electrodes based on charged ligands

The solvents used to dissolve the ion-exchanger must fulfil the following requirements.

(a) A low vapor pressure in order to avoid excessive loss of solvent by evaporation.

(b) A sufficiently high viscosity to prevent rapid loss by flow across the membrane. This is particularly important for electrodes provided with porous plugs.

(c) A low solubility in the (aqueous) sample solution. Most of the solvents used have a dielectric constant which is so low that ion association takes place to a considerable extent. Hence, eqns. (119) and (121) should be applicable. However, it should also be noticed that the derivation of these equations starts from the assumption of a steady state, a situation that usually is not attained during measurements. Therefore, the values of the selectivity coefficients expected from the theory represent only a rough estimation of the values obtained in real experiments.

Although a great number of ion-exchange systems are available and so ISEs may be developed for almost any cation or anion, only a small number of systems show a satisfactory behaviour. In particular, the selectivity leaves much to be desired. This is the reason why only a few electrodes are commercially available at present. The most important are the calcium and the nitrate electrodes, whereas the perchlorate, the fluoroborate and the chloride electrodes have also found some application.

a. Calcium selective electrode. The ion-exchangers used for the preparation of calcium selective membranes are derived from diesters of phosphoric acid. The choice of the phosphate ester was made because it was well-known that calcium forms stable complexes with phosphate and polyphosphate ions. By the introduction of long alkyl chains, the solubility in water is decreased, whereas the advantage of the diester over the monoester is that problems of mixed complex formation (e.g. calcium hydrogen complexes) are almost completely eliminated.

R = C_8H_{17} to $C_{16}H_{33}$
Dialkyl phosphate

Di-*n*-octylphenyl phosphate

The use of a solvent with very polar substituents such a di-*n*-octylphenyl phosphate furthers the selectivity for calcium relative to magnesium and other alkaline earth metals [152]. This selectivity of calcium relative to magnesium is almost completely nullified by replacement of the di-*n*-octylphenyl phosphate by the less polar 1-decanol. In that case, the electrode senses the total calcium plus magnesium concentration. The electrode, advertised as a "divalent cation electrode", is especially suitable in the determination of water

346

hardness. The electrode is also sensitive to zinc(II) and other divalent cations.

The calcium selective electrode exhibits a linear response from 10^{-1} M down to about 5×10^{-5} M of calcium(II). The slope of the calibration curve is nearly Nernstian. The lower limit of detection (ca. 10^{-5} M) is related to the solubility of the calcium dialkyl phosphate. This limit can be lowered by increasing the length of the aliphatic chain of the alkyl group, but the ion-exchanger will become less soluble and precipitation or gelling may occur. A decrease of the concentration of the calcium phosphate would also be favourable, but the electrical resistance of the membrane will increase, attended by longer response times.

The upper limit of the linear response range is set by the increase of calcium ion transport across the membrane. This transport can be restricted by a better match of the calcium concentration in the inner solution (usually 10^{-1} to 10^{-2} M) to the concentration in the sample solution. In that case the electrode can show a Nernstian response to calcium concentrations of more than 1 M.

The selectivity to calcium is reasonable with respect to the alkali metals ($k_{Ca,alkali}^{pot} < 10^{-3}$) and acceptable with respect to other alkaline earth metals ($k_{Ca,Mg}^{pot} \sim 10^{-1}$ to 10^{-2}; $k_{Ca,Sr}^{pot} \sim 10^{-2}$; $k_{Ca,Ba}^{pot} \sim 10^{-2}$). Iron(II), lead(II), copper(II), zinc(II) interfere, the selectivity coefficients being ~ 0.8, ~ 0.6, ~ 0.3, and ~ 3, respectively. After exposure to solutions containing zinc(II), a rather long recovery time must be observed. At concentration levels above 10^{-3} M, iodide and perchlorate interfere owing to their solubility in the organic phase.

The E vs. pH curves at a constant activity level of calcium(II) in the solution show characteristic dips below pH 5. Above pH 11, a change can be observed in the potential due to the formation of $CaOH^+$ and $Ca(OH)_2$ complexes. In the range $5 < pH < 11$, the response to calcium is virtually independent of pH. The peculiar dips at $pH < 5$ have been attributed to changes in the distribution constant, k_{Ca}, cause by the formation of species such as $Ca(HS_2)_2$ in which S denotes the ion-exchanging site [153].

For concentration changes from 10^{-4} to 10^{-1} M, response times (t_{95} values) of several seconds have been reported for vigorously stirred solutions, but in the presence of many interfering ions or moderate stirring, a marked increase of the response times has been observed.

b. Nitrate selective electrode. For the preparation of nitrate selective membranes, two types of ion-exchanger have been proposed. Ross [154] has suggested tris (substituted 1,10-phenanthroline)-nickel(II) nitrate dissolved in solvents such as nitrobenzene or *p*-nitrocymene, whereas other authors have introduced the use of quaternary ammonium salts such as tridodecylhexadecyl ammonium nitrate dissolved in *n*-octyl-2-nitrophenylether [155], methyltricapryl ammonium nitrate (aliquat 336S in the nitrate form) dissolved in 1-decanol [156] and tetraheptyl ammonium nitrate in nitrobenzene or chloroform [157].

The phenanthroline-based nitrate electrode exhibits a linear response in the range of 10^{-1} to about 10^{-4} M with a slope of ca. 57 mV/decade [158]. The lower limit of detection is about 10^{-5} M. The electrode is applicable in the pH range 3—8. Below pH 3, protonation of the ligand will be the limiting factor; above pH 8, hydroxyl ions will start to interfere. Approximately the same performance is observed for the quaternary ammonium ion-based nitrate electrodes.

Although the selectivity coefficients determined by various investigators are not always unambiguous, a general trend can be observed in the order of increasing values of $k_{NO_3,X}^{pot}$: $SO_4^{2-} \sim F^- < Cl^- < NO_2^- \sim Br^- < NO_3^- \sim ClO_3^- < I^- \sim BF_4^- < ClO_4^- < OH^-$ ($k_{NO_3,Cl}^{pot} \sim 10^{-3}$; $k_{NO_3,ClO_3}^{pot} \sim 2$; $k_{NO_3,ClO_4}^{pot} \sim 10^3$). This sequence parallels the order of decreasing anion hydration energy suggesting that aqueous solvation processes play an important role in determining the selectivity for these ions. The implication of this selectivity order is that the nitrate electrode is, in fact, more selective to I^-, BF_4^- and especially to ClO_4^- ions than to NO_3^- ions. Response times are reported to be in the order of many seconds to several minutes.

Despite the rather poor performance of the nitrate electrode in comparison with many other ISEs, the electrode has gained a great popularity because other methods for the determination of nitrate are rather cumbersome.

c. Perchlorate selective electrode. The nitrate electrode described above is about a thousand times more selective towards ClO_4^- than to NO_3^- and so this electrode can be converted to a perchlorate electrode by replacement of the nitrate anion in the organic phase by the perchlorate anion. The Nernstian response range of the perchlorate electrode is somewhat larger than for nitrate, 10^{-1} to

348

5×10^{-5} M; the limit of detection is about 10^{-6} M. The upper pH limit is shifted to about pH 10 ($k^{pot}_{ClO_4,OH} \sim 1$). Ross [154] has recommended the use of the iron(II) phenanthroline instead of the nickel(II) complex. Because the perchlorate electrode has found little application, almost no literature about its performance is available.

d. Tetrafluoroborate selective electrode. This electrode was prepared by Carlson and Paul [159] by shaking the nickel(II) phenanthroline nitrate with several portions of a solution containing the BF_4^- ion. The electrode exhibits a Nernstian response between 10^{-1} and 10^{-4} M; the limit of detection is approximately 5×10^{-6} M. The suitable pH range is $3 < pH < 12$. The electrode can be used for the determination of boron after its conversion in the sample to the fluoroborate anion by means of hydrofluoric acid.

e. Chloride selective electrode. The liquid membrane chloride selective electrode is almost completely superseded by the solid-state chloride electrode based on the silver halide (−silver sulfide) membranes, because this latter electrode is more stable, has a longer lifetime and shows a better selectivity for chloride with respect to perchlorate and nitrate. However, the solid-state electrode cannot be used in the presence of iodide and small amounts of sulfide. In that case, the liquid membrane electrode can offer some advantages. The ion-exchanger is dimethyldistearyl ammonium chloride. The linear range is rather limited, 10^{-1} to 5×10^{-4} M. The limit of detection is about 10^{-5} M. The pH range extends from pH 3 to pH 10.

(2) Liquid membrane electrodes based on neutral carriers

As in the case of liquid membranes with charged ligands, the solvents used for the neutral carriers should have a low vapor pressure, a sufficiently high viscosity, and a low solubility in order to prevent a too rapid loss of active material from the membrane. As was pointed out in Sect. 3.(D)(2)b, the selectivity coefficient is independent of the kind of solvent when the interfering ion and the primary ion form complexes with the neutral carriers of the same stoichiometry. However, when the valencies differ, the selectivity will depend on both the polarity of the solvent as reflected in the value of the dielectric constant and the concentration of the carrier. Preference to monovalent cations is observed in solvents of low

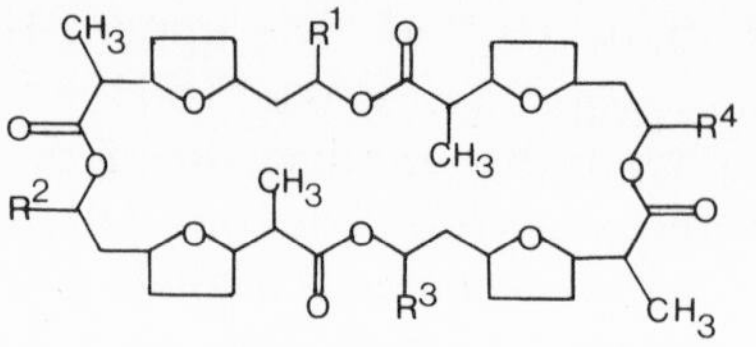

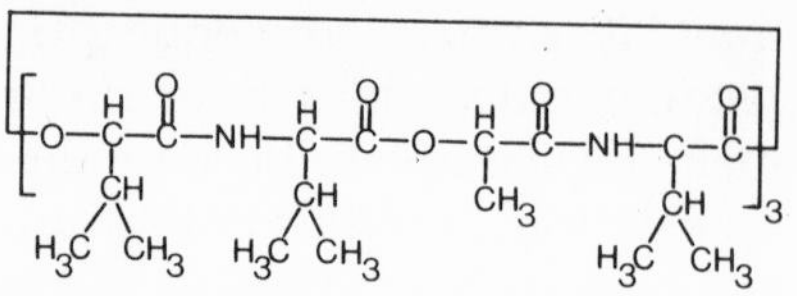

Fig. 17. Structure of some macrotetrolides. $R^1 = R^2 = R^3 = R^4 = CH_3$, nonactin; $R^1 = R^2 = R^3 = CH_3$; $R^4 = C_2H_5$, monactin; $R^1 = R^3 = CH_3$; $R^2 = R^4 = C_2H_5$, dinactin; $R^1 = CH_3$; $R^2 = R^3 = R^4 = C_2H_5$, trinactin.

Fig. 18. Structure of valinomycin.

polarity whereas an increase of the polarity will favour the selectivity for divalent cations [160]. This is in good agreement with theoretical predictions.

The use of neutral carriers started with the work of Stefanac and Simon [161] who were the first to realize that macrocyclic antibiotics which show a large selectivity for potassium might be suitable for the preparation of potassium selective membranes.

Several classes of compounds can be distinguished. At first the macrotetrolides (tetralactone of nonactinic acid and derivatives) such as nonactin, monactin, dinactin, and trinactin, which have eight oxygen atoms in a flexible ring, can be mentioned (Fig. 17). These compounds form complexes with potassium which are far more stable than the complexes with sodium ions.

Another group of macrocyclic compounds that has been suggested is formed by the cyclic depsipeptides (α-amino acids and α-hydroxy aliphatic acids alternatingly bound in a ring) such as valinomycine, which also shows a large selectivity for potassium (Fig. 18). This compound plays an important role in the transport of potassium across the membranes of living cells.

Apart from the natural compounds mentioned so far, an interesting group of substances is formed by the "crown ethers" dis-

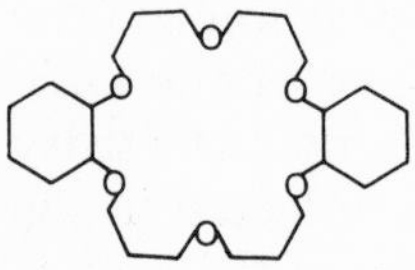

Fig. 19. Structure of the cyclopolyether dicyclohexyl-18-crown-6.

covered by Pedersen [162]. These cyclopolyethers (Fig. 19) form cavities with the oxygen atoms directed to the centre. The selectivity for alkali metals depends on the ring size, which determines the diameter of the cavity, and on the number of oxygen atoms.

Morf and Simon [163] (see also ref. 22, p. 430) have summarized the requirements governing the suitability of electrically neutral ligands to function as selective carriers for alkali or alkaline earth metals including ammonium ions.

(a) A carrier molecule should possess polar and non-polar groups.

(b) There should be preferably 5—8 but not more than 12 coordinating sites such as oxygen atoms.

(c) The carrier should be able to adopt a stable conformation in which a cavity is formed with the polar groups directed inwards and the non-polar groups forming a lipophilic shell around the coordination sphere. The cation should just fit in the cavity.

(d) High selectivities are achieved by the presence of bridged structures or hydrogen bonds making the arrangement more rigid. On the other hand, the ligand should be flexible enough to allow a sufficiently fast exchange.

The slow exchange reactions of most macrocyclic compounds have led Morf and his co-workers (see ref. 16) to the synthesis of acyclic compounds that otherwise fulfil the requirements mentioned above. Some of these ligands are presented in Fig. 20. The high flexibility of these structures ensures a rapid complexation and decomplexation, which make them very suitable for the preparation of membranes for

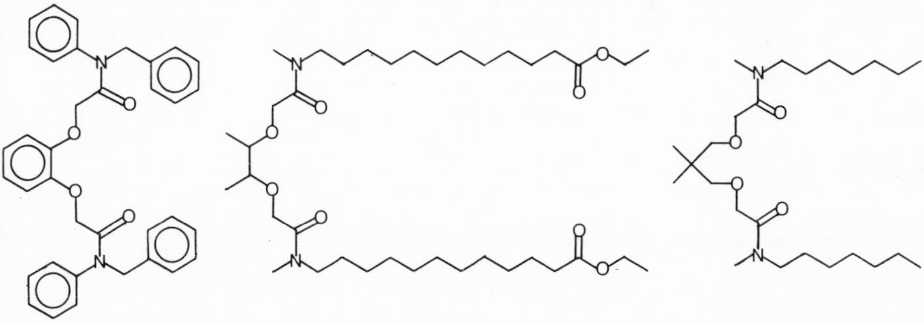

(a)   (b)   (c)

Fig. 20. Structures of some synthetic neutral carriers. (a) Selective for Na$^+$; (b) selective for Ca^{2+}; (c) selective for Li$^+$.

References pp. 385—392

ISEs. Although, for instance, the selectivity of the Ca^{2+} sensor based on the synthetic neutral carrier shown in Fig. 20(b) is superior to the "classical" one based on diesters of phosphoric acid with respect to Mg^{2+}, Zn^{2+}, and H^+, electrodes of this type are not yet commercially available. The same applies, for the moment, to all of the synthetic neutral carrier-based electrodes. The brief discussion of the performance of the various electrodes will therefore be restricted to the potassium and the ammonium selective liquid membrane electrodes.

a. Potassium selective liquid membrane electrode. The liquid membrane potassium selective electrodes commercially available are based on valinomycine usually dissolved in diphenyl ether. The electrode response is linear in the range of about 10^{-1} to 10^{-5} M with a slope of 58.5—59.0 mV/decade. The practical limit of detection is $\sim 10^{-6}$ M. The response time depends on the stirring rate for those electrodes in which the valinomycine is fixed in a PVC matrix or in a porous plug. In vigorously stirred solutions, response times are of the order of seconds; in slowly stirred solutions, several minutes are necessary to obtain the final value. Lindner et al. [164] have shown that silicone rubber-based electrodes are rather insensitive to flow rates. They attributed this behaviour to the formation of a swollen silica layer comparable with the gel layers formed with glass electrodes. This layer can be considered to form a fixed diffusion layer. On the other hand, the response times are much larger due to the lower values of the diffusion constants in the layer.

Liquid membrane potassium electrodes have a great advantage over potassium selective glass electrodes with respect to the interference by sodium ions; this selectivity is highly desirable for the direct determination of potassium in blood. Also, the selectivity in comparison with alkaline earth metals is acceptable. Selectivity coefficients as determined by Pioda et al. [151] are $k_{K,Na}^{pot} \sim 2 \times 10^{-4}$; $k_{K,Li}^{pot} \sim 2 \times 10^{-4}$; $k_{K,Cs}^{pot} \sim 0.5$; $k_{K,Rb}^{pot} \sim 5$; $k_{K,Ca}^{pot} \sim 2 \times 10^{-4}$; $k_{K,Mg}^{pot} \sim 2 \times 10^{-4}$; $k_{K,Ba}^{pot} \sim 10^{-4}$; $k_{K,NH_4}^{pot} \sim 10^{-2}$. The electrode response is influenced by iodide, hydroxide, chromate, and oxalate and very seriously by tetraphenylborate [165]. Furthermore, surface active compounds with a net positive charge, such as cetyltrimethyl ammonium bromide, affect the response [166].

b. Ammonium selective liquid membrane electrode. A limited number of papers has been published on the applications of liquid

352

membrane ammonium selective electrodes. All these electrodes were based on the macrotetrolides nonactin and monactin [167,168]. The selectivity sequence is reported to be $NH_4^+ > K^+ > Rb^+ > H^+ > Cs^+ > Li^+ > Na^+ > Ca^{2+}$ ($k_{NH_4,K}^{pot} \sim 10^{-1}$; $k_{NH_4,Na}^{pot} \sim 2 \times 10^{-3}$). A Nernstian response is observed in the range 10^{-1} to 10^{-5} M. The limited success of this electrode is certainly partly due to the availability of two alternatives: the ammonium selective glass electrode and the very selective ammonia gas-sensing electrode to be discussed in Sect. 4.(D).

During the last decade, considerable progress has been made in the synthesis of ligands tailor-made for one single ion. The present prospect is that many of them will be applied for the preparation of commercially available liquid membrane electrodes in the near future.

(D) GAS-SENSING ELECTRODES

Potentiometric gas-sensing electrodes are, in fact, complete electrochemical cells consisting of an ion selective indicating electrode in conjunction with a reference electrode. The complete device is separated from the sample solution either by a hydrophobic gas-permeable membrane or by an air-gap of several millimeters (Fig. 21). The operating principle is based on the diffusion through the membrane or air-gap of the gaseous species to be determined until the

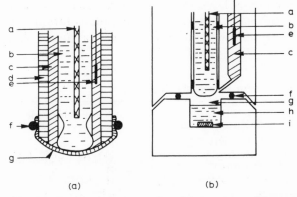

(a)   (b)

Fig. 21. Gas-sensing electrodes. (a) gas-permeable membrane electrode; (b) air-gap electrode. a, inner reference electrode; b, filling solution of ISE; c, bulk internal electrolyte; d, shaft; e, reference electrode; f, "O"-ring; g, gas-permeable membrane/air-gap; h, sample solution; i, stirring bar.

References pp. 385–392   353 partial pressure (or, actually, the chemical potential) on both sides of the membrane has become equal. If the permeating species evokes a change in the activity of the ion for which the indicating electrode is sensitive, the total EMF observed will depend on the partial pressure of the gas in the sample solution. This concept was first introduced by Stow et al. [169] for the determination of carbon dioxide. Shortly thereafter, Severinghaus and Bradley [170] improved the construction. Their device comprises a pH-sensing glass electrode and a sodium hydrogen carbonate solution as the internal electrolyte in the thin layer between the glass surface and the gas-permeable membrane. The CO_2 diffusing through the membrane affects the pH of the internal electrolyte in the film according to the reactions

$$CO_2 + H_2O \rightleftharpoons H_2CO_3 \rightleftharpoons H^+ + HCO_3^-; HCO_3^- \rightleftharpoons H^+ + CO_3^{2-}$$
$$(K_{a1}) \qquad\qquad\qquad\qquad (K_{a2})$$

Ross et al. [171] have summarized several other equilibria which might be suitable for gas-sensing probes together with possible electrodes as indicators (Table 7). The same authors have also presented a steady-state model in order to describe the behaviour of gas-sensing electrodes with respect to time response. Although their quantitative result is obtained after the introduction of some simplifying restrictions, it seems to be allowable to make some general predictions. The response time will be faster the thinner the membrane and the electrolyte layer, and the larger the diffusion coefficient of the gaseous species inside the membrane. A large distribution coefficient of this species between the membrane phase and the sample solution will also improve the response time. Furthermore, it was predicted that

TABLE 7

Possible equilibria and electrodes for gas-sensing electrodes

Species	Equilibrium	Sensing electrode
NH_3	$NH_3 + H_2O \rightleftharpoons NH_4^+ + OH^-$	H^+
	$x\,NH_3 + M^{n+} \rightleftharpoons M(NH_3)^{n+}$	$M = Ag^+, Cu^{2+}, Cd^{2+}$
SO_2	$SO_2 + H_2O \rightleftharpoons H^+ + HSO_3^-$	H^+
NO_2	$2\,NO_2 + H_2O \rightleftharpoons NO_3^- + NO_2^- + 2\,H^+$	H^+, NO_3^-
H_2S	$H_2S \rightleftharpoons HS^- + H^+$	S^{2-}
HCN	$Ag(CN)_2 \rightleftharpoons Ag^+ + 2\,CN^-$	Ag^+
Cl_2	$Cl_2 + H_2O \rightleftharpoons 2\,H^+ + ClO^- + Cl^-$	H^+, Cl^-
CO_2	$CO_2 + H_2O \rightleftharpoons H^+ + HCO_3^-$	H^+

going from a very low to a high concentration, the response time is independent of the magnitude of the concentration change. This was supported by the experiments of Bailey and Riley [172]. For the reverse process, the response time will vary with the ratio of the concentrations before and after the concentration change. On the basis of their theory, Ross et al. predict that the response time for a concentration change from 10^{-1} to 10^{-5} M can be approximately thirteen times greater than for the reverse change.

The diffusion coefficients in air are, in general, more than 10^4 times larger than in the various types of membranes. This more than counterbalances the increase in membrane thickness (gas-permeable membranes are $\sim 10^{-2}$ cm thick; the air-gap is $\sim 10^0-10^{-1}$ cm). Hence, a shorter response time can be expected for air-gap electrodes in comparison with gas-sensing electrodes provided with membranes [173]. With appropriate wetting procedures, the electrolyte layer on the surface can be kept very thin which also furthers the quick response. Air-gap electrodes are also very satisfactory for the determination of sample solutions containing compounds that might clog the membrane micropores or attack the membrane. The electrolyte layer of an air-gap electrode has to be renewed after each measurement. This limits the application in continuous analysis, but it also implies, on the other hand, that the electrolyte can be easily replaced by another kind of electrolyte. So, the same device can be used for all gases that can be detected with the same sensing electrode, e.g. NH_3, CO_2, or SO_2 in conjunction with a pH electrode.

With membrane-based gas-sensors, there is no need for renewal of the electrolyte film after each measurement. Therefore, this type of electrode is more suitable for continuous monitoring. Moreover, there is less danger of drying out of the active surface of the indicating electrode, thereby impairing the response.

Although gas-sensing electrodes are widely applied nowadays, comparatively little has been published on their analytical evaluation. It is only recently that a number of papers have appeared dealing with the measuring range, the response time, and the selectivity. As for the limit of detection, Hansen and Larsen [174] have assumed that this limit is due to the fact that the species to be determined are naturally present in the internal electrolyte because of protolysis of the corresponding acid or base, e.g.

$$NH_4^+ + H_2O \rightleftharpoons NH_3 + H_3O^+$$

$$HCO_3^- + H_2O \rightleftharpoons H_2CO_3 + OH^-; H_2CO_3 \rightleftharpoons H_2O + CO_2$$

If the sample solution does not contain the gaseous species, diffusion from the electrolyte film to the sample solution will occur, which tends to lower the concentration in the film. Now, two possibilities can be considered with respect to the relation between the composition of the electrolyte film and that of the bulk of the internal electrolyte. Hansen and Larsen have implicitly assumed that the film is continuously renewed from the bulk keeping the composition effectively constant. Bailey and Riley [175], correctly, question this assumption. They have pointed out that, in most cases, the lower limit of detection is much lower than can be expected from Hansen and Larsen's theory. They explain this fact by supposing that the renewal is negligible. Mascini and Cremisini [176] have suggested that both the diffusion rate of the gaseous species across the membrane and the rate of supply from the bulk to the electrolyte film have to be taken into account at the same time. Van der Pol [177] has shown that, for an ammonia-sensing electrode, the limit of detection cannot be explained by equilibrium considerations alone, but that the limit is due to the fact that, below a certain ammonia level in the sample solution, the velocity with which the system tends to reach equilibrium rapidly decreases to zero.

Interferences can be expected from other gaseous species in the solution, particularly if they can produce variations in the activity of the ion sensed by the indication electrode as well. Mascini and Cremisini have considered the case of two volatile acidic species HA and HB in the sample in conjunction with an internal electrolyte containing B^-, i.e. an electrode intended for the determination of HB. The possible interference of HA is governed by the reaction

$$HA + B^- \rightleftharpoons HB + A^-$$

with an equilibrium constant $k = k_a^A/k_a^B$, where k_a^A and k_a^B are the dissociation constants of HA and BH, respectively. Provided that the sample is acidic enough to ensure the complete conversion of both species to their volatile forms and that the membrane is an air-gap or a membrane that does not discriminate between the two acidic species, the ratio $[HB]/[HA] = r$ in the electrolyte film can be taken equal to the ratio in the sample solution. As long as $r > 10k$, the equilibrium is shifted to the left-hand side and HA does not interfere. The system $HB-B^-$ will determine the pH in the electrolyte film. On the other hand, if $r < 10^{-1}k$, the reaction is shifted to the right and the $HA-A^-$ system will determine the pH in the film. By changing

356

the pH of the sample solution, the selectivity can be influenced. On raising the pH, the stronger one of the two acidic species will be converted to the involatile anionic form. Of course, an analogous argument can be given for alkaline species like ammonia and related alkaline compounds.

At the moment, gas-sensing electrodes for NH_3, CO_2, SO_2, NO_x, and H_2S are commercially available. The ammonia and the carbon dioxide selective electrodes can be used for concentrations up to about 1 M; for the other electrodes, this upper limit is approximately 10^{-2} M; at higher concentrations a continuous potential drift is observed. The lower limit of Nernstian response depends on the kind of electrode and its construction. For most electrodes, a value of 10^{-6} to 10^{-5} M has been reported. The practical limit of detection is about 10 times lower. To get a maximum response, the pH of the sample solution has to be adjusted. For the various sensors the pH ranges are NH_3 pH > 12; CO_2 pH < 3.3; SO_2 pH < 0.7; NO_x pH < 2; H_2S pH < 5.

(E) ENZYME ELECTRODES

Enzyme electrodes consist of an ion selective electrode covered by a layer of an immobilized or insolubilized enzyme. The enzyme reacts with the compound to be assayed highly selectively or even completely specifically. The reaction products formed diffuse in the enzyme layer towards the ISE which responds to these products either directly or indirectly in the way the gas-sensing membrane electrode responds. Because Guilbault has already written a complete chapter on enzyme electrodes in this series [178], only some general aspects will be recapitulated here.

The advantage of enzyme electrodes over other enzymatic determinations is the fact that there is no need for reagent preparation and the fact that there is almost no consumption of the often very expensive enzymes: the same electrode can be used for many assays. Furthermore, the performance of the determination is very simple once the electrode is ready for operation. For the immobilization of enzymes, several techniques have been proposed, e.g. micro encapsulation within semipermeable thin walls; adsorption on inert supports; covalent cross-linking by bifunctional reagents into macroscopic particles; physical entrapment into gel lattices; and covalent binding to water-insoluble matrices. The latter technique is now the most widely used method and well-suited for the construction of enzyme electrodes. It offers the advantage of an increased stability which is not irreversibly affected by changes in pH, ionic strength, solvents, or temperature. The immobilized enzyme is spread over the surface of the base membrane and the coating preferably protected by a dialysis membrane.

Enzyme electrodes, when in operation, should be calibrated daily because there is always some loss in activity. The lifetime depends strongly on the way the enzyme is fixed as well as on the purity of the enzyme. Physically entrapped enzymes will keep their function for several weeks or during 50—100 assays; chemically bound enzymes can be used for much longer periods and many more determinations if the synthesis is properly effected.

The overall response time depends on the rate of diffusion of the substrate, the reaction rate with the enzyme in the membrane, diffusion of the reaction products towards the ISE surface and, of course, the response time of the ISE itself. The rate of diffusion is increased by reduction of the Nernst diffusion layer in the solution, and so by the rate of stirring. The reaction rate is governed, according to the Michaelis—Menten equation, by the enzyme activity and the concentration of substrate and is also affected by factors such as pH, temperature, and the presence of inhibitors. To reach the steady state situation inside the membrane as quickly as possible, the membrane should be very thin. This requires highly active enzymes in order to maintain an acceptable reaction rate of the substrate with the enzyme. Response times are usually of the order of a few minutes. Washing is necessary between two consecutive assays and, especially with thick membranes, wash times of more than 10 min have to be part of the procedure. All enzyme electrodes sense substrates in the range of 10^{-2} to 10^{-4} M. For more details and a discussion of several applications the reader is referred to the contribution by Guilbault, Vol. VIII, Chap. 1 of this series.

A major problem is the lack of commercially available enzyme electrodes and even the lack of the purified enzymes themselves. In this connection, an interesting development has to be mentioned. Recently, it has been shown that living bacterial cells can be used in conjunction with a gas-sensing electrode to form *bio-selective sensors* [179—181]. These bacterial electrodes offer a number of advantages over the conventional enzyme electrode. These advantages include a longer electrode lifetime, the possibility of a regeneration of the response by storage of the electrode in the appropriate growth medium, elimination of the often tedious and expensive process of enzyme purification as well as the possibility of performing assays of compounds for which appropriate enzymes have not yet been isolated.

(F) ION SENSITIVE FIELD EFFECT TRANSISTORS (ISFETs)

In 1970, Bergveld [71] introduced a new concept for the construction of ISEs. His intention was to develop a device that would overcome some of the problems associated with the use of conventional ion selective micro-electrodes. The very high resistivity of such conventional electrodes necessitates the use of high-impedance voltmeters and thoroughly shielded connecting cables. The pick-up of noise in the cables would be greatly reduced if a high-to-low impedance conversion could be accomplished directly after the ion selective membrane, i.e. in situ. An extra advantage would be that the low-impedance output signal thus obtained would allow the use of inexpensive low-impedance measuring instruments. Bergveld used a metal oxide semiconductor field effect transistor (MOSFET) to accomplish the impedance conversion on the spot. Such a MOSFET is schematically depicted in Fig. 22. It consists of a p-type silicon wafer in which two n-type regions exist, the so-called source and drain. The metal gate is separated from the silicon by an insulating layer of SiO_2. Application of voltage between gate (positive) and the

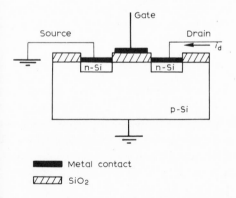

Fig. 22. Metal oxide semiconductor field effect transistor.

bulk p-type silicon (negative) will cause an accumulation of electrons in the silicon at the Si—SiO$_2$ interface forming a conductive channel between drain and source. Hence, the drain—source current, i_d, depends on the applied voltage. When the metal gate is replaced by an ion selective membrane, the potential difference across the membrane will contribute to the total potential difference between solution and the bulk silicon and so influence the drain—source current. Such a device was given the name ISFET. At present, ISFETs sensitive to hydrogen [71,182], potassium [182,183], and calcium [183] have been reported. Although, under certain circumstances, a reference electrode can be omitted, as was shown by de Rooy [184], other authors always prefer the use of reference electrodes. However, it only makes sense to develop miniature ISFETs if the reference electrodes are at least of the same size or smaller. An elegant solution of this problem was recently presented by Comte and Janata [185]. They connected another source and drain on the same silicon wafer together with a pH-sensitive membrane as the gate. A small compartment above this gate was filled with a pH buffer and connected by a small capillary with sample solution. This part of the chip functions as a reference FET. The difference between the two drain—source currents is correlated with the activity of the ion the membrane is selective for. An attractive feature of this differential device is that noise, and fluctuations due to temperature variations, are largely compensated. The performance of this type of electrode is not yet such that they can compete with the macro ISEs, but it can be expected that good progress will be made in the near future. Applications will be found in the biomedical field.

5. Instrumentation

(A) VOLTMETER

For the accurate measurement of the EMF of an electrochemical cell, the voltmeter has to meet certain requirements. The input impedance of the instrument should be at least a thousand times larger than the impedance of the cell in order to be able to measure the EMF with an error of less than 0.1%. The total impedance of the cells is generally mainly the result of the electrical resistance of the ion selective membrane, but special attention has also to be paid to the reference electrode. In particular, the clogging of the liquid junction can be the cause of a very high impedance. Another factor of importance can be the resistance of the solution, especially if it does not contain sufficient background electrolyte or if non-aqueous solvents are used. The low conductivity of the solution can be the cause of irreproducible and drifting potentials. It can often be overcome by the addition of a large amount of an indifferent electrolyte. An increase in the amount of electrolyte has the additional advantage that the double layer is thinner, thus reducing the contribution of the electokinetic potential to the total EMF. This makes the potential less sensitive to variations in the stirring rate or, more generally, of the hydrodynamic conditions in the solution.

Since a large number of ISEs exhibit an impedance of over 10^9 Ω, the voltmeter should have an input impedance of at least 10^{12} Ω. Such instruments are commercially available as mV-meters, pH-meters or pIon-meters. Most modern instruments operate according to the same principles. At first, a high to low impedance transformation is achieved by means of a field effect transistor (FET) followed by a low impedance amplifier. If an a.c. amplifier is used, the signal must be modulated before the amplification step and demodulated thereafter. The second part of the meter consists of the measuring device which can be either analog or digital. Furthermore, instruments that are suitable for direct reading of the pH or pIon must include the possibility to adjusting the slope as well as compensating for deviations of the temperature of the sample solution from that of the solutions used for the calibration. Some instruments even allow the direct reading of concentrations, for instance by using an anti-log scale with proper calibration.

Sometimes, reference electrodes which also have a high impedance are preferable. For instance, a pH glass electrode can be used as a reference electrode in solutions of constant pH. The advantage of the use of such an ISE as the reference electrode is that any liquid junction can be avoided. The input of the reference electrode has then also to be provided with a FET.

Special care must be taken of the metal contacts and particularly of the sockets. Connections between different metals cause a contact potential that is sensitive to temperature variations. For most metal—metal contacts, these variations are restricted to about 10^{-3} to 10^{-4} mV/°C, but corrosion can significantly enhance this value, e.g. the CuO/Cu contact potential changes approximately 1 mV per °C.

Another important aspect to be considered is the electrical shielding. To avoid serious problems arising from the pick-up of noise and to protect the total device against influences of electrostatic origin, the high impedance lead must be carefully shielded. Therefore, ISEs are provided with low-noise coaxial cables. The isolation resistance between the central conducting lead and the shielding braid forms an impedance parallel to the impedance of the voltmeter. The effective impedance of the cable, including the plug—socket connection, should be more than 10^{13} Ω. In particular, the isolation of the connectors often forms a weak point in the whole operation; frequent cleaning is absolutely necessary. To avoid condensation on the contacts under humid conditions, it is often advisable to keep the instrument switched on, even if it is not in operation.

The instrument must always be grounded but special attention has to be paid to avoiding so-called ground loops. Such loops can easily be formed if two instruments that are electrically connected (e.g. voltmeter and recorder) are separately grounded.

(B) REFERENCE ELECTRODES

A reference electrode is required in conjunction with an indicating electrode (e.g. an ISE) to complete an electrical circuit. The two electrodes form the electrochemical cell. In addition, many ISEs have an internal reference electrode. Generally, any electrode which responds selectively to a species in the sample solution that is not sensed by the indicating electrode can function as a reference electrode as long as the concentration of this species is kept constant. However, only "real" reference electrodes will be considered in this section. This discussion is not meant to give a full account of all the types available or described in the literature (a comprehensive review is given in a monograph by Ives and Janz [186]), but it has to be emphasized that the reproducibility of the EMF measurement is equally affected by the performance of the indicating electrode and the reference electrode.

The reference electrodes commonly used are the calomel electrode (Hg/Hg_2Cl_2 electrode) and the silver/silver chloride electrode. The latter is often applied as the internal reference electrode on many ISEs (e.g. glass electrodes). Both electrodes are secondary reference electrodes, the primary reference electrode being the standard hydrogen electrode. The calomel and the silver/silver chloride electrodes

362

are electrodes of the second kind. In contrast to electrodes of the first kind, which exhibit a potential that is primarily fixed by the concentration of ions of the metal of which the electrode is made, electrodes of the second kind have a potential that is fixed by the concentration of an anion with which the metal ion can form a sparingly soluble salt. To ensure stable equilibrium conditions, the sparingly soluble salt must be present on the surface of the metallic electrode. The quality of the electrode strongly depends on the preparation procedure, but it can be safely stated that most of the commercial devices meet the required analytical standards. Reference electrodes should be allowed to carry only a very small current in order to avoid an irreversible alteration of the surface structure and hence of the potential.

The calomel and the silver/silver chloride electrodes are usually filled with a saturated potassium chloride solution. Temperature changes will therefore affect not only the Nernstian response (slope factor) and the solubility product of the Hg_2Cl_2 or $AgCl$, but also the activity of the chloride ion due to the solubility of KCl. The saturated calomel electrode has a potential of 247.7 mV at 20°C and of 244.4 mV at 25°C. The corresponding figures for the Ag/AgCl electrode are 201.9 mV at 20°C and 197.0 mV at 25°C [187].

Substances such as sulfides and powerful oxidizing or reducing agents will affect the potential. Should the case arise, the internal electrolyte must be protected against the penetration of such components. On the other hand, it is also possible that the potassium chloride which leaks from the reference electrode in the sample solution will interact with the ion to be determined or with the surface of the indicating electrode. In both cases, an intermediate solution must be used. Double junction electrodes are available for this purpose.

The most frequently occurring cause of problems with reference electrodes is the obstruction of the free diffusion of the ions across the phase boundary, or liquid junction, between the internal solution and the sample solution. Several types of liquid junction have been described. For analytical applications, mainly junctions of the "restrained diffusion" type are used, the restriction being either a ceramic plug, an asbestos wick, or a sleeve. The first two junctions have small leakage rates; the latter has a much larger rate of loss of the internal electrolyte. A high leakage rate is especially important when the sample solution contains compounds that might clog the junction. The disadvantage of fast leakage is that the internal electrolyte must be replenished very frequently.

The potentials at the ISE as well as at the reference electrode depend on the temperature and, in many cases, on the stirring rate. In the case of the latter aspect, it is advisable to take care of constant hydrodynamic conditions at the surface of the ISE and at the liquid junction of the reference electrode. This means that the stirring rate should not only be kept constant, but it should also be the same during the calibration procedure and the determination. So, the electrodes should preferably be kept in a fixed position in the beaker. When the electrode assembly is used for the indication of a titration reaction, only relative potential changes are of importance to locate the end-point and it is only necessary to fix the conditions during one titration.

Many magnetic stirrers produce some heat which makes them stir faster. Therefore, it is better to use synchronous stirrers. To avoid warming the sample solution, the beaker should be insulated from the stirrer. When a very high accuracy is required, the cell should be thermostatted. Although, most voltmeters are provided with an ability to compensate for a difference in the temperature at which the determination is performed and the temperature at which the calibration was made, this compensation usually applies to the slope factor (RT/zF) only. For high accuracy, calibration and determination should be performed at the same temperature.

6. Methods of analysis with ion selective electrodes

(A) INTRODUCTION

One of the basic problems in the application of ISEs in analytical chemistry is connected with the fact that ion activities instead of ion concentrations are measured. Activities might be of more importance for the physical chemist or the physiologist, but the analyst is generally more interested in the concentration. A relation exists between activity and concentration which can be expressed as

$$a_i = \gamma_i C_i \tag{129}$$

in which γ_i is the activity coefficient of the ion i. When the amounts of all ionic species in the solution tend to zero ($\Sigma C_j \rightarrow 0$), the interaction between the ions in the solution will become negligible and the behaviour of the solution is said to be "ideal". In that case, γ_i becomes equal to unity. It is not possible to determine single ion activities experimentally because each kind of ion in a solution is accompanied by counter ions. However, it is possible to calculate the single ion activities on the basis of a theoretical model introduced by Debye and Hückel. They derived the relationship

$$\log \gamma_i = - \frac{A z_i^2 I^{1/2}}{1 + B \overset{0}{a} I^{1/2}} \tag{130}$$

where $I = 0.5 \sum_j C_j z_j^2$, I is called the *ionic strength* of the solution, A and B are constants which depend on the solvent and on the temperature, $\overset{0}{a}$ is a parameter related to the distance up to which the two ions forming an ion pair can approach each other. At room temperature, $A \simeq 0.51$ and $B \simeq 3.3 \times 10^{-7}$. Typical values for $\overset{0}{a}$ are in the order of 0.3—0.5 nm. Therefore, the product $B\overset{0}{a}$ does not differ very much from unity and below $I \simeq 0.01$, the whole term $B\overset{0}{a}I^{1/2}$ can be neglected. This means that, at low ionic strength, the activity coefficient is independent of the kind of ion apart from its electrical charge. At larger ionic strengths, the individual character of each kind of ion becomes important.

The Debye—Hückel relationship holds for ionic strengths in the range $0 < I < 0.1$. In this region, long-range ion—ion interactions are of major importance, whereas at values $I > 0.1$ the ions approach each other so closely that specific short-range interactions begin to play a role. Because the distance of closest approach depends on the hydration (solvation) number, this number becomes a parameter in extended equations for the single ion activity. At the same time, the equations lose their universality.

(B) CALIBRATION CURVE; THE USE OF METAL BUFFERS

If a calibration curve is made by measuring the electrode response to standard solutions prepared by serial dilution without the addition of extra indifferent salt, the ionic strength will increase linearly with increase of the concentration. This will lead to a gradual decrease of the activity coefficient and the calibration curve [E vs. log(concentration)] will show a negative departure from the straight line for concentrations above 10^{-3} to 10^{-2} M. This curvature of the calibration curve can be precluded by working at constant ionic strength. This is commonly achieved by adding a large excess of an indifferent electrolyte, the *ionic strength buffer*. In that case, the activity coefficient is constant and the E vs. log(concentration) curve is linear over the entire concentration range. The line will be parallel to the E vs. log(activity) curve, the shift being $(RT/z_i F) \ln \gamma_i$. In contrast to the preparation of concentration standards, the preparation of activity standards is accompanied with the fundamental problem of the choice of an appropriate standard activity scale. These problems have been extensively discussed by Bates [188,189].

Serial dilution of standards is a satisfactory procedure for calibration down to about 10^{-6} M. For calibration at lower concentrations, it is advisable to use buffer solutions. Metal ion buffers are prepared by adding an excess of complexing agent, L, to the metal ion solution. According to the formula for the stability constant

$$K_{ML} = \frac{[ML]}{[M][L]}$$

the concentration of M in the solution can be represented by

$$[M] = \frac{[ML]}{K_{ML}[L]}$$

or

$$pM = \log K_{ML} + \log \frac{[L]}{[ML]}$$

Hence, at a fixed ratio [ML]/[L], the value of [M] is also fixed. Of course, if L is subject to a protolytic side reaction, the conditional stability constant must be used [190]. The ligand to be used for the preparation of a buffer has to fulfil two requirements: first, the conditional stability constant should be accurately known; second, the ligand itself should not affect the electrode response.

Calibration without buffers can be facilitated in various ways. To a relatively large volume of the sample background electrolyte, if necessary mixed with any reagent or ionic strength buffer, small increments of the ion to be standardized are added. When the solution is concentrated, it is possible to use a microburette (maximum volume 1 ml) so that any correction for dilution is unnecessary. Bailey and Pungor [191] have used the elegant technique of coulometric generation to increase the concentration to be standardized. A simple continuous method for the performance of a calibration of ISEs was proposed by Horvai et al. [192]. They started with a vessel containing the concentrated standard solution. The vessel was equipped with an overflow to keep the volume of the solution constant. Background electrolyte is pumped into the vessel at a constant rate in order to dilute the original standard solution, decreasing the concentration of the ion to be standardized. The whole procedure can easily be automated.

Sometimes, the pH will change on serial dilution. This can lead to the situation in which the ISE is no longer responding properly to the ion of interest. Clearly, the use of a pH buffer will improve the results [193].

(C) DIRECT MEASUREMENTS

The direct potentiometric method with ISEs is based on single measurements and fitting the EMF value found for a sample of unknown concentration on the ordinate of the calibration curve and reading the corresponding concentration from the abscissa. The highest accuracy is obtained when the reading falls in the linear Nernstian or near-Nernstian part of the curve. Near the limit of detection, the sensitivity ($dE/d\log c$) is much lower.

Since it is extremely difficult to obtain reproducible measurements within ± 0.1 mV, the error in the concentration measurement is, at best, about 0.4% for a monovalent ion and 0.8% for a divalent ion. More generally, the error can be given by the expression

$$\% \text{ error} \approx 4|z_i|\Delta E \tag{131}$$

where ΔE is the error in the EMF measurement in mV.

The direct method can be used for the analysis of samples in which the ion to be analyzed is only partly in the "free" state, whereas the remaining part is complexed. In that case, only the concentration of free ion is determined. By a suitable pretreatment, the complexes can be destroyed allowing the determination of the total amount. A good example forms the determination of fluoride in the presence of iron(III) or aluminum(III), which both form strong fluoride complexes. Without any pretreatment, only the free fluoride is measured. On the addition of a mixture containing a pH buffer and a complexing agent like DCTA, which forms stable complexes with both iron(III) and aluminum(III), the complexed fluoride will be set free and the total fluoride is measured.

When measurements have to be made in special well-defined matrices, it is desirable to calibrate the ISE with standard solutions which contain the same amounts of accompanying compounds, e.g. for the determination of calcium in blood serum, it is preferable to use standard calcium solutions containing sodium chloride at a concentration level of about 0.15 M [193].

Many matrix problems can be overcome by using standard addition/subtraction methods or by the performance of a potentiometric titration. These methods will be briefly discussed in the next two sections.

(D) STANDARD ADDITION/SUBTRACTION METHOD

Both the standard addition and the standard subtraction method are based on a change in the concentration of the species to be determined. This is achieved either by the addition of a known amount of standard solution of this species or by the addition of a standard solution of a compound that reacts stoichiometrically, preferably in a $1 : 1$ ratio, with the species of interest. The determination is performed as follows. First, the EMF, E_1, of V_1 ml of the sample solution whose concentration is C_1 is measured.

$$E_1 = \text{constant} + S \log[x_1 \gamma_1 C_1] \tag{132}$$

where S is the slope of the calibration curve, γ_1 is the activity coefficient, and x_1 is the fraction of uncomplexed species. Subsequently, a small volume, ΔV, of a standard solution (concentration C_s; $C_s \gg C_1$) is added and the EMF, E_2, is again measured.

$$E_2 = \text{constant} + S \log \left[x_2 \gamma_2 \left(\frac{V_1 C_1 + \Delta V C_s}{V_1 + \Delta V} \right) \right] \tag{133}$$

When ΔV is small compared with V_1, eqn. (133) can be transformed to

$$E_2 = \text{constant} + S \log \left[x_2 \gamma_2 \left(C_1 + \frac{\Delta V}{V_1} C_s \right) \right] \tag{134}$$

Subtraction of E_1 from E_2 leads to

$$E_2 - E_1 = \Delta E = S \log \left[\frac{x_2 \gamma_2}{x_1 \gamma_1} \frac{C_1 + \dfrac{\Delta V}{V_1} C_s}{C_1} \right] \tag{135}$$

368

If $x_2 \simeq x_1$ and $\gamma_1 \simeq \gamma_2$, which is essential for the method, this equation can be reduced to

$$\Delta E = S \log \left[1 + \frac{\Delta V}{V_1} \frac{C_s}{C_1} \right] \tag{136}$$

or

$$C_1 = C_s \frac{\Delta V}{V_1} \left[\frac{1}{(\text{antilog } \Delta E/S) - 1} \right] \tag{137}$$

For the standard subtraction method, E_2 is subtracted from E_1 and the equation corresponding to eqn. (136) reads

$$E_1 - E_2 = \Delta E^* = S \log \frac{1}{1 - (\Delta V/V_1)(C_s/C_1)} \tag{138}$$

This equation can be rewritten as

$$C_1 = C_s \frac{\Delta V}{V_1} \frac{1}{1 - (1/\text{antilog } \Delta E^*/S)} \tag{139}$$

A variant on the addition techniques discussed above is the analate addition method proposed by Durst [194]. This method, which implies the addition of a small volume of the sample solution to the standard solution, is especially useful when only small sample volumes are available.

The accuracy of the standard addition method can be improved by adding several increments to the same sample solution and measurement after each addition: the so-called *multiple standard addition method*. The method is based on a generalization of eqn. (133), viz.

$$E_x = \text{constant} + S \log x_x \gamma_x \left[\frac{V_1 C_1 + \Delta V_x C_s}{V_1 + \Delta V_x} \right] \tag{140}$$

If x_x and γ_x are constant

$$E_x = \text{constant} + S \log[x_x \gamma_x] + S \log \left[\frac{V_1 C_1 + \Delta V_x C_s}{V_1 + \Delta V_x} \right]$$

$$= E' + S \log \left[\frac{V_1 C_1 + \Delta V_x C_s}{V_1 + \Delta V_x} \right] \tag{141}$$

This equation can be transformed to

$$(V_1 + \Delta V_x) \, \text{antilog}[(E_x - E')/S] = V_1 C_1 + \Delta V_x C_s$$

or

$$(V_1 + \Delta V_x) \, \text{antilog}(E_x/S) = (V_1 C_1 + \Delta V_x C_s) \, \text{antilog}(E'/S) \qquad (142)$$

A plot of $(V_1 + \Delta V_x) \, \text{antilog}(E_x/S)$ on the ordinate versus ΔV_x on the abscissa should yield a straight line which intersects the abscissa at $(\Delta V_x)_0$. Then the equality

$$V_1 C_1 + (\Delta V_x)_0 = 0 \qquad (143)$$

holds. So

$$C_1 = -\frac{(\Delta V_x)_0 C_s}{V_1} \qquad (144)$$

The procedure can be simplified by keeping ΔV_x small with respect to V and by using special antilog paper provided by the Orion Research Inc. [195]. Multiple standard subtraction is, in fact, part of a linear titration procedure and will be discussed in the next section.

(E) TITRATIONS

The inherent advantages of potentiometric titrations over the direct evaluation of the EMF measurements are better accuracy and higher reliability. This is particularly true for the determinations with those ISEs which do not exhibit long-term stability or which do not show a constant sensitivity over the entire concentration range. The improvement is due to the fact that titrations are absolute and no calibration of the ISE is necessary. The accuracy of titrations is primarily determined by the accuracy of standardization and volume measurement of the titrant and so errors of less than 0.5% can be attained. Moreover, it is possible to increase the selectivity of the analysis through adjustment of such conditions so that only the relevant species is titrated.

It is beyond the scope of this chapter to give a full discussion of the theory of potentiometric titrations. Only some aspects of general importance associated with titrations indicated with ISEs will be considered. A separate chapter on potentiometric titrations has been published in Vol. IIA of this series [252]. A recent monograph [253] on automatic potentiometric titrations also contains details of recent development.

When ISEs are used for the indication of direct or back titrations, it is not necessary that the ion to be titrated is the one that is sensed by the electrode; it is also possible that the electrode is selectively sensitive to the ion of the titrant. So, the lead electrode has been used for the indication of the titration of sulfate by using lead(II) as the titrant.

The end-point of the potentiometric titration is mostly located at the steepest point of an S-shaped titration curve. This is only correct if the titration curve is symmetrical. Asymmetry can occur in the presence of interfering substances or by the fact that the total variation in the concentration during the titration does not fit in the linear range of the calibration curve. A drawback of this method of locating the end-point is that measurements in the neighbourhood of the equivalence point are particularly important, while in this region the concentration of the two reactants is relatively small and the equilibrium state is sometimes slowly attained. Furthermore, when the concentration to be determined is very low or the value of the equilibrium constant is rather small, the S-shape is not very pronounced and, consequently, it will be difficult to locate the end-point. This difficulty can be overcome to some degree by using a linear instead of a logarithmic indication method.

In a logarithmically indicated titration, the quantity plotted on the ordinate is proportional to the logarithm of the concentration (activity), whereas in a linearly indicated titration, the quantity plotted is directly proportional to the concentration itself. In that case, each titration curve consists of two linear branches connected by a curved portion. On extrapolation, these branches intersect at the end-point. The degree of curvature near the end-point depends on the concentration to be determined and the value of the equilibrium constant. A potentiometric titration curve, which is a logarithmic one in origin, can be linearized according to Gran [196] by application of the relationship

$$E = E' + S \log \frac{V_1 C_1 - \Delta V_x C_t}{V_1 + \Delta V_x} \tag{145}$$

where C_t is the concentration of the titrant. This equation can be transformed to

$$(V_1 + \Delta V_x) 10^{(E-E')/S} = V_1 C_1 - \Delta V_x C_t \tag{146}$$

Hence, a plot of the left-hand side of eqn. (146) vs. ΔV_x yields a straight line which intersects the abscissa at the end-point for which

$$V_1 C_1 = (\Delta V_x)_{ep} C_t \tag{147}$$

holds. Equations (145)—(147) apply to the case where the ion to be determined is sensed. Then, the measurements before the equivalence point have to used for the linearization. When the titrant ion is sensed, the points measured after the equivalence point have to be used. Then the relationship

$$E = E' + S \log \frac{\Delta V_x C_t - V_1 C_1}{\Delta V_x + V_1} \tag{148}$$

holds and a plot of $(\Delta V_x + V_1)10^{(E-E')/S}$ vs. ΔV_x will intersect the abscissa again at

$$(\Delta V_x)_{ep} = \frac{V_1 C_1}{C_t} \tag{149}$$

Notice that in the subtraction method, the slope of the straight line is the opposite of that in the standard addition method.

When microburettes are used and the sample volume V_1 is not too small, the factor $(\Delta V_x + V_1)$ due to the dilution can be taken as constant and only the term $10^{(E-E')/S}$ need be plotted against ΔV_x.

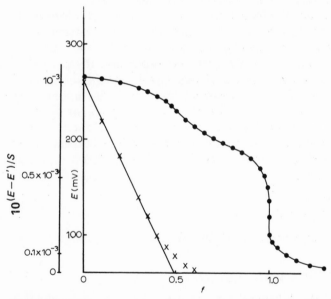

Fig. 23. Titration curve of a mixture of 10^{-3} M Cd(II) and 10^{-3} M Cu(II) with EDTA indicated by means of a CuISE (experimental results taken from ref. [197]). ●, E vs. f; ✕, $10^{(E-E')/S}$ vs. f.

372

In Fig. 23, an example is given to illustrate the usefulness of the linearization. The titration curve for a mixture of Cu(II) and Cd(II) with EDTA consists of two parts; in the first part, the more stable Cu(II)EDTA complex is formed followed by the complexation of cadmium. The second end-point can be clearly located in the logarithmic E vs. f curve (f is the titration parameter), but the first end-point is not detectable. Apparently, the difference between the stability constants of the Cu(II)EDTA and the Cd(II)EDTA complexes is too small. However, when the measurements are accurate enough, the first part of the curve can be linearized yielding the end-point by intersection of the line with the abscissa.

Figure 23 illustrates another aspect of importance: in the presence of Cu(II)EDTA, the copper selective electrode can be used for the indication of the titration of Cd(II) with EDTA. In practice, the amount of Cu(II)EDTA can be very small; it has to function only as an indicator species. This principle can be applied to many other determinations where titration reactions have to be indicated for which no ISE is available. For other aspects of locating the end-point, see ref. 254.

(F) EXPERIMENTAL DETERMINATION OF SELECTIVITY COEFFICIENTS

Knowledge of the values of selectivity coefficients is very important for the estimation of the interference of the other compounds in a sample. If not available, it is good analytical practice to determine the coefficients in the course of the development of the analytical procedure. The determination of these selectivity coefficients, $k_{A,B}^{pot}$, can be based on potential measurements either in separate or in mixed solutions.

In the separate solution method, the EMF response is measured with each of two separate solutions, one containing the ion $A^{z_A\pm}$ (but no B), the other containing the ion B^{+z_B} (but no A) at the same activity ($a_A = a_B$). If the measured values are E_A and E_B, respectively, the value of $k_{A,B}^{pot}$ may be calculated from the equation

$$\log k_{A,B}^{pot} = \frac{E_B - E_A}{2.303(RT/z_A F)} + \left(1 - \frac{z_A}{z_B}\right) \log a_A \tag{150}$$

It should be noted that the selectivity coefficient depends on a_A and so on the activity (concentration) level at which the determination is performed unless $z_A = z_B$. This implies that, in the case $z_A \neq z_B$, the value of $k_{A,B}^{pot}$ should always be accompanied with an indication of the concentration level.

This separate solution procedure is not recommended by the IUPAC commision [41,41a] except in those cases where the mixed solution method is inconvenient or not feasible.

The mixed solution or fixed interference method is based on EMF measurements with solutions of a constant level of interference, a_B, and variable activity of the primary ion, a_A. The use of mixed solutions is expected to give a more realistic representation of the electrode behavior.

The curve represented by the general equation

$$E = \text{constant} + \frac{2.303RT}{z_A F} \log[a_A + k_{A,B}^{pot}(a_B)^{z_A/z_B}]$$

consists of two linear portions. The first branch is represented by

$$E = \text{constant} + \frac{2.303RT}{z_A F} \log a_A$$

and the second by

$$E = \text{constant} + \frac{2.303RT}{z_A F} \log k_{A,B}^{pot}(a_B)^{z_A/z_B}$$

At the point of intersection, the value of a_A has to fulfil the relationship

$$k_{A,B}^{pot} = \frac{(a_A)_{\text{intersection}}}{(a_B)^{z_A/z_B}} \tag{151}$$

Srinivisan and Rechnitz [198] have proposed another mixed solution method for the determination of $k_{A,B}^{pot}$, but the procedure is arithmetically rather cumbersome and has no advantages over the fixed interference method.

7. Applications of ion selective electrodes

(A) INTRODUCTION. A BRIEF SURVEY OF THE DIFFERENT FIELDS

The great versatility of ion selective electrodes can be illustrated by the large number of papers published during the last decade; over 2500 publications on ISEs have appeared during the last two decades.

374

They cover a very wide variety of applications ranging, for instance, from the determination of bromide in photographic baths to calcium in beer, from the determination of cyanide in plating baths to potassium in serum. Comprehensive surveys are available in the biennial reviews by Buck [25] and the extensive reviews by Koryta [18,19]. Also, some of the manufacturers of ISEs have a bibliography available. Although a great many of the applications are related to rather specific analytical problems, some general fields of application are soil analysis, biological fluids, including serum, blood, and urine, water analysis, including drinking water, sea water, and waste water, and all kinds of beverages and foodstuffs.

Without having the intention of being complete, some references of the applications of the most important ISEs in the various fields are presented in Table 8.

(B) ION SELECTIVE MICRO ELECTRODES

Ion selective micro electrodes are especially designed for in vivo measurements. Walker and Brown [239,240] introduced such electrodes for intracellular measurements with a design based on a glass micropipette. Tip diameters as small as 1 μm have been obtained; an example is shown in Fig. 24 [241].

In a recent review article, Koryta et al. give a summary of the state of the art [242].

Potassium selective electrodes based on the ion carrier valinomycin have a large electrical resistance, therefore Baum and Lynn [243] have used tetra(p-chlorophenyl)borate. A chloride selective micro electrode has been described by Coetzee and Freiser [244]. Brown et al. [241] have constructed a calcium selective micro electrode and a similar electrode was described by Oehme et al. [245]. Recently, a lithium selective micro electrode has been made and used for the measurement of the accumulation of lithium in neurons [246].

There are problems inherent to the use of micro electrodes in in vivo measurements. Spreading electrical fields can influence the potential and, therefore, have to be avoided. This can be achieved by using double-barreled electrodes, one barrel filled with the ion-exchanger solution and the other with NaCl (for the case of K^+ determination) or with Na_2SO_4 or sodium glutamate solution (for Cl^- determinations). Another difficulty is the large electrical resistance

TABLE 8

Applications of the most important ISEs

Species	Determination of	Refs.
Soil and soil extracts		
Ca^{2+}	Calcium in soil	199
Cl^-	Chloride in soil extracts	200
F^-	Fluoride in soil	201
	Total fluoride in soil and vegetation	202
NH_3, NH_4^+	Ammonia in Kjeldahl analysis of soils	203
	Ammonia in soil extracts	204
NO_3^-	Nitrate in plants, soil extracts, and water	205
	Nitrate nitrogen in soil extracts	206—209
S^{2-}	Sulfide in submerged soils	210
SO_4^{2-}	Sulfate in soil; titration using a Pb-electrode	211
Biological fluids		
Br^-	Bromide in biological fluids, serum or plasma	212,213
Ca^{2+}	Free calcium in serum or plasma	214—216
Cl^-	Chloride in biological fluids	217
F^-	Fluoride in plasma	218
NH_3, NH_4^+	Ammonia in plasma	219
Na^+, K^+	Sodium and potassium in whole blood	220
Water analysis		
Br^-	Bromide, chloride, and fluoride in atmospheric precipitation	221
Ca^{2+}	Calcium in river and sea water	222
Cl^-	Chloride in sea water	223
F^-	Fluoride in potable and natural waters	224,225
NH_3, NH_3^+	Ammonia in natural waters	204
	Ammonia in water and wastes	226
	Ammonia in sea water	227, 228
NO_3^-	Nitrate in natural water	205, 229
S^{2-}	Sulfide in water	230
CN^-	Cyanide in waste water	231
Cu^{2+}	Copper in water	235
Beverages and foodstuffs		
Ca^{2+}	Calcium in skim milk	232
Cl^-	Chloride in cheese	233
	Chloride for the detection of abnormal milk	234
F^-	Fluoride in beverages	236
	Fluoride in wines	237
NO_3^-	Nitrate in baby food	238

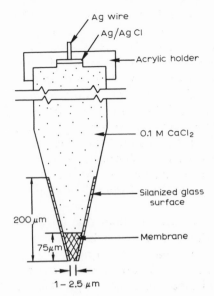

Fig. 24. Construction features of the calcium sensitive micro electrode [241].
(By kind permission of *Analytica Chimica Acta*.)

of micro electrodes. This demands a careful shielding of the high impedance cables as discussed in Sect. 4.(F) and 5.(A).

The response time of micro electrodes is of crucial importance in cases where the ionic activity changes in cells or tissues due to some external stimulation has to be measured. Response times of the order of seconds, as observed for many ISEs, are much too long in many such excitation studies.

(C) ION SELECTIVE FLOW-THROUGH ELECTRODES

Flow-through detectors are of special importance for the continuous monitoring in, for instance, process streams, effluent streams, rivers, etc. The reproducible way in which the sample is presented to the sensor in continuous flow systems also makes these systems very attractive for the measurement of discrete samples injected in some carrier stream. In many cases, a substantial increase in the precision together with a high sampling rate can be achieved.

The demands for these two applications are, in general, somewhat different. For the monitoring of continuous sample streams, the systems must run unattended for long periods, it should be easy to operate, it must be robust, and it must resist chemical attack. In principle, ISEs meet these requirements. Although it is sometimes possible to immerse the ISE—reference electrode assembly directly into the sample stream (if necessary together with a temperature compensation device), normally, a small part of the main stream is bled off. This allows some manipulation of the sample solution such as the adjustment to a constant temperature and pressure, and a pretreatment of the sample by adding buffers, to adjust the proper pH, or reagents, e.g. agents to mask interferents. Furthermore, it is easier to maintain constant hydrodynamic conditions at the surface of the ion sensitive membrane and the liquid junction of the reference electrode.

Because sample streams often flow through pipelines that are good electrical conductors, e.g. steel, and these lines have a direct connection to the earth, the sample solution will become grounded in such cases. To avoid undesirable loops, care must be taken in the careful insulation of the grounding of the voltmeter and the sample stream. Problems of this kind can also be overcome by an electrolytic separation of the main stream from the side stream passing the electrodes. This can be accomplished by making the side stream drop in a vessel from which the sample side stream is led to the sensor. This will cause a certain extra delay in the indication of concentration changes but, in general, a delay time of some seconds or even minutes will be acceptable in the majority of industrial applications.

For use in continuous analysis, such as continuous flow analysis (CFA) or the flow injection analysis (FIA), the response time of the monitor is of paramount importance because it is the main parameter that determines the sampling capacity. Therefore, the hold up and so the cell volume must be very small (of the order of μl) and the stagnant layer should be thin to ensure a rapid establishment of the equilibrium between the concentration at the surface and in the bulk. Moreover, the carry-over should be minimized by an appropriate design of the cell.

Several types of ion selective flow-through electrode are commercially available. These devices generally consist of a cap provided with an inlet and an outlet that can be fitted to an ISE [Fig. 25(a)]. Thompson and Rechnitz [247] have suggested another construction suitable for solid-state membranes. They have simply drilled a hole through the ion sensitive pellet. A somewhat similar approach was

378

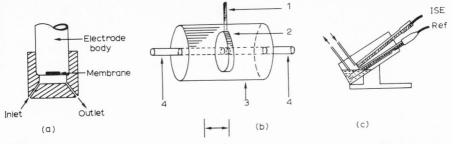

Fig. 25. Three types of flow-through electrodes. (a) Electrode provided with flow-through cap. (b) Disposable flow-through electrode. 1, Silver wire; 2, ion sensitive pellet; 3, polymer body; 4, inlet/outlet. (c) Flow-through electrode assembly according to Růžička et al. [250].

used by Blaedel and Dinwiddie [248] and by van der Linden and Oostervink [249] [Fig. 25(b)]. The latter authors have embedded a pellet in a polymer body before drilling a hole through the pellet. The response time, t_{90}, was found to be about 50 s which is much too long for CFA of FIA but which makes the electrode fast enough to monitor, for instance, the copper content of tap water. They have also designed an electrode which can easily be dismounted. The response of this type is much faster ($t_{90} \sim 6$ s) and can be used in FIA.

A very elegant way of using ISEs in FIA has been developed by Růžička et al. [250]. In their set-up, an ordinary ISE is placed above the solution level in a vessel in such a position that the active surface just makes contact with the solution. A reference electrode is immersed in the solution [Fig. 25(c)]. The sample stream flows across the surface of the membrane in a very thin layer (<1 mm). The response time of the device was found to be very short and a sample rate of over two hundred samples per hour is feasible.

Flow-through ion selective electrodes have also been applied as detectors in chromatography. Recently, Kojima et al. [251] have described the construction and performance of a dual ion selective electrode allowing the simultaneous detection of chlorine- and bromine-containing compounds in gas chromatography.

8. Specifications of commercially available electrodes

The specifications presented in Table 9 are the manufacturers' data. For more details the reader should consult the instruction manuals.

TABLE 9

Details of commercially available ion selective electrodes

Electrode	Manufacturer	Model number	Sensitivity (mV/decade)	Measuring range	Optimum pH range	Temp. range (°C)	Type
Ammonia/ ammonium	EIL	8002-8	-59	$5\times10^{-2}-5\times10^{-6}$			Gas-sensing
	HNU	ISE-10-10-00	-58	$10^0-5\times10^{-7}$			Gas-sensing
	Orion	95-10	-57	10^0-10^{-6}		0-50	Gas-sensing
	Philips	IS 561-NH4	+56 ± 3	$10^{-1}-10^{-6}$	4— 7	0-50	Liquid
	Radelkis	OP-NH3-07113	Theoretical	$10^{-1}-10^{-6}$			Gas-sensing
	Tacussel	PNH3-1	Theoretical	$10^{-1}-10^{-5}$			Gas-sensing
Barium	Corning	003 59 124M	+24—29	$10^0-5\times10^{-6}$			Liquid
	EDT	EE-Ba	+23 ± 3	$10^0-5\times10^{-6}$			Liquid
	Philips	IS 561-Ba		10^0-10^{-5}	1.7—7	0-50	Liquid
Bromide	Corning	003 59 094L	-52—59	$10^0-5\times10^{-6}$			Solid-state
	EDT	EE-Br		$10^0-5\times10^{-6}$			Solid-state
	Ingold	157201		$10^0-5\times10^{-6}$	2—12	0-80	Solid-state
	HNU	ISE-30-35-00	-56	$10^0-5\times10^{-7}$			Solid-state
	Metrohm	EA 306-Br	-56	$10^0-5\times10^{-6}$	0—14	0-50	Solid-state
	Orion	94-35	-56	$10^0-5\times10^{-6}$		0-80	Solid-state
	Philips	IS 550-Br	-56 ± 3	10^0-10^{-6}	1—11	0-50	Solid-state
	Radiometer	F1022Br	Theoretical	10^0-10^{-6}	0—14	-5-60	Solid-state
	Radelkis	OP-Br-07113	-56—59	10^0-10^{-6}	2—12		Solid-state
	Tacussel PBR	2PBR 2		10^0-10^{-6}	2—14		Solid-state
Cadmium	Ingold	157202		10^0-10^{-7}	3— 7	0-80	Solid-state
	Metrohm	EA 306-Cd	+28	10^0-10^{-7}	2—14	0-80	Solid-state
	Orion	94-48	+28	10^0-10^{-7}		0-80	Solid-state
	Philips	IS 550-Cd	+27 ± 2	10^0-10^{-6}	1—14	0- 50	Solid-state
	Tacussel	PCD 1		$10^{-1}-10^{-6}(10^{-10})$	1— 7		Solid-state
Calcium	Corning	003 59 114H		10^0-10^{-5}			Liquid
	EDT	EE-Ca	+24—29	10^0-10^{-5}	4— 9		Liquid
	HNU	ISE-20-20-00	+28	$10^0-5\times10^{-7}$			Liquid
	Metrohm	EA 301-Ca	+28	$10^0-8\times10^{-6}$	5.5-11	0- 50	Liquid
	Orion	93-20	+28	$10^0-5\times10^{-7}$	2.5-11	0-40	Liquid

Ion	Make	Type	Temp.	Detection limit	pX	Temp. range (°C)	Electrode type
	(HNO) Orion	ISE10-2200	+56	10^{-2}–10^{-4}			
	Orion	95-20	+56	10^{-2}–10^{-4}		0–50	Gas-sensing
Carbon dioxide	Radelkis	OP-9353	+56	Detn. lim. 0.13–32.00 kPa		0–50	Gas-sensing
Chloride	Corning	003 59 092J		10^0–10^{-4}			Solid-state
	EDT	EE-Cl	−52—59	10^0–10^{-4}	3–10		Solid-state
	EIL	8004-2	−59	10^0–5×10^{-5}	2–11	0–80	Solid-state
	Ingold	157203		10^0–5×10^{-5}			Solid-state
	HNU	ISE-30-17-00	−57	10^0–5×10^{-5}	0–14	0–50	Solid-state
	Metrohm	EA 306-Cl	−56	10^0–5×10^{-5}		0–80	Solid-state
	Orion	94-17	−56	10^0–5×10^{-5}		0–50	Solid-state
	Orion	93-17	−56	10^0–8×10^{-6}		0–50	Solid-state
	Philips	IS 550-Cl	−56 ± 3	10^0–10^{-5}	1–10		Solid-state
	Radiometer	F1012Cl	Theoretical	10^0–10^{-5}	0–14		Solid-state
	Radelkis	OP-Cl-07113	−55—59	10^0–10^{-5}	3–10	−5—60	Solid-state
	Tacussel	PCL 3		10^0–10^{-5}	2–14		Solid-state
Copper	Corning	003 59 112E		10^0–10^{-6}			Solid-state
	EDT	EE-Cu		10^0–10^{-6}			Solid-state
	Ingold	157207	+27	10^0–10^{-8}	3–7		Solid-state
	HNU	ISE-30-29-00	+26	10^0–5×10^{-7}		0–80	Solid-state
	Metrohm	EA 306-Cu	+26	10^0–10^{-7}	0–14	0–80	Solid-state
	Orion	94-29	+27 ± 2	Sat.–10^{-8}		0–80	Solid-state
	Philips	IS 550-Cu	Theoretical	10^0–10^{-6}	1–14	0–50	Solid-state
	Radiometer	F1112Cu	+26—29	10^0–10^{-6}	0–14		Solid-state
	Radelkis	OP-Cu-07113		10^0–10^{-7}	3–7	−5—60	Solid-state
	Tacussel	PCU 2		10^{-1}–10^{-6} (10^{-10})	2–8		Solid-state
Cyanide	Corning	003 59 100A		10^{-2}–10^{-6}	≥ 12		Solid-state
	EDT	EE-CN	−52—59	10^{-2}–10^{-6}	≥ 12		Solid-state
	Ingold	157204		10^{-2}–10^{-6}	11–13	0–80	Solid-state
	HNU	ISE-30-13-00	−63	10^{-2}–5×10^{-7}		0–100	Solid-state
	Metrohm	EA 306-CN	−58	10^{-2}–10^{-6}	0–14	0–80	Solid-state
	Orion	94-06	−58	10^{-1}–10^{-7}		0–50	Solid-state
	Philips	IS 550-CN	−56 ± 3	10^{-1}–10^{-7}	1–12		Solid-state
	Radiometer	F1042CN	Theoretical	10^{-3}–10^{-6}	0–14		Solid-state
	Radelkis	OP-CN-07113	−55—59	10^{-2}–10^{-6}	11–13	−5—60	Solid-state
	Tacussel	PCN 2		10^0–10^{-7}	9–14		Solid-state

TABLE 9 (continued)

Electrode	Manufacturer	Model number	Sensitivity (mV/decade)	Measuring range	Optimum pH range	Temp. range (°C)	Type
Fluoride	Corning	003 59 106K		10^0–10^{-6}			Solid-state
	EDT	EE-F	−52—59	10^0–10^{-5}	4— 8		Solid-state
	Ingold	157205	−56	10^0–10^{-6}	5— 8	0—80	Solid-state
	Metrohm	EA 306-F	−56	10^0–10^{-6}	0— 7/0—11	0—80	Solid-state
	Orion	94-09	−56	Sat.–10^{-6}		0—80	Solid-state
	Philips	IS 550-F	−56 ± 3	10^{-1}–10^{-6}	4— 8	0—80	Solid-state
	Radiometer	F1052F	Theoretical	10^0–3×10^{-7}	3—13		Solid-state
	Radelkis	OP-F-07113	−57—59	10^0–10^{-7}	5— 5.5	0—60	Solid-state
	Tacussel	PF 4		10^0–10^{-7}	5— 8		Solid-state
Fluoroborate	Metrohm	EA 301-BF$_4$	−57	Sat.–3×10^{-6}	2-12		Liquid
	Orion	93-05	−57	Sat.–3×10^{-6}		0—50	Liquid
Iodide	Corning	003 59 096N		10^0–10^{-6}			Solid-state
	EDT	EE-I	−52—59	10^0–5×10^{-6}	2—12		Solid-state
	Ingold	157206		10^0–2×10^{-7}	3—12	0—80	Solid-state
	HNU	ISE-30-53-00	−59	10^0–5×10^{-8}			Solid-state
	Metrohm	EA 306-I	−56	10^0–5×10^{-6}	0—14	0—50	Solid-state
	Orion	94-53	−56	10^0–5×10^{-6}		0—80	Solid-state
	Philips	IS 550-I	−56 ± 3	10^0–10^{-7}	1—12	0—50	Solid-state
	Radiometer	F10321	Theoretical	10^0–10^{-6}	0—14		Solid-state
	Radelkis	OP-I-07113	−57—59	10^{-1}–10^{-7}	3—12	−5—60	Solid-state
	Tacussel	PI 2		10^0–10^{-7}	1—14		Solid-state
Lead	Ingold	157200		10^0–10^{-7}	4— 7	0—80	Solid-state
	Metrohm	EA 306-Pb	+25	10^0–10^{-7}	2—14	0—80	Solid-state
	Orion	94-82	+25	10^0–10^{-6}		0—80	Solid-state
	Tacussel	PPB 1		10^{-1}–10^{-5} (10^{-10})			Solid-state
Lithium	Philips	IS 561-Li	+56 ± 3	10^0–10^{-5}	3—12	0—50	Liquid
Nitrate	Corning	003 59 116K		10^0–10^{-5}			Liquid
	EDT	EE-N	−55—59	10^{-1}–10^{-5}	4—11/2—12		Liquid
	EIL	8006-2	−59	10^{-1}–10^{-5}		5—40	Liquid

Ion	Manufacturer	Type no.	Slope	Concentration range	pH range	Temp. (°C)	Electrode type
	Philips	IS 561-NO3	-56 ± 3	$10^0\text{--}10^{-5}$	3–12	0–50	Liquid
	Tacussel	PNO3-1	$+58$	$10^0\text{--}5 \times 10^{-6}$	2–10	0–50	Liquid
Nitrite	HNU	ISE-10-38-00		$2 \times 10^{-2}\text{--}5 \times 10^{-8}$			Gas-sensing
Perchlorate	Metrohm	EA 301-ClO$_4$	-58	$10^0\text{--}2 \times 10^{-6}$	4–11	0–50	Liquid
	Orion	93-81	-58	$10^0\text{--}2 \times 10^{-6}$		0–50	Liquid
Potassium	Corning	003 59 118M		$10^0\text{--}10^{-5}$			Liquid
	EDT	EE-K	$+50\text{--}59$	$10^0\text{--}5 \times 10^{-5}$	4– 9	0–50	Liquid
	Metrohm	EA 301-K	$+56$	$10^0\text{--}10^{-5}$	1–12	0–50	Liquid
	Orion	93-19	$+56$	$10^0\text{--}10^{-6}$		0–50	Liquid
	Philips	IS 561-K	$+56 \pm 3$	$10^0\text{--}10^{-6}$	2–12	0–60	Liquid
	Radiometer	F2312K	Theoretical	$10^0\text{--}10^{-6}$	1.5–12.5		Liquid
	Radelkis	OP-K-07113	$+55\text{--}59$	$10^0\text{--}10^{-6}$	3–10		Liquid
	Tacussel	PKV		$10^0\text{--}10^{-6}$	6–14		Glass
Silver	Corning	003 59 104H		$10^0\text{--}10^{-7}$			Solid-state
	EDT	EE-Ag		$10^0\text{--}10^{-7}$			Solid-state
	Ingold	157208		$10^0\text{--}10^{-7}$	2– 9	0–80	Solid-state
	HNU	ISE-30-47-00	$+59$	$10^0\text{--}10^{-7}$		0–80	Solid-state
	Metrohm	EA 306-S/Ag	$+56$	$10^0\text{--}10^{-7}$	0–14	0–80	Solid-state
	Orion	94-16		$10^0\text{--}10^{-7}$		0–50	Solid-state
	Philips	IS 550-Ag	$+56 \pm 3$	$10^0\text{--}10^{-6}$	1–14		Solid-state
	Radiometer	F1212S	Theoretical	$10^0\text{--}10^{-6}$	0–14		Solid-state
	Radelkis	OP-Ag-07113	$+55\text{--}59$	$10^0\text{--}10^{-7}$	2– 9	$-5\text{--}60$	Solid-state
	Tacussel	PAG 2		$10^0\text{--}10^{-7}\ (10^{-10})$	2– 8		Solid-state
Sodium	Corning	003 59 122P		$10^{+2}\text{--}10^{-6}$	5–12		Glass
	EDT	EE-Na	$+59$	$10^{+2}\text{--}10^{-6}$	5–12		Glass
	EIL	1048-2		$10^{-1}\text{--}5 \times 10^{-6}$	5–12	0–80	Glass
	Ingold	pNa-20		$10^0\text{--}10^{-7}$	5– 9 (pH $\geqslant$ pNa + 3)	0–70	Glass
	Ingold	pNa-1000		$10^0\text{--}10^{-5}$		0–80	Glass
	Metrohm	EA 109-Na	$+56$	Sat.–10^{-6}		0–50	Glass
	Orion	94-11	$+55 \pm 3$	$10^0\text{--}10^{-5}$	4–12		Liquid
	Philips	IS 561-Na	$+59$	$10^0\text{--}10^{-5}$	pH $\geqslant$ pNa + 3		Glass
	Radiometer	G502Na	$+57\text{--}59$	$10^0\text{--}10^{-5}$	pH $\geqslant$ pNa + 3	10–60	Glass
	Radelkis	OP-Na-07113		$10^0\text{--}10^{-5}$			Glass
	Tacussel	PNAV		$10^0\text{--}10^{-6}$	6–14		Glass

TABLE 9 (continued)

Electrode	Manufacturer	Model num	Sensitivity (mV/decade)	Measuring range	Optimum pH range	Temp. range (°C)	Type
Sulfide	Corning	003 59 098A	-25—29	10^0—10^{-7}	≥ 12		Solid-state
	EDT	EE-S	-27	10^0—10^{-6}	≥ 12		Solid-state
	Ingold	157208	-27	10^0—10^{-7}	13—14	0—80	Solid-state
	HNU	ISE-30-47-00	-28	10^0—10^{-7}			Solid-state
	Metrohm	EA 306-S/Ag	-28	10^0—10^{-7}	0—14	0—80	Solid-state
	Orion	94-16	-28	10^0—10^{-7}		0—80	Solid-state
	Philips	IS 550-S	-27 ± 2	10^0—10^{-6}	1—14	0—50	Solid-state
	Radiometer	F1212S	Theoretical	10^0—10^{-6}	8—14	-5—60	Solid-state
	Radelkis	OP-S-07113	-29—31	10^0—10^{-6}	13—14		Solid-state
	Tacussel	PS 3		10^0—10^{-7} (10^{-10})	11—14		Solid-state
Sulfur dioxide	EIL	8010-8	59	5×10^{-2}—5×10^{-5}		0—40	Gas-sensing
Thiocyanate	Corning	003 59 102E		10^0—10^{-5}			Solid-state
	EDT	EE-T		10^0—10^{-5}			Solid-state
	Ingold	157209		10^0—5×10^{-6}	2—12	0—80	Solid-state
	Metrohm	EA 306-SCN	-57	10^0—5×10^{-6}	2—10	0—50	Solid-state
	Orion	94-58	-57	10^0—5×10^{-6}		0—80	Solid-state
	Radelkis	OP-SCN-07113	-55—59	10^{-1}—5×10^{-6}	2—12		Solid-state
	Tacussel	PSCN 1		10^{-1}—10^{-5} (10^{-6})	0—14		Solid-state
Water hardness	Corning	003 59 120P		10^0—10^{-5}			Liquid
	EDT	EE-W	$+22$—29	10^{-1}—5×10^{-5}	4.5—8		Liquid
	HNU	ISE-20-32-00	$+27$	10^{-2}—2×10^{-6}			Liquid
	Metrohm	EA 301-Me^{2+}	$+24$	10^0—4×10^{-6}	5.5—11	0—50	Liquid
	Orion	93-32	$+24$	10^{-2}—4×10^{-6}		0—50	Liquid
Ortho-phosphate	Tacussel	PP04-1	-15	10^{-1}—10^{-8}			Solid-state

References

1 W. Nernst, Z. Phys. Chem., 2 (1888) 613; 4 (1889) 129.
2 M. Planck, Ann. Phys. Chem., 39 (1890) 161; 40 (1890) 561.
3 W. Ostwald, Z. Phys. Chem., 6 (1890) 71.
4 M. Cremer, Z. Biol., 47 (1906) 562.
5 F. Haber and Z. Klemensiewicz, Z. Phys. Chem. (Leipzig), 67 (1909) 385.
6 F.G. Donnan, Z. Elektrochem., 17 (1911) 572.
7 T. Teorell, Trans. Faraday Soc., 33 (1937) 1053.
8 K.H. Meyer and J.F. Sievers, Helv. Chim. Acta, 19 (1936) 649.
9 B. Lengyel and E. Blum, Trans. Faraday Soc., 30 (1934) 461.
10 B.P. Nicolsky and T.A. Tolmacheva, Zh. Fiz. Khim., 10 (1937) 495 (see ref. 26).
11 I.M. Kolthoff and H.L. Sanders, J. Am. Chem. Soc., 59 (1937) 416.
12 H.J.C. Tendeloo, Proc. Acad. Sci. (Amsterdam), 38 (1935) 434; J. Biol. Chem., 113 (1936) 333.
13 G. Eisenman, in R.A. Durst (Ed.), Ion-Selective Electrodes, Natl. Bur. Stand. Spec. Publ. 314, Washington, 1969, Chap. 1.
14 A.K. Covington, in R.A. Durst (Ed.), Ion-Selective Electrodes, Natl. Bur. Stand. Spec. Publ. 314, Washington, 1969, Chap. 3.
15 J.W. Ross, Science, 156 (1967) 1378.
16 W.E. Morf and W. Simon, Hung. Sci. Instrum., 41 (1977) 1.
17 M.S. Frant and J.W. Ross, Science, 154 (1966) 1553.
18 J. Koryta, Anal. Chim. Acta, 61 (1972) 329; see also J. Koryta, Ion Selective Electrodes, Cambridge University Press, Cambridge, 1975.
19 J. Koryta, Anal. Chim. Acta, 91 (1977) 1.
20 R.A. Durst (Ed.), Ion-Selective Electrodes, Natl. Bur. Stand. Spec. Publ. 314, Washington, 1969.
21 E. Pungor (Ed.), Proc. Symp. Ion Selective Electrodes, Mátrafüred, Hungary, 1972, Akadémiai Kiadó, Budapest, 1973.
22 IUPAC Int. Symp. Selective Electrodes, 1973, University of Wales, Institute of Science and Technology, Cardiff.
23 E. Pungor (Ed.), Proc. Symp. Ion Selective Electrodes, Mátrafüred, Hungary, 1976, Akadémiai Kiadó, Budapest, 1977.
24 E. Pungor and I. Buzas (Eds.), Ion Selective Electrodes. Conference held at Budapest, Hungary 5—9 September, 1977, Elsevier, Amsterdam, 1978.
25 R.P. Buck, Anal. Chem., 44 (1972) 270R; 46 (1974) 28R; 48 (1976) 23R; 50 (1978) 1R.
26 G. Eisenman (Ed.), Glass Electrodes for Hydrogen and other Cations, Marcel Dekker, New York, 1967.
27 G.J. Moody and J.D.R. Thomas, Selective Ion-Sensitive Electrodes, Merrow, Watford, Herts., Gt. Britain, 1971.
28 K. Camman, Das Arbeiten mit ionenselektiven Elektroden, Springer-Verlag, Berlin, 1973.
29 P.L. Bailey, Analysis with Ion-Selective Electrodes, Heyden, London, 1976.
30 N. Lakshminarayahnaiah, Membrane Electrodes, Academic Press, New York, 1976.

31 F. Helfferich, Ion Exchange, McGraw-Hill, New York, 1962, p. 340.
32 Manual for symbols and terminology for physicochemical quantities and units, Pure Appl. Chem., 21 (1970) 1.
33 J.O'M. Bockris and A.K.N. Reddy, Modern Electrochemistry, Plenum Press, New York, 1970.
34 R.P. Buck, Crit. Rev. Anal. Chem., 5 (1975) 323.
35 R.P. Buck, in E. Pungor and I. Buzas (Eds.), Ion Selective Electrodes. Conference held at Budapest, Hungary, 5—9 September, 1977, Elsevier, Amsterdam, 1978, p. 21.
36 R.P. Buck, Anal. Chem., 40 (1968) 1432.
37 A. Hulanicki and A. Lewenstam, Talanta, 23 (1976) 661; 24 (1977) 171.
38 J.O'M. Bockris and A.K.N. Reddy, Modern Electrochemistry, Plenum Press, New York, 1970, p. 1056.
39 W. Jaenicke, Z. Elektrochem., 55 (1951) 648; W. Jaenicke and M. Haase, Z. Elektrochem., 63 (1959) 521.
40 G.J.M. Heijne and W.E. van der Linden, Anal. Chim. Acta, 96 (1978) 13.
41 Commission on Analytical Nomenclature of the IUPAC Analytical Chemistry Division, Pure Appl. Chem., 48 (1976) 129.
41a IUPAC Inf. Bull., (1) (1978) 70.
42 A. Shatkay, Anal. Chem., 48 (1976) 1039.
43 G.A. Rechnitz and H.F. Hameka, Z. Anal. Chem., 214 (1965) 252.
44 G. Johansson and K. Norberg, J. Electroanal. Chem., 18 (1968) 239.
45 P.L. Markovic and J.O. Osburn, AIChE J., 19 (1973) 504.
46 W.E. Morf, E. Lindner and W. Simon, Anal. Chem., 47 (1975) 1596.
47 J. Mertens, P. van den Winkel and D.L. Massart, Anal. Chem., 48 (1976) 272.
48 E. Lindner, K. Tóth and E. Pungor, Anal. Chem., 48 (1976) 1071.
49 W.E. Morf, Anal. Lett., 10 (1977) 87.
50 J. Buffle and N. Parthasarathy, 93 (1977) 111; 93 (1977) 121.
51 M.J.D. Brand and G.A. Rechnitz, Anal. Chem., 42 (1970) 478.
52 K. Camman and G.A. Rechnitz, Anal. Chem., 48 (1976) 856.
53 K. Camman, in E. Pungor and I. Buzas (Eds.), Ion Selective Electrodes, Conference held at Budapest, Hungary, 5—9 September, 1977, Elsevier, Amsterdam, 1978, p. 297.
54 K. Camman, Anal. Chem., 50 (1978) 936.
55 J.O'M. Bockris and A.K.N. Reddy, Modern Electrochemistry, Plenum Press, New York, 1970, Sect. 11.2.10.
56 J. Koryta, J. Dvořák and V. Boháčková, Electrochemistry, Methuen, London, 1970, p. 193.
57 P. Henderson, Z. Phys. Chem., 59 (1907) 118; 63 (1908) 325.
58 W.E. Morf, Anal. Chem., 49 (1977) 810.
59 H.-R. Wuhrmann, W.E. Morf and W. Simon, Helv. Chim. Acta, 56 (1973) 1011.
60 F. Helfferich, Ion Exchange, McGraw-Hill, New York, 1962, p. 351.
61 W.E. Morf and W. Simon, Hung. Sci. Instrum., 41 (1977) 1.
62 W.E. Morf, G. Kahr and W. Simon, Anal. Lett., 7 (1974) 9.
63 H.S. Carslaw and J.C. Jaeger, Conduction of Heat in Solids, Oxford University Press, London, 1959.

64 J.P. Sandblom, G. Eisenman and J.L. Walker, Jr., J. Phys. Chem., 71 (1967) 3862.
65 M. Sato, Electrochim. Acta, 11 (1966) 361.
66 M. Koebel, Anal. Chem., 46 (1974) 1559.
67 R.P. Buck and V.R. Sheppard, Anal. Chem., 46 (1974) 1559.
68 T. Hepel, M. Hepel and M. Lesko, Analyst, 102 (1977) 132.
69 T. Hepel and M. Hepel, Electrochim. Acta, 22 (1977) 295.
70 J. Růžička and C.G. Lamm, Anal. Chim. Acta, 53 (1971) 206; 54 (1971) 1.
71 P. Bergveld, IEEE Trans. Bio-Med. Electron., 19 (1972) 342.
72 H. James, G. Carmack and H. Freiser, Anal. Chem., 44 (1972) 856.
73 G.A. Rechnitz, IUPAC Int. Symp. Selective Electrodes, 1973, University of Wales, Institute of Science and Technology, Cardiff, p. 468.
74 E. Pungor and K. Tóth, IUPAC Int. Symp. Selective Electrodes 1973, University of Wales, Institute of Science and Technology, Cardiff, p. 441.
75 R.M. van de Leest and A. Geven, J. Electroanal. Chem., 90 (1978) 97.
76 J. Veselý, Chem. Listy, 65 (1971) 86.
77 R. Bock and S. Strecker, Z. Anal. Chem., 235 (1968) 322.
78 L.G. Sillén and A.E. Martell, Stability constants of Metal—Ion Complexes, Spec. Publ. No. 17, The Chemical Society, London, 1964.
79 M.S. Frant and J.W. Ross, Anal. Chem., 40 (1968) 1169.
80 J.J. Lingane, Anal. Chem., 39 (1967) 881.
81 A. Hulanicki, M. Trojanowicz and M. Cichy, Talanta, 23 (1976) 47.
82 J. Veselý, Collect. Czech. Chem. Commun., 36 (1971) 3364.
83 J.W. Ross, in R.A. Durst (Ed.), Ion-Selective Electrodes, Natl. Bur. Stand. Spec. Publ. 314, Washington, 1969, Chap. 2.
84 H. Adametzova and R. Vadura, J. Electroanal. Chem., 55 (1974) 53.
85 H.J.C. Tendeloo and A. Krips, Rec. Trav. Chim. Pays-Bas, 76 (1957) 703, 946.
86 E. Pungor, K. Tóth and J. Havas, Hung. Sci. Instrum., 3 (1965) 2.
87 G. Kahr, Diss. Nr. 4927, ETH, Zürich, 1972.
88 G.J. Moody and J.D.R. Thomas, Selective Ion-Sensitive Electrodes, Merrow, Watford, Herts., Gt. Britain, 1971, p. 62.
89 C. Kittel, Introduction to Solid State Physics, Wiley, New York, 1976, 5th edn., Chap. 17.
90 E. Pungor and K. Tóth, Analyst, 95 (1970) 625.
91 E. Pungor and K. Tóth, Pure Appl.Chem., 34 (1972) 105; 36 (1973) 441.
92 K. Tóth and E. Pungor, Anal. Chim. Acta, 51 (1970) 221.
93 W.E. Morf, G. Kahr and W. Simon, Anal. Chem., 46 (1974) 447.
94 D.J. Crombie, G.J. Moody and J.D.R. Thomas, Anal. Chim. Acta, 80 (1975) 1.
95 K. Camman, Das Arbeiten mit ionenselektiven Elektroden, Springer-Verlag, Berlin, 1973, p. 77.
96 J. Veselý, O.J. Jensen and B. Nicolaisen, Anal. Chim. Acta, 62 (1972) 1.
97 J. Gulens and B. Ikeda, Anal. Chem., 50 (1978) 782.
98 Yu.G. Vlasov and S.B. Kocheregin, in E. Pungor and I. Buzas (Eds.), Ion Selective Electrodes. Conference held at Budapest, Hungary, 5—9 September, 1977, Elsevier, Amsterdam, 1978, p. 597.

99 R.E. van de Leest, Analyst, 101 (1976) 433.
100 M.M. Sorrentino and G.A. Rechnitz, Anal. Chem., 46 (1974) 943.
101 E. Pungor, K. Tóth and G. Nagy, Mikrochim. Acta, (1978) 531.
102 M.S. Frant and J.W. Ross, Ger. Off., 1,942,379, 1970.
103 M. Mascini and A. Liberti, Anal. Chim. Acta, 53 (1971) 202.
104 E.H. Hansen, C.G. Lamm and J. Růžička, Anal. Chim. Acta, 59 (1972) 403.
105 J.C. Czaban and G.A. Rechnitz, Anal. Chem., 45 (1973) 471.
106 H. Thompson and G.A. Rechnitz, Chem. Instrum., 4 (1973) 239.
107 G.J.M. Heijne, W.E. van der Linden and G. den Boef, Anal. Chim. Acta, 89 (1977) 287.
108 G.J.M. Heijne, W.E. van der Linden and G. den Boef, Anal. Chim. Acta, 93 (1977) 99.
109 G.J.M. Heijne and W.E. van der Linden, Anal. Chim. Acta, 96 (1978) 13.
110 A. Jyo, T. Hashizume and N. Ishibashi, Anal. Chem., 49 (1977) 1869.
111 G.J.M. Heijne, W.E. van der Linden and G. den Boef, Anal. Chim. Acta, 98 (1978) 221.
112 D.J. Crombie, G.J. Moody and J.D.R. Thomas, Talanta, 21 (1974) 1094.
113 P. Lanza, Euroanalysis III, Dublin, 1978, Abstr. No. 109.
114 D. Midgley, Anal. Chim. Acta, 87 (1976) 19.
115 G.B. Oglesby, W.C. Duer and F.J. Millero, Anal. Chem., 49 (1977) 877.
116 M.F. El-Taras, E. Pungor and G. Nagy, Anal. Chim. Acta, 82 (1976) 285.
117 I. Sekerka and J.F. Lechner, Anal. Lett., 11 (1978) 415.
118 W.E. van der Linden and R. Oostervink, Anal. Chim. Acta, 108 (1979) 169.
119 A.F. Zhukov, A.V. Vishnya, Ya.L. Kharif, Yu.I. Urusov, F.K. Volynets, E.I.R. Ryzhikov and A.V. Gordievskii, Zh. Anal. Khim., 30 (1975) 1761.
120 H. Hirata and K. Higashiyama, Z. Anal. Chem., 257 (1971) 104.
121 M.J.D. Brand, J.J. Militello and G.A. Rechnitz, Anal. Lett., 2 (1969) 523.
122 G.J.M. Heijne, W.E. van der Linden and G. den Boef, Anal. Chim. Acta, 100 (1978) 193.
123 M. Mascini and A. Liberti, Anal. Chim. Acta, 60 (1972) 405.
124 P. Kivalo, R. Virtanen, K. Wickström, M. Wilson, E. Pungor, G. Horvai and K. Tóth, Anal. Chim. Acta, 87 (1976) 401.
125 Orion Research Inc. Instruction Manual Lead Electrode, Model 94-82.
126 A.V. Gordievskii, V.S. Shterman, A.Ya. Syrchenkov, N.I. Savvin, A.F. Zhukov and Yu.I. Urusov, Zh. Anal. Khim., 27 (1972) 2170.
127 T. Anfält and D. Jagner, Anal. Chim. Acta, 55 (1971) 477.
128 R.E. Van de Leest, Analyst, 102 (1977) 509.
129 M. Dole, The Glass Electrode, Wiley, New York, 1941.
130 L. Kratz, Die Glaselektrode und ihre Anwendungen, D. Steinkopf, Frankfurt a.M., 1950.
131 K. Schwabe, pH Messtechnik, Th. Steinkopf, Dresden, Leipzig, 1963.
132 G. Eisenman, Glass Electrodes for Hydrogen and Other Cations, Marcel Dekker, New York, 1967.
133 G. Eisenman, R.G. Bates, G. Mattock and S.M. Friedman, The Glass Electrode, Wiley-Interscience, New York, 1966.
134 R.G. Bates, Determination of pH, Theory and Practice, Wiley, New York, 1973, 2nd edn.

135 W.H. Zachariasen, J. Am. Chem. Soc., 54 (1932) 3841.
136 G. Eisenman, D.O. Rudin and J.U. Casby, Science, 126 (1957) 831.
137 G. Eisenman, Biophys. J., 2 (1962) 159.
138 A.A. Belyestin and A.A. Lev, Chemistry in Natural Science, Leningrad State University, 32 (1965) (quoted in ref. 28).
139 G. Eisenman, R.G. Bates, G. Mattock and S.M. Friedman, The Glass Electrode, Wiley-Interscience, New York, 1966, Chap. 1.
140 K. Schwabe and H. Dahms, Monatsber. Dtsch. Akad. Wiss. Berlin, 1 (1959) 279.
141 D. Hubbard et al., J. Res. Natl. Bur. Stand., 37 (1946) 223; 41 (1948) 273; 44 (1950) 247; 45 (1950) 430; 46 (1951) 168.
142 Z. Boksay, G. Bouquet and S. Dobas, Phys. Chem. Glasses, 8 (1967) 140; 9 (1968) 69.
143 A. Wikby, J. Electroanal. Chem., 33 (1971) 145; 38 (1972) 429, 441; 39 (1972) 103.
144 F.G.K. Baucke, in E. Pungor and I. Buzas (Eds.), Ion Selective Electrodes. Conference held at Budapest, Hungary, 5—9 September, 1977, Elsevier, Amsterdam, 1978, p. 215.
145 M.F. Wilson, E. Haikala and P. Kivalo, Anal. Chim. Acta, 74 (1975) 395, 411.
146 N. Lakshminarayahnaiah, Transport Phenomena in Membranes, Academic Press, New York, 1969.
147 S.B. Tuwiner, Diffusion and Membrane Technology, Van Nostrand—Reinhold, Princeton, New Jersey, 1962.
148 F.P. IJsseling and E. van Dalen, Anal. Chim. Acta, 36 (1966) 166; 40 (1968) 421; 43 (1968) 77; 45 (1969) 121.
149 P.L. Bailey, Analysis with Ion-Selective Electrodes, Heyden, London, 1976, p. 119.
150 L.A.R. Pioda and W. Simon, Chimia, 23 (1969) 72.
151 L.A.R. Pioda, V. Stanková and W. Simon, Anal. Lett., 2 (1969) 665.
152 J.W. Ross, in R.A. Durst (Ed.), Ion-Selective Electrodes, Natl. Bur. Stand. Spec. Publ. 314, Washington, 1969, Chap. II, pp. 65 and 66.
153 J. Růžička, E.H. Hansen and J.C. Tjell, Anal. Chim. Acta, 67 (1973) 155.
154 J.W. Ross, U.S. Pat. 3,483,112, 1969; see also ref. 83.
155 J.W. Davies, G.J. Moody and J.D.R. Thomas, Analyst, 97 (1972) 87.
156 C.J. Coetzee and H. Freiser, Anal. Chem., 40 (1968) 2071; 41 (1969) 1128.
157 R.E. Reinsfelder and F.A. Schultz, Anal. Chim. Acta, 65 (1973) 425.
158 S. Potterton and W.D. Shults, Anal. Lett., 1 (1967) 11.
159 R.M. Carlson and J.L. Paul, Anal. Chem., 40 (1968) 1292.
160 U. Fiedler, Anal. Chim. Acta, 89 (1977) 111.
161 Z. Stefanac and W. Simon, Microchem. J., 12 (1967) 125.
162 C.J. Pedersen, J. Am. Chem. Soc., 89 (1967) 7017; 92 (1970) 386, 391.
163 W.E. Morf and W. Simon, Helv. Chim. Acta, 54 (1971) 2683.
164 E. Lindner, K. Tóth, E. Pungor, W.E. Morf and W. Simon, Anal. Chem., 50 (1978) 1627.
165 S. Lal and G.D. Christian, Anal. Lett., 3 (1970) 11.
166 S.M. Hammond and P.A. Lambert, J. Electroanal. Chem., 53 (1974) 155.
167 R.P. Scholer and W. Simon, Chimia, 24 (1970) 372.

168 R.E. Cosgrove, C.A. Mask and I.H. Krull, Anal. Lett., 3 (1970) 457.
169 R.W. Stow, R.F. Baer and B.F. Randall, Arch. Phys. Med. Rehabil., 38 (1957) 646.
170 J.W. Severinghaus and A.F. Bradley, J. Appl. Physiol., 13 (1958) 515.
171 J.W. Ross, J.H. Riseman and J.A. Krueger, Pure Appl. Chem., 36 (1973) 473.
172 P.L.Bailey and M. Riley, Analyst, 100 (1975) 145.
173 J. Růžička and E.H. Hansen, Anal. Chim. Acta, 69 (1974) 129.
174 E.H. Hansen and N.R. Larsen, Anal. Chim. Acta, 78 (1975) 459.
175 P.L. Bailey and M. Riley, Analyst, 102 (1977) 213.
176 M. Mascini and C. Cremisini, Anal. Chim. Acta, 97 (1979) 237.
177 F. van der Pol, Anal. Chim. Acta, 97 (1978) 245.
178 G.G. Guilbault, in G. Svehla (Ed.), Comprehensive Analytical Chemistry, Vol. VIII, Elsevier, Amsterdam, 1977, Chap. I.
179 P.D'Orazio and G.A. Rechnitz, Anal. Chem., 49 (1977) 2083.
180 P.D'Orazio, M.E. Meyerhoff and G.A. Rechnitz, Anal. Chem., 50 (1978) 1531.
181 M.A. Jensen and G.A. Rechnitz, Anal. Chim. Acta, 101 (1978) 125.
182 S.D. Moss, J. Janata and C.C. Johnson, Anal. Chem., 47 (1975) 2238.
183 P.T. McBride, J. Janata, P.A. Comte, S.D. Moss and C.C. Johnson, Anal. Chim. Acta, 101 (1978) 239.
184 N.F. de Rooy, Thesis, Technical Univ. Enschede, The Netherlands, 1978.
185 P.A. Comte and J. Janata, Anal. Chim. Acta, 101 (1978) 247.
186 D.J.G. Ives and G.D. Janz, Reference Electrodes. Theory and Practice, Academic Press, New York, 1969.
187 F.G.K. Baucke, J. Electroanal. Chem., 67 (1976) 291.
188 R.G. Bates, Pure Appl. Chem., 36 (1973) 407.
189 R.G. Bates, in E. Pungor and I. Buzas, Ion Selective Electrodes, Conference held at Budapest, 5—9 September, 1977, Elsevier, Amsterdam, 1978, p. 3.
190 A. Ringbom, Complexation in Analytical Chemistry, Interscience, New York, 1963.
191 P.L. Bailey and E. Pungor, Anal. Chim. Acta, 64 (1973) 423.
192 G. Horvai, K. Tóth and E. Pungor, Anal. Chim. Acta, 82 (1976) 45.
193 J.D.R. Thomas, in E. Pungor and I. Buzas, Ion Selective Electrodes, Conference held at Budapest, 5—9 September, 1977, Elsevier, Amsterdam, 1978, p. 175.
194 R.A. Durst (Ed.), Ion-Selective Electrodes, Natl. Bur. Stand. Spec. Publ. 314, Washington, 1969, Chap. 11.
195 Orion Research Inc., News Lett., 2 (1970) 49.
196 G. Gran, Analyst, 77 (1952) 661.
197 J.M. van der Meer, G. den Boef and W.E. van der Linden, Anal. Chem. Acta, 76 (1975) 261.
198 K. Srinivisan and G.A. Rechnitz, Anal. Chem., 41 (1969) 1203.
199 E.A. Woolson, J.H. Axley and P.C. Kearney, Soil Sci., 109 (1970) 279.
200 B.W. Hill and G.W. Langdale, Soil Sci. Plant Anal., 2 (1971) 237.
201 S. Larsen and A.E. Widowson, J. Soil Sci., 22 (1971) 210.
202 N.R. McQuaker and M. Gurney, Anal. Chem., 49 (1977) 45.
203 J.M. Bremner and M.A. Tabatabai, Soil Sci. Plant Anal., 3 (1972) 159.

204 W.L. Banwart, M. Tabatabai and J.M. Bremner, Soil Sci. Plant Anal., 3 (1972) 449.
205 P.J. Milham, A.S. Awad, R.E. Paul and J.H. Bull, Analyst, 95 (1970) 751.
206 A. Øien and A.R. Selmer-Olsen, Analyst, 94 (1969) 888.
207 G.R. Smith, Anal. Lett., 8 (1975) 503.
208 R.J.K. Myers and E.A. Paul, Can. J. Soil Sci., 48 (1968) 369.
209 A.R. Mack and R.B. Sanderson, Can. J. Soil Sci., 51 (1971) 95.
210 A.I. Allam, G. Pitts and J.P. Hollis, Soil Sci., 114 (1972) 456.
211 J.D. Goertzen and J.D. Oster, Soil Sci. Soc. Am. Proc., 36 (1972) 391.
212 H.J. Degenhart, G. Abeln, B. Bevaart and J. Baks, Clin. Chim. Acta, 38 (1972) 217.
213 S. Poser, W. Poser and B. Müller-Oerlinghausen, Z. Klin. Chem. Klin. Biochem., 12 (1974) 350.
214 J.H. Ladenson and G.N. Bowers Jr., Clin. Chem., 19 (1973) 565.
215 H. Husdan, M. Leung, D. Oreopoulos and A. Rapoport, Clin. Chem., 23 (1977) 1775.
216 S. Madsen and K. Ølgaard, Clin. Chem., 23 (1977) 690.
217 L. Szabo, M.A. Kenny and W. Lee, Clin. Chem., 19 (1973) 727.
218 C. Fuchs, Clin. Chim. Acta, 60 (1975) 157.
219 A.F. Attilis, D. Autizi and L. Capocaccia, Biochem. Med., 14 (1975) 109.
220 J.H. Ladenson, Clin. Chem., 23 (1977) 1912.
221 R.C. Harris and H.H. Williams, J. Appl. Meteorol., 8 (1969) 299.
222 M. Mascini, Anal. Chim. Acta, 56 (1971) 312.
223 N. Ogata, Jpn. Anal., 21 (1972) 780.
224 N.T. Crosby, A.L. Dennis and J.G. Stevens, Analyst, 93 (1968) 643.
225 J.E. Harwood, Water Res., 3 (1969) 273.
226 R.F. Thomas and R.L. Booth, Environ. Sci. Technol., 7 (1973) 523.
227 T.R. Gilbert and A.M. Clay, Anal. Chem., 45 (1973) 1757.
228 A.G.A. Merks, Neth. J. Sea Res., 9 (1975) 371.
229 J. Mertens and D.L. Massart, Bull. Soc. Chim. Belg., 82 (1973) 179.
230 E.W. Baumann, Anal. Chem., 46 (1973) 1345.
231 J.H. Riseman, Am. Lab., 4 (1972) 63.
232 P.J. Muldoon and B.J. Liska, J. Dairy Sci., 52 (1969) 460.
233 A.W. Randell and P.M. Linklater, Aust. J. Dairy Technol., 27 (1972) 51.
234 P.J. Muldoon and B.J. Liska, J. Dairy Sci., 54 (1971) 117.
235 M.J. Smith and S.E. Manahan, Anal. Chem., 45 (1973) 836.
236 W.P. Ferne and N.A. Shane, J. Food Sci., 34 (1969) 317.
237 G. DeBaenst, J. Mertens, P. van de Winkel and D.L. Massart, J. Pharm. Belg., 28 (1973) 188.
238 M.A. Liedtke and C.E. Meloan, J. Agric. Food Chem., 24 (1976) 410.
239 J.L. Walker, Anal. Chem., 43 (1971) 89A.
240 J.L. Walker and A.M. Brown, Science, 167 (1970) 1502.
241 H.M. Brown, J.P. Pemberton and J.D. Owen, Anal. Chim. Acta, 85 (1976) 261.
242 J. Koryta, M. Brezina, J. Pradác and J. Pradácová, in A. Bard (Ed.), Electroanalytical Chemistry, Vol. 11, Marcel Dekker, New York, 1979, Chap. 2.
243 G. Baum and M. Lynn, Anal. Chim. Acta, 65 (1973) 393.
244 C.J. Coetzee and H. Freiser, Anal. Chem., 40 (1968) 2071.

245 M. Oehme, M. Kessler and W. Simon, Chimia, 30 (1976) 204.
246 R.C. Thomas, W. Simon and M. Oehme, Nature (London), 258 (1976) 754.
247 H. Thompson and G.A. Rechnitz, Chem. Instrum., 4 (1972) 239.
248 W.J. Blaedel and D.E. Dinwiddie, Anal. Chem., 47 (1975) 1070.
249 W.E. van der Linden and R. Oostervink, Anal. Chim. Acta, 101 (1978) 419.
250 J. Růžička, E.H. Hansen and E.A. Zagatto, Anal. Chim. Acta, 88 (1977) 1.
251 T. Kojima, M. Ichise and Y. Seo, Anal. Chim. Acta, 101 (1978) 273.
252 D.G. Davies in C.L. Wilson and D.W. Wilson (Eds.), Comprehensive Analytical Chemistry, Vol. IIA, Elsevier, Amsterdam, 1964, pp. 65—173.
253 G. Svehla, Automatic Potentiometric Titrations, Pergamon Press, Oxford, 1978, pp. 31 ff.
254 G. Svehla, Automatic Potentiometric Titrations, Pergamon Press, Oxford, 1978, pp. 187 ff.

Index

395

396

398

standard for mass spectrometry, 211
Permissible tolerance, between laboratories, 52
Permselectivity, 280
Petroleum analysis, application of mass spectrometry to, 246
pH, effect of, on LaF_3 membrane electrode, 323
pH (glass) electrodes, *see also*: Glass electrode 338, 340, 342
Phosphate, I.S.E. for determination of, 384
Phosphate esters, application of, to construction of I.S.E.s, 346
Phosphorus, abundance of natural isotopes of, 207, 258
Photographic emulsions, as ion-detectors, 263
Photographic-plate detector, sensitivity and precision of, 265
Photographic recording of mass spectrum, 202, 261—265
Photoionisation sources, for mass spectrometry, 181—182
Photometry, application of law of propagation of errors to, 87—90
Platinum, isotopic composition of, 260
Poiseuille equation, 175
Poisson distribution, 13
Polarograms, statistical aspects of interpretation of, 86
Polarographic analysis, application of law of propagation of errors to, 86—87
Polyaminocarboxylic acids, effect of, on behaviour of mixed-sulphide membranes, 291
Polycrystalline membranes, application of, to I.S.E.s, 277
Polynomials of a single variable, 122—138
Population mean, 4
Potassium, determination of, in blood, by I.S.E., 352
—, isotopic composition of, 258, 259
—, selective liquid membrane electrode for, 352
—, selectivity of antibiotics for, 350
Potassium ions, microelectrode selective for, 375

—, selective electrons for, 339, 345, 383
Potentiometric analysis, statistical aspec of, 84—86, 126
Potentiometric recorders, for mass spe trometry, 197
Potentiometric selectivity coefficien 289
Potentiometric titrimetry, with I.S.E. 370—373
Precipitates, formation of ternary con pounds in mixed, 318
Precision, 2
Probability density function, 11
Probability distributions, 12—25
—, discrete, 12—13
Propagation of errors, applications of la of, 73—100
Pseudo molecular ions, 239
—, fragmentation of, 240

Q-test, description of, 31—32
Quadrupole mass spectrometer, 191 200—201
Quartile, 5
Quasi-equilibrium theory, 203, 305
Q-values, subroutine for calculation o critical, 33—34

Radio frequency spark, application of, t mass spectrometry, 255
Radiometric data, application of law o propagation of errors to, 99—100
Random errors, 1
Regression, linear, 104—107, 109—113
Regression analysis, by linearizing trans formations, 139—141
Relative frequencies, 7
Relative isotopic abundances, tables of 257
Repeatability, definition of, 51
Reproducibility, of mean value, 70—73
Resolving power, of mass analysers, 191 193—194
Response time, of liquid membranes 307—309
—, —, with dissolved charged ligands, 315

404

406